ENCYCLOPEDIA OF BIOCHEMISTRY

ENCYCLOPEDIA OF BIOCHEMISTRY

Vol. 5

By

Sananda Chatterjee

DAE, DCA, CMCNet

Scientific Consultant

Institute for Natural Sciences

Kolkata

(West Bengal)

DISCOVERY PUBLISHING HOUSE PVT. LTD.

NEW DELHI-110 002

Published by:
Namit Wasan
DISCOVERY PUBLISHING HOUSE PVT. LTD.
4383/4B, Ansari Road, Darya Ganj
New Delhi-110 002 (India)
Phone : +91-11-23279245; 23253475; 43596065
E-mail : discoverybooksindia@gmail.com
discoverypublishinghouse@gmail.com
namitwasan9@gmail.com
web : www.discoverypublishinggroup.com

Reprinted: 2020
First Edition: 2012

ISBN: 978-81-8356-876-0 (Set)

Encyclopedia of Biochemistry

Printed at:
Infinity Imaging Systems
Delhi

Preface

It is always an ultimate outlook to give students the best information in a good package. So I took up the job of writing a series of books in chemistry starting from a very introductory level to this advance level of biochemistry.

Biochemistry, to express something about it, in one word, would be difficult for me. It should be ones part of life; the reactions of protein, amino-acids, the DNA RNA—double helix pattern. The nynhydrin reactions are all the wonders of biochemistry—man's anato-biochemical linkages.

Students who want to become doctors or medical practitioners must be a good biochemist first because biochemistry is the theory behind and the medicine is the practical application.

The effects and side effects of the particular medicine, whether it is allergic or non-allergic, to the patients are well understood by the doctors but are taught in biochemistry.

The chemical functions of the different human organs and systems; the kidney, the respiratory system, the blood, the chemical functions of hormones, vitamins, the liver etc are all well covered by this book.

The subject of this book is therefore deeply intended for those upcoming doctors who keep themselves for the real service of the mankind.

The object of this book is not only meant for the doctors but it is the last bangle of that long garland which has started all along from the introduction chapter of The Architecture of Chemistry, The Mans Scientific Attitude.

Biochemistry also comes in use in the field of agriculture, the biochemical configuration of the soil. The biochemical configuration of the plants to be cultivated, the application of the pesticides, and the relation of the pesticides with the human population are well studied under biochemistry.

The Agricultural Scientist, the Pathological Scientist, Forensic Scientist, is definitely inter-joined with each other by a single buffer of the Biochemistry.

This book is dedicated to those promising scientific readers and interested students of biochemistry and advanced chemical science, who wants to make a fruitful attempt in the field of modern scientific world.

The author will be grateful and surely thankful to see the best utilization of this book in any of the field of application.

Sananda Chatterjee

Preface

Contents

Vitamin and Hormones

Introduction

As we have read in this chapter about the vitamins and hormones we shall discuss about the metabolic implementation of those vitamins and hormones in this chapter only.

VITAMINS

As we all know that the vitamins are broadly divided into two groups the fat soluble and water soluble. The table below shows them properly.

	Vitamin	Biological Name	Chemical Name	Fat-soluble or water-soluble
1.	A	Antixerophthalmic	Axerophthol	Fat-Soluble
2.	D	Antirachitic	Calciferol	
3.	E	Antisterility	IX-Tocopherol	
4.	K	Antihaemorrhagic	Phylloquinone	"
5.	C	Antiscorbutic -	L-Ascorbic Acid	Water-Soluble
6.				

Vitamin B. Complex

(i)	BI	Antiberiberi/anti-neuritic factor	AneurinThiamine	Water-Soluble	
(ii)	B2	Growth factor	Riboflavin: Lacto, Ovo, Verdo-orhepatoflavin	"	"

(iii)	Niacin	Antiperllagra factoror anti-black-tonguefactor	Nicotinic acid	"	"
(iv)	B6	Anti dermatitis factoror Anti-acrodynia factor	Pyridoxine	"	"
(v)	Pantothenic Acid (B_5)	Chick anti dermatitis factor or anti-arcodynia factor	Pantothenic acid	"	"
(vi)	H	Skin factor Anti-egg-white injury factor	Biotin	"	"
(vii)	Folic Acid (B_9)	Anti-anaemia factor	Folic acid (petroylglutamic acid)	"	"
(viii)	Inositol	Mouse anti-alopecia factor	Inositol	"	"
(ix)	PABA	Chromotrichia factor Anti-grey-hair factor	Para-aminobenzoic acid	"	"
(x)	B12	Anti-perniciousanæmia factor	Rubramin; cobione; cyano-cobalamin		
(xi)	APF	Animal Protein factor			

The table below gives the information about the diseases caused by the deficiency of these vitamins:

Vitamins	Symptoms of deficiency	Sources
1	2	3
vitamin A	Susceptibility to infection; poor vision in twilight, Retarded growth	Milk, butter, eggs, liver, margarine, tomatoes, carrots, fish liver oils
vitamin B1 thiamin	Loss of appetite; nerve disorders; fatigue; poor digestion Retarded growth	Meat, especially pork, wholemeal bread and cereals, milk, vegetables
vitamin B2 riboflavine	Sores at corners of the mouth; other skin and membrane disorders	Meat, milk, green vegetables, eggs, poultry
vitamin C	Slow healing; tendency to bruise and bleed easily; sore gums Scurvy	Many raw fruits, especially the citrus group, and vegetables, tomatoes, melon

1	2	3
vitamin D	Poor bone and tooth development; dental decay; Rickets	Fish liver oils, liver, fortified milk and baby cereals, irradiated margarines, etc., sunshine
vitamin E	Infertility	Seed germs, green vegetables
vitamin K	Faulty clotting of the blood Bleeding	Meat, poultry, fish, potatoes, peanuts; whole grain cereals

We shall now discuss in detail the natural availability sources and normal proportion of vitamins must be present in the human body in order to function properly.

Recommended Dietary Allowances (RDA) and Adequate Intakes (AI) for Vitamine

Age (yr)	Thiamin RDA (mg/day)	Riboflavin RDA (mg/day)	Niacin RDA (mg/day)[a]	Biotin AI (µg/day)	Pantothenic aicd AI (mg/cay)	Vitamin B_6 RDA (mg/day)	Folate RDA (mg/day)[b]	Vitamin B_{12} RDA (mg/day)	Choline AI (mg/day)	Vitamin C RDA (mg/day)	Vitamin A RDA (mg/day)[e]	Vitamin D AI (mg/day)[3]	Vitamin E RDA (mg/day)[e]	Vitamin K AI (mg/day)
1	2	3	4	5	6	7	8	9	10	11	12	13	14	15
Infats														
0-0.5	0.2	0.3	2	5	1.7	0.1	65	0.4	125	40	400	5	4	2.0
0.5-1	0.3	0.4	4	6	1.8	0.3	80	0.5	150	50	500	5	5	2.5
Children														
1-3	0.5	0.5	6	8	2	0.5	150	0.9	200	15	300	5	6	30
4-8	0.6	0.6	8	12	3	0.6	200	1.2	250	25	400	5	7	55
Males														
9-13	0.9	0.9	12	20	4	1.0	300	1.8	375	45	600	5	11	60
14-18	1.2	1.3	16	25	5	1.3	400	2.4	550	75	900	5	15	75
19-30	1.2	1.3	16	30	5	1.3	400	2.4	550	90	900	5	15	120
31-50	1.2	1.3	16	30	5	1.3	400	2.4	550	90	900	5	15	120
51-70	1.2	1.3	16	30	5	1.3	400	2.4	550	90	900	5	15	120
>70	1.2	1.3	16	30	5	1.3	400	2.4	550	90	900	5	15	120
Females														
9-13	0.9	0.9	12	20	4	1.0	300	1.8	375	45	600	5	11	60
14-18	1.0	1.0	14	25	5	1.2	400	2.4	400	65	700	5	15	75
19-30	1.1	1.1	14	30	5	1.3	400	2.4	425	75	700	5	15	90

1	2	3	4	5	6	7	8	9	10	11	12	13	14	15
31-50	1.1	1.1	14	30	5	1.3	400	2.4	425	75	700	5	15	90
51-70	1.1	1.1	14	30	5	1.3	400	2.4	425	75	700	5	15	90
>70	1.1	1.1	14	30	5	1.3	400	2.4	425	75	700	5	15	90
Pregnancy														
<18	1.4	1.4	18	30	6	1.9	600	2.6	450	80	750	5	15	75
19-30	1.4	1.4	18	30	6	1.9	600	2.6	450	80	750	5	15	75
31-50	1.4	1.4	18	30	6	1.9	600	2.6	450	80	750	5	15	75
Lactation														
<18	1.4	1.6	17	35	7	2.0	500	2.8	550	115	1200	5	19	75
19-30	1.4	1.6	17	35	7	2.0	500	2.8	550	120	1300	5	19	90
31-50	1.4	1.6	17	35	7	2.0	500	2.8	550	120	1300	5	19	90

We shall now discuss the detail functions of these vitamins in human metabolism.

VITAMIN A

The discovery of vitamin A may have stemmed from research dating back to 1906, indicating that factors other than carbohydrates, proteins, and fats were necessary to keep cattle healthy. By 1917 one of these substances was independently discovered by Elmer McCollum at the University of Wisconsin-Madison, and Lafayette Mendel and Thomas Burr Osborne at Yale University. Since "water-soluble factor B" (Vitamin B) had recently been discovered, the researchers chose the name "fat-soluble factor A" (vitamin A). Vitamin A was first synthesized in 1947 by two Dutch chemists, David Adriaan van Dorp and Jozef Ferdinand Arens.

Vitamin A is a vitamin that is needed by the retina of the eye in the form of a specific metabolite, the light-absorbing molecule retinal. This molecule is absolutely necessary for both scotopic and color vision. Vitamin A also functions in a very different role, as an irreversibly oxidized form retinoic acid, which is an important hormone-like growth factor for epithelial and other cells.

In foods of animal origin, the major form of vitamin A is an ester, primarily retinyl palmitate, which is converted to an alcohol (retinol) in the small intestine. The retinol form functions as a storage form of the vitamin, and can be converted to and from its visually active aldehyde form, retinal. The associated acid (retinoic acid), a metabolite that can be irreversibly synthesized from vitamin A, has only partial vitamin A activity, and does not function in the retina or some essential parts of the reproductive system.

All forms of vitamin A have a beta-ionone ring to which an isoprenoid chain is attached, called a *retinyl group*. This structure is essential for vitamin activity. The orange pigment of carrots - beta-carotene - can be represented as two connected retinyl groups, which are used in the body to contribute to vitamin A levels. Alpha-carotene and gamma-carotene also have a single retinyl group, which give them some vitamin activity. None of the other carotenes have vitamin activity. The carotenoid beta-cryptoxanthin possesses an ionone group and has vitamin activity in humans.

Vitamin A can be found in two principal forms in foods:

- retinol, the form of vitamin A absorbed when eating animal food sources, is a yellow, fat-soluble substance. Since the pure alcohol form is unstable, the vitamin is found in tissues in a form of retinyl ester. It is also commercially produced and administered as esters such as retinyl acetate or palmitate.
- The carotenes alpha-carotene, beta-carotene, gamma-carotene; and the xanthophyll beta-cryptoxanthin (all of which contain beta-ionone rings), but no other carotenoids, function as vitamin A in herbivores and omnivore animals, which possess the enzyme required to convert these compounds to retinal. In general, carnivores are poor converters of ionine-containg carotenoids, and pure carnivores such as cats and ferets lack beta-carotene 15,15'-monooxygenase and cannot convert any carotenoids to retinal (resulting in *none* of the carotenoids being forms of vitamin A for these species).

METABOLIC FUNCTIONS

Vitamin A plays a role in a variety of functions throughout the body, such as:

- Vision the anatobiochemsitry
- Gene transcription
- Immune function
- Embryonic development and reproduction
- Bone metabolism
- Hæmatopoiesis
- Skin health
- Antioxidant activity

VISION THE ANATOBIOCHEMISTRY

The role of vitamin A in the vision cycle is specifically related to the retinal form. Within the eye, 11-*cis*-retinal is bound to rhodopsin (rods) and iodopsin (cones) at conserved lysine residues. As light enters the eye, the 11-*cis*-retinal is isomerized to the all-"trans" form. The all-"trans" retinal dissociates from the opsin in a series of steps called bleaching. This isomerization induces a nervous signal along the optic nerve to the visual center of the brain. Upon completion of this cycle, the all-"trans"-retinal can be recycled and converted back to the 11-"cis"-retinal form via a series of enzymatic reactions. In addition, some of the all-"trans" retinal may be converted to all-"trans" retinol form and then transported with an interphotoreceptor retinol-binding protein (IRBP) to the pigment epithelial cells. Further esterification into all-"trans" retinyl esters allow this final form to be stored within the pigment epithelial cells to be reused when needed. The final conversion of 11-*cis*-retinal will rebind to opsin to reform rhodopsin in the retina. Rhodopsin is needed to see black and white as well as see at night. It is for this reason that a deficiency in vitamin A will inhibit the reformation of rhodopsin and lead to night blindness.

The Structure of the Eye

The area within the eye that detects light and colour is the called the Retina. The two types of detection cell present, rods and cones, process information coming through the Lens and send it down the optic nerve to the brain. Rod cells (of which there are around 100 million) detect the degree of lightness entering the eye and their sensitivity is dependent on the amount of Rhodopsin present which is itself generated within the cells. However, Rhodopsin is destroyed by bleaching on exposure to light and therefore rod cells only work in low light as at high illumination the reduced level of this photosensitive pigment leads to a very low sensitivity. Cone cells (of which there are around three million) are also sensitive to light levels but retain their function upto high illumination via use of the pigment Iodopsin. Detection of colour is a function of the three types of cone cell present within the retina: between them they cover the visible spectrum. This is because each type is sensitive to a different range of wavelengths with maximums corresponding to red (long), green (medium) or blue (short).

Rod Cells, or Rods

are photoreceptor cells in the retina of the eye that can function in less intense light than can the other type of photoreceptor, cone cells. Named for their cylindrical shape, rods are concentrated at the outer edges of the retina and are used in peripheral vision. On average, there are approximately 92 million rod cells in the human retina. More sensitive than cone cells, rod cells are almost entirely responsible for night vision.

Structure and Function

Rods are a little narrower than cones but have the same structural basis. The pigment is on the outer side, lying on the pigment epithelium. This end contains many stacked disks, probably from the folding inward of the limiting membrane surrounding this section. Rods disks have three layers separated by two 3.5 nm spaces; the outer layers are 2 nm, and the middle is 4 nm. The disks are 10-20 nm apart. Frogs are believed to have 1,000 disks in their outer segment. Rods have a high area for visual pigment and thus substantial efficiency of light absorption. Because they have only one type of light-sensitive pigment, rather than the three types that human cone cells have, rods have little, if any, role in color vision.Like cones, rod cells have a synaptic terminal, an inner segment, and an outer segment. The synaptic terminal forms a synapse with another neuron, for example a bipolar cell. The inner and outer segments are connected by a cilium, which lines the distal segment. The inner segment contains organelles and the cell's nucleus, while the rod outer segment (abbreviated to ROS), which is pointed toward the back of the eye, contains the light-absorbing materials.

Sensitivity

A rod cell is sensitive enough to respond to a single photon of light, and is about 100 times more sensitive to a single photon than cones. Rods require less light to function than cones, they are therefore the primary source of visual information at night (scotopic vision). Cone cells, on the other hand, require tens to hundreds of photons to become activated. Additionally, multiple rod cells converge on a single interneuron, collecting and amplifying the signals. However, this convergence comes at a cost to visual acuity (or image resolution) because the pooled information from multiple cells is less distinct than it would be if the visual system received information from each rc 1 cell individually. The convergence of rod cells also tends to make peripheral vision very sensitive to movement, and is responsible for the phenomenon of an individual seeing something vague occur out of the corner of his or her eye. Rod cells also respond more slowly to light than cones do, so stimuli they receive are added over about 100 milliseconds. While this makes rods more sensitive to smaller amounts of light, it also means that their ability to sense temporal-changes, such as quickly changing images, is less accurate than that of cones. Experiments by George Wald and others showed that rods are most sensitive to wavelengths of light around 498 nm (green-blue), and are completely insensitive to wavelengths longer than about 640 nm (red). This fact is responsible for the Purkinje effect, in which blue colors appear more intense relative to reds at twilight, when rods take over as the cells responsible for vision.

Response to Light

In vertebrates, activation of a photoreceptor cell is actually a hyperpolarization (inhibition) of the cell. When they are not being stimulated, such as in the dark, rod cells and cone cells depolarize and release a neurotransmitter spontaneously. This neurotransmitter hyperpolarizes the bipolar cell. Bipolar cells exist between photoreceptors and ganglion cells and act to transmit signals from the photoreceptors to the ganglion cells. As a result of the bipolar cell being hyperpolarized, it does not release its transmitter at the bipolar-ganglion synapse and the synapse is not excited.

Activation of photopigments by light sends a signal by hyperpolarizing the rod cell, leading to the rod cell not sending its neurotransmitter, which leads to the bipolar cell then releasing its transmitter at the bipolar-ganglion synapse and exciting the synapse.

Deplorization of Rod Cells

Depolarization of rod cells (causing release of their neurotransmitter) occurs because in the dark, cells have a relatively high concentration of cyclic guanosine 3'-5' monophosphate (cGMP), which opens ion channels (largely sodium channels, though calcium can enter through these channels as well). The positive charges of the ions that enter the cell down its electrochemical gradient change the cell's membrane potential, cause depolarization, and lead to the release of the neurotransmitter glutamate. Glutamate can depolarize some neurons and hyperpolarize others, allowing photoreceptors to interact in an antagonistic manner.When light hits photoreceptive pigments within the photoreceptor cell, the pigment changes shape. The pigment, called rhodopsin (photopsin is found in cone cells) comprises a large protein called opsin (situated in the plasma membrane), attached to which is a covalently-bound prosthetic group: an organic molecule called retinal (a derivative of vitamin A). The retinal exists in the 11-*cis*-retinal form when in the dark, and stimulation by light causes its structure to change to all-trans-retinal. This structural change causes a series of changes in the opsin that ultimately lead it to activate a regulatory protein called transducin (a type of G protein), which leads to the activation of cGMP phosphodiesterase, which breaks cGMP down into 5'-GMP. Reduction in cGMP allows the ion channels to close, preventing the influx of positive ions, hyperpolarizing the cell, and stopping the release of neurotransmitters. Though cone cells primarily use the neurotransmitter substance acetylcholine, rod cells use a variety. The entire process by which light initiates a sensory response is called visual phototransduction.Activation of a single unit of rhodopsin, the photosensitive pigment in rods, can lead to a large reaction in the cell because the signal is amplified. Once activated, rhodopsin can activate hundreds of transducin molecules, each of which in turn activates a phosphodiesterase molecule, which can break down over a thousand cGMP molecules per second. Thus, rods can have a large response to a small amount of light.As the retinal component of rhodopsin is derived from vitamin A, a deficiency of vitamin A causes a deficit in the pigment needed by rod cells. Consequently, fewer rod cells are able to sufficiently respond in darker conditions, and as the cone cells are poorly adapted for sight in the dark, blindness can result. This is night-blindness.

Revert to the Resting State

Rods make use of three inhibitory mechanisms (negative feedback mechanisms) to allow a rapid revert to the resting state after a flash of light.

Firstly, there exists a rhodopsin kinase (RK) which would phosphorylate the cytosolic tail of the activated rhodopsin on the multiple serines, partially inhibiting the activation of transducin. Also, an inhibitory protein - arrestin then binds to the phosphorylated rhodopsins to further inhibit the rhodopsin's activity.

While arrestin shuts off rhodopsin, an RGS protein [functioning as a GTPase-activating proteins(GAPs)] drives the transducin (G-protein) into an "off" state by increasing the rate of hydrolysis of the bounded GTP to GDP. Also as the cGMP sensitive channels allow not only the influx of sodium ions, but also calcium ions, with the decrease in concentration of cGMP, cGMP sensitive channels are then closed and reducing the normal influx of calcium ions. The decrease in the concentration of calcium ions stimulates the calcium ion-sensitive proteins, which would then activate the guanylyl cyclase to replenish the cGMP, rapidly restoring its original concentration. The restoration opens the cGMP sensitive channels and causes a depolarization of the plasma membrane.

Desensitization

When the rods are exposed to a high concentration of photons for a prolonged period, they become desensitized (adapted) to the environment.

As rhodopsin is phosphorylated by rhodopsin kinase [a member of the GPCR kinases (GRKs)], it binds with high affinity to the arrestin. The bound arrestin can contribute to the desensitization process in at least two ways. First, it prevents the interaction between the G protein and the activated receptor. Second, it serves as an adaptor protein to aid the receptor to the clathrin-dependent endocytosis machinery (to induce receptor-mediated endocytosis).

Cone Cells

Cone cells, or cones, are photoreceptor cells in the retina of the eye that function best in relatively bright light. The cone cells gradually become sparser towards the periphery of the retina.

A commonly cited figure of six million in the human eye was found by Osterberg in 1935. Researchers indicating an average closer to 4.5 million cone cells and 90 million rod cells in the human retina.

Cones are less sensitive to light than the rod cells in the retina (which support vision at low light levels), but allow the perception of color. They are also able to perceive finer detail and more rapid changes in images, because their response times to stimuli are faster than those of rods. Because humans usually have three kinds of cones with different photopsins, which have different response curves and thus respond to variation in color in different ways, they have trichromatic vision. Being color blind can change this, and there have been reports of people with four or more types of cones, giving them tetrachromatic vision.

Types

Humans normally have three kinds of cones. The first responds most to light of long wavelengths, peaking in the yellow region; this type is designated L for long. The second type responds most to light of medium-wavelength, peaking at green, and is abbreviated M for medium. The third type responds most to short-wavelength light, of a violet color, and is designated S for short. The three types have peak wavelengths near 564-580 nm, 534-545 nm, and 420-440 nm, respectively. The difference in the signals received from the three cone types allows the brain to perceive all possible colors, through the opponent process of color vision. The color yellow, for example, is perceived when the L cones are stimulated slightly more than the M cones, and the color red is perceived when the L cones are stimulated significantly more than the M cones. Similarly, blue and violet hues are perceived when the S receptor is stimulated more than the other two. The S cones are most sensitive to light at wavelengths around 420 nm. However, the lens and cornea of the human eye are increasingly absorptive to smaller wavelengths, and this sets the lower wavelength limit of human-visible light to approximately 380 nm, which is therefore called 'ultraviolet' light. People with aphakia, a condition where the eye lacks a lens, sometimes report the ability to see into the ultraviolet range. At moderate to bright light levels where the cones function, the eye is more sensitive to yellowish-green light than other colors because this stimulates the two most common of the three kinds of cones almost equally. At lower light levels, where only the rod cells function, the sensitivity is greatest at a blueish-green wavelength.

Structure

Cone cells are somewhat shorter than rods, but wider and tapered, and are much less numerous than rods in most parts of the retina, but greatly outnumber rods in the fovea. Structurally, cone cells have a cone-like shape at one end where a pigment filters incoming light, giving them their different response curves. They are typically 40-50 µm long, and their diameter varies from .50 to 4.0 µm, being smallest and most tightly packed at the center of the eye at the fovea. The S cones are a little larger than the others.

Photobleaching can be used to determine cone arrangement. This is done by exposing dark-adapted retina to a certain wavelength of light that paralyzes the particular type of cone sensitive to that wavelength for up to thirty minutes from being able to dark-adapt making it appear white in contrast to the grey dark-adapted cones when a picture of the retina is taken. The results illustrate that **S** cones are randomly placed and appear much less frequently than the **M** and **L** cones. The ratio of **M** and **L** cones varies greatly among different people with regular vision (e.g. values of 75.8% **L** with 20.0% **M** versus 50.6% **L** with 44.2% **M** in two male subjects). Like rods, each cone cell has a synaptic terminal, an inner segment, and an outer segment as well as an interior nucleus and various mitochondria. The synaptic terminal forms a synapse with a neuron such as a bipolar cell. The inner and outer segments are connected by a cilium. The inner segment contains organelles and the cell's nucleus, while the outer segment, which is pointed toward the back of the eye, contains the light-absorbing materials. Like rods, the outer segments of cones have invaginations of their cell membranes that create stacks of membranous disks. Photopigments exist as transmembrane proteins within these disks, which provide more surface area for light to affect the pigments. In cones, these disks are attached to the outer membrane, whereas they are pinched off and exist separately in rods. Neither rods nor cones divide, but their membranous disks wear out and are worn off at the end of the outer segment, to be consumed and recycled by phagocytic cells.

The response of cone cells to light is also directionally nonuniform, peaking at a direction that receives light from the center of the pupil; this effect is known as the Stiles-Crawford effect.

THE RHODOPSIN

Rhodopsin, also known as visual purple, is a pigment of the retina that is responsible for both the formation of the photoreceptor cells and the first events in the perception of light. Rhodopsins belong to the G-protein coupled receptor family and are extremely sensitive to light, enabling vision in low-light conditions. Exposed to light, the pigment immediately photobleaches, and it takes about 30 minutes to regenerate fully in humans.

Structure

Rhodopsin consists of the protein moiety opsin and a reversibly covalently bound cofactor, retinal. Opsin, a bundle of seven transmembrane helices, binds retinal, a photoreactive chromophore, in a central pocket. Retinal is produced in the retina from Vitamin A. Isomerization of 11-*cis*-retinal into all-*trans*-retinal by light induces a conformational change in opsin that activates the associated G protein and triggers a second messenger cascade.Rhodopsin of the rods most strongly absorbs green-blue light and therefore appears reddish-purple, which is why it is also called "visual purple". It is responsible for *monochromatic* vision in the dark. Several closely related opsins exist that differ only in a few amino acids and in the wavelengths of light that they absorb most strongly. Humans have four different other opsins beside rhodopsin. The photopsins are found in the different types of the cone cells of the retina and are the basis of color vision. They have absorption maxima for yellowish-green (photopsin I), green (photopsin II), and bluish-violet (photopsin III) light. The remaining opsin (melanopsin) is found in photosensitive ganglion cells and absorbs blue light most strongly.The structure of rhodopsin has been studied in detail via X-ray crystallography on rhodopsin crystals. The photoisomerization dynamics has been investigated with time-resolved IR spectroscopy and UV/Vis spectroscopy. A first photoproduct called photorhodopsin forms within 200 femtoseconds after irradiation followed within picoseconds by a second one called bathorhodopsin with distorted all-trans bonds. This intermediate can be trapped and studied at cryogenic temperatures. Several models (e.g. the *bicycle-pedal mechanism*, *hula-twist mechanism*) attempt to explain how the retinal group can change its conformation without clashing with the enveloping rhodopsin protein pocket.

Recent data supports that it is a functional monomer as opposed to a dimer, which was the paradigm of G-coupled protein receptors for many years.

Rhodopsin and Retinal Disease

Mutation of the rhodopsin gene is a major contributor to various retinopathies such as retinitis pigmentosa. The disease-causing protein generally aggregates with ubiquitin in inclusion bodies, disrupts the intermediate filament network and impairs the ability of the cell to degrade non-functioning proteins which leads to photoreceptor apoptosis. Other mutations on rhodopsin lead to X-linked congenital stationary night blindness, mainly due to constitutive activation, when the mutations occur around the chromophore binding pocket of rhodopsin. Several other pathological states relating to rhodopsin have been discovered including poor post-Golgi trafficking, dysregulative activation, rod outer segment instability and arrestin binding.

THE MECHANISM OF RHODOPSIN WITHIN THE ROD CELLS

The protein Rhodopsin contains the protonated retinal-Schiff's base complex which naturally lies in the inter-membrane pocket formed by the seven trans-membrane a-helical receptors. There are many flat

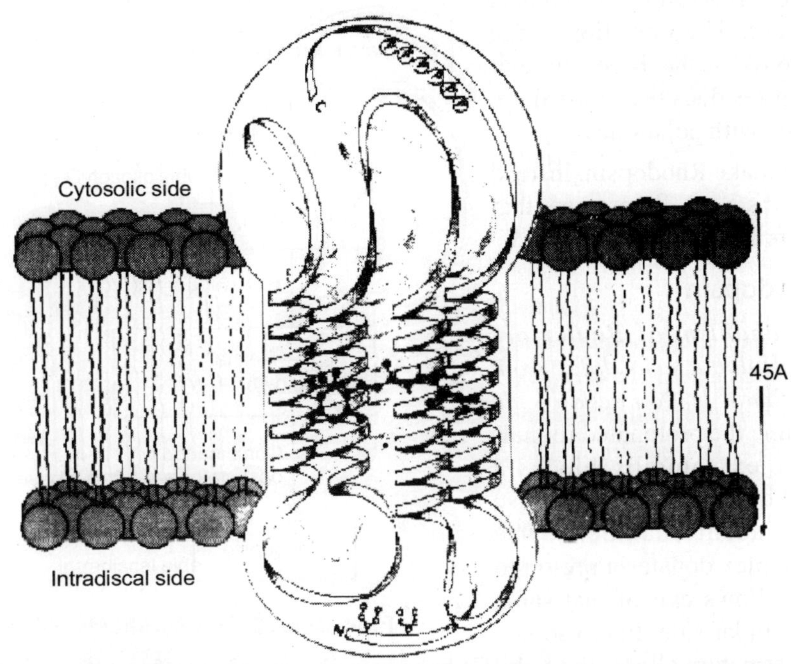

Fig. 12.1 : The Position of 11-*cis* retinal within Rhodopsin

discs of rhodopsin within the outer segment of a rod cell which upon light detection undergo a photo-isomeric change from Rhodopsin (11-*cis*) to all-*trans* retinal. After the photoisomerisation cascade which occurs via 5 shortlived intermediates (flowchart 1), *trans* retinal diffuses away and is converted back into 11-*cis* retinal before re-entry into the cycle. This process occurs via reduction to all-trans retinol followed by oxidation/isomerisation in the dark. Photoexcited Rhodopsin (4th of the 5 intermediates) triggers an enzymatic cascade process resulting in the hydrolysis of GMP. This in turn closes cation-specific channels within the rod cell membrane which are naturally open to influx of Na^+ in the dark, and due to the effect of hyper-polarisation, the inner synatic body sends a nerve signal to other neurons in the Retina. Finally the light-induced lowering of calcium levels aids recovery of excited neurons to a passive, "dark" state and the cycle starts again upon detection of light. The photoreceptors of cone cells are

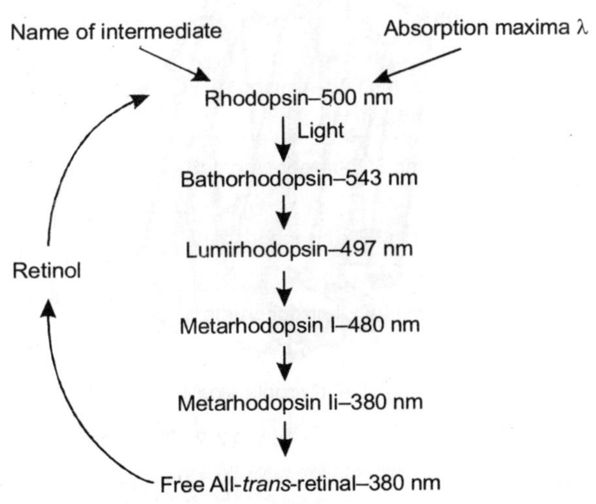

Flor Chart 1: The 5 Intermediates of the Rhodopsin Cycle

also seven a-helical receptors with 11-*cis*-retinal as their chromophore. The detection range varies from green to red as the three nonpolar hydroxyl-containing residues near retinal are sequentially replaced with polar ones.

Humans cannot make Rhodopsin, instead they use and external source, β-carotene, that is found in food in order to synthesis it:

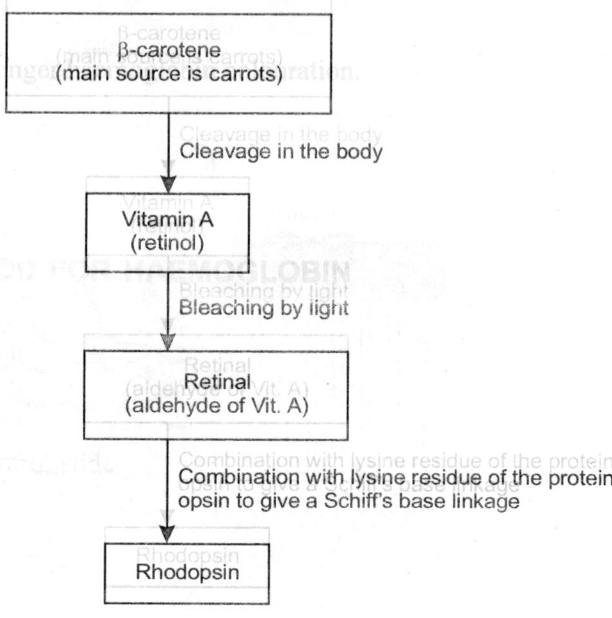

Flow Chart 2: The Synthesis of Rhodopsin

Synthesis of Rhodopsin

How does the mechanism of Bacteriorhodopsin found in Halobacteria differ from that of Rhodopsin found in Rod cells of the human eye?: This time the protonated retinal-Schiff's base complex naturally blocks a channel through the membrane otherwise formed by two adjacent chambers. The protonated trans complex donates a proton to Asp-85 which then allows exit of that same proton to the extracellular side. Photoisomerisation to the 13-*cis* structure allows the Schiff's base to pick up a proton from the Asp-96 residue on the cytosolic side. Upon reorientation of the *cis* form to the *trans*, the cycle of isomerisation and proton pumping continues.

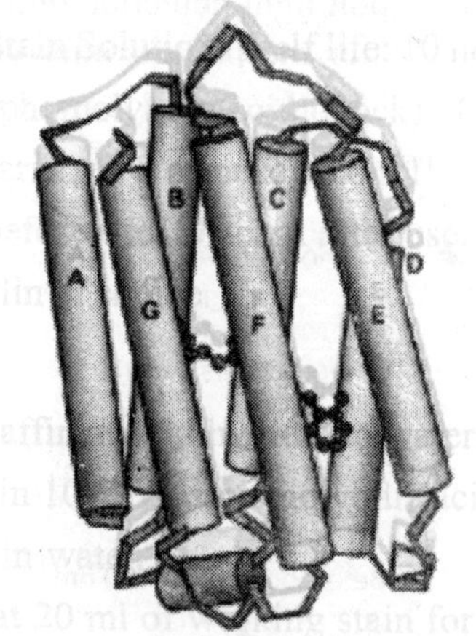

A. Cytosolic side

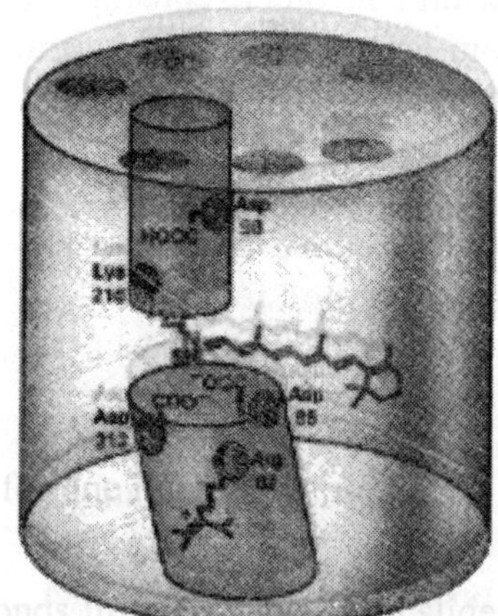

B. Extracellular side

Fig. 12.2 : The Structure of Bacteriodopsin

SUB-STHE RHODOPSIN CYCLE

The visual pigment **rhodopsin** (sometimes called "visual purple") is bound to the plasma membrane of the rod and forms transmembrane protein complexes within it. Rhodopsin undergoes a cyclic decomposition and reconstitution in response to the presence of light. This rather complicated cycle is the basis for absorption of light and its transduction into a nervous signal.

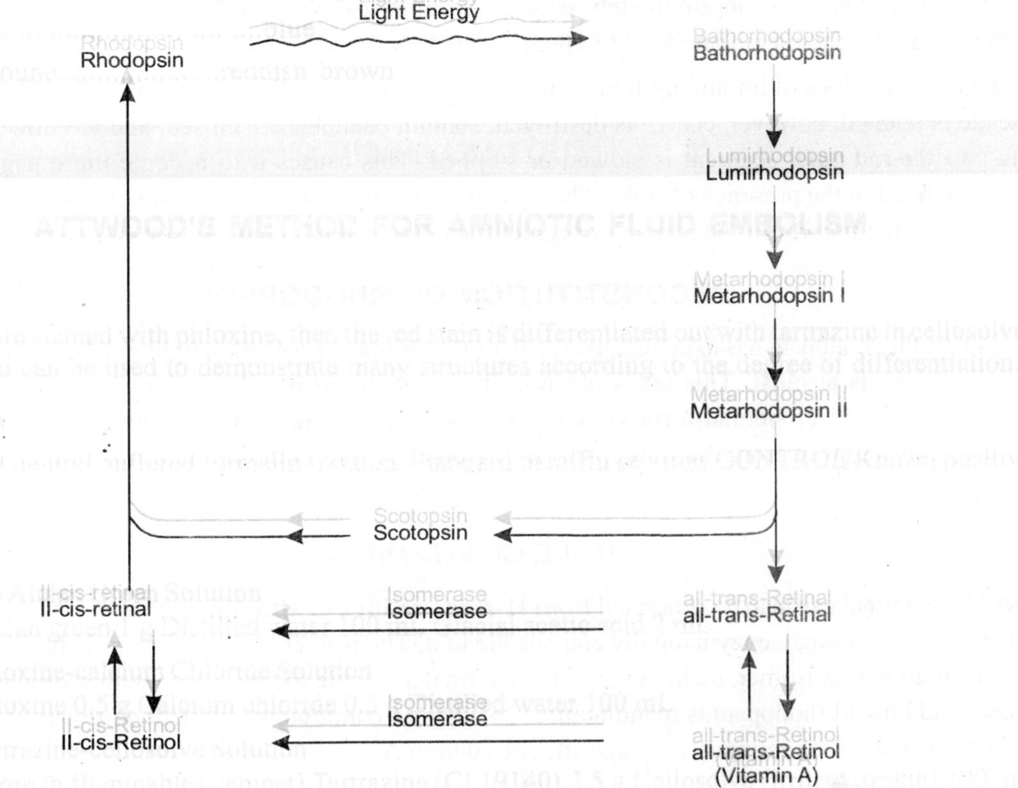

Fig. 12.3 : The Rhodopsin Cycle

Rhodopsin has two components: **scotopsin**, a protein moiety; and **11-*cis*-retinal**, a carotene derivative. When combined, these two subunits create the conjugated rhodopsin molecule.

Energy from impinging light excites the electrons in the 11-cis-retinal subunit and converts it to a different configuration, **11-*trans*-retinal**. Because this is conformationally incompatible with the scotopsin moiety, it begins to detach from it, and the rhodopsin conjugate begins to break up into its component parts.

The disintegration of rhodopsin into retinal and scotopsin is progressive, with a series of short-lived intermediate compounds formed, as shown in the diagram to the right. The eventual result is release of the two components of rhodopsin from each other completely. One of the breakdown products, **metarhodopsin II**, is the agent that ultimately effects the change in the rod membrane's charge.

Metarhodopsin II is an enzyme. It acts to activate a second membrane-bound protein in the rod, **transducin**. Transducin is in its turn an enzyme activating rod-resident **phosphodiesterase**, a third enzyme in the cascade, capable of hydrolyzing cyclic GMP.

Cyclic GMP's role is to keep sodium channels in the membrane of the rod open, so that sodium flux is facilitated. In dark conditions, this is the normal situation: channels are open, sodium flux from the extracellular space is approximately equal to sodium loss via the pump system of the inner rod segment, and the rod membrane is not hyperpolarized.

Under conditions of impinging light, when the metarhodopsin II—transducin—phosphodiesterase cascade is initated, however, cGMP is destroyed, sodium channels are closed, and the flow of sodium ions into the rod outer segment is slowed or stopped. This causes it to become more negative, *i.e.*, hyperpolarized in the presence of light. The cascade is reversed by the presence of **rhodopsin kinase**, another enzyme present in the rod outer segment, and the sodium channels are re-opened.

RECONSTITUTION OF RHODOPSIN

Obviously, the rhodopsin has to be reconstituted, or the ability to respond to light will be lost completely in a few seconds at most. This takes place by two side pathways. First, the 11-trans-retinal is re-converted to the 11-*cis*-retinal form via an isomerase enzyme. Since the scotopsin moiety is present (having been removed from the rhodopsin) it immediately will combine with this to regenerate new rhodopsin.

ROLE OF VITAMIN A

New 11-*cis*-retinal can also be generated from **11-*trans*-retinol**, or **vitamin A**. Vitamin A is a derivative form of 11-*trans*-retinal, enzymatically convertible to it. The isomerase reaction can in turn convert the *trans* form to the *cis* isomer, making new 11-*cis*-retinal available to recombine with scotopsin. By this pathway additional rhodopsin is manufactured to adapt to continuously dark conditions. The pigment epithelium layer of the retina is a storage site for vitamin A.

IMPLICATIONS

The vast amplification of the plasma membrane in the rod outer segment, and the catalytic properties of the enzymes in the cascade from metarhodopsin II—transducin—phosphodiesterase—cGMP breakdown mean that a single photon of light has a great effect on the rod membrane. One photon, the minimum quantity of light possible, will cause the movement of millions of sodium ions, because of the catalytic nature of the enzymes and the large surface area provided for them to work. Hence the rods are extremely sensitive to light. Animals that live in dark environments (such as deep-water fishes and cave-dwelling creatures) always have far more rods than cones, because it is important to them to be able to see in the minimum amount of light. Other adaptations of the eye structure, such as a large pupil diameter, are also important; but the sensitivity of the chemical cascade is the most important factor in determining the ability of an animal to respond to dim light. Adaptation to dark in most animals is a matter of generating more 11-*cis*-retinal from vitamin A, and conjugating it to scotopsin to make more rhodopsin. Similarly, reduction of sensitivity to light means a reduction of the availability of rhodopsin, and hence a conversion of the 11-*cis*-retinal to the inactive trans form, and conversion of trans retinal back to vitamin A, making it unavailable for conjugation to scotopsin. Anything which interferes with the rhodopsin cycle or the enzyme cascade it triggers will obviously affect vision, especially in the dark. An individual on a diet deficient in vitamin A can have reserves so low as to make him incapable of producing enough 11-*cis*-retinal to see effectively in dim light. Drugs or chemicals that affect vitamin A metabolism may lead to similar problems. Hereditary deficiencies in these pathways are also possible sources of this "night blindness."

THE VISUAL PHOTOTRANSDUCTION

It is a process by which light is converted into electrical signals in the rod cells, cone cells and photosensitive ganglion cells of the retina of the eye.

The visual cycle is the biological conversion of a photon into an electrical signal in the retina. This process occurs via G-protein coupled receptors called opsins which contain the chromophore 11-*cis* retinal. 11-cis retinal is covalently linked to the opsin receptor via a Schiff base forming a retinylidene protein. When struck by a photon, 11-cis retinal undergoes photoisomerization to all-trans retinal which changes the conformation of the opsin GPCR leading to signal transduction cascades which causes closure of a cyclic GMP-gated cation channel, and hyperpolarization of the photoreceptor cell.

Following isomerization and release from the opsin protein, all-*trans* retinal is reduced to all *trans* retinol and travels back to the retinal pigment epithelium to be "recharged". It is first esterified by lecithin-retinol acyltransferase (LRAT) and then converted to 11-*cis* retinol by the isomerohydrolase RPE65. The isomerase activity of RPE65 has been shown, but it is still uncertain whether it also acts as a hydrolase. Finally, it is oxidized to 11-*cis* retinal before traveling back to the rod outer segment where it can again be conjugated to an opsin to form a new, functional visual pigment (rhodopsin).

Photoreceptors

The process of phototransduction is a complicated one, and in order to understand it, one must have an understanding of the structure of the photoreceptor cells involved in vision: the rods and cones. These cells contain a chromophore (11-*cis*-retinal, the aldehyde of Vitamin A1 and light-absorbing portion) bound to a cell membrane protein, opsin. Rods deal with low light level and do not mediate color vision. Cones, on the other hand, can code the color of an image through comparison of the outputs of the three different types of cones. Each cone type responds best to certain wavelengths, or colors, of light because each type has a slightly different opsin. The three types of cones are L-cones, M-cones and S-cones that respond optimally to long wavelengths (reddish color), medium wavelengths (greenish color), and short wavelengths (bluish colour) respectively.

Process

To understand the photoreceptor's behaviour to light intensities, it is necessary to understand the roles of different currents. There is an ongoing outward potassium current through nongated K^+-selective channels. This outward current tends to hyperpolarize the photoreceptor at around -70 mV (the equilibrium potential for K^+). There is also an inward sodium current carried by cGMP-gated sodium channels. This so-called 'dark current' depolarizes the cell to around -40 mV. Note that this is significantly more depolarized than most other neurons. A high density of Na^+-K^+ pumps enables the photoreceptor to maintain a steady intracellular concentration of Na^+ and K^+.

In the Dark

Photoreceptor cells are strange cells because they are depolarized in the dark, i.e. light hyperpolarizes and switches off these cells, and it is this 'switching off' that activates the next cell and sends an excitatory signal down the neural pathway. In the dark, cGMP levels are high and keep cGMP-gated sodium channels open allowing a steady inward current, called the dark current. This dark current keeps the cell depolarized at about +40 mV.

The depolarization of the cell membrane opens voltage-gated calcium channels. An increased intracellular concentration of Ca^{2+} causes vesicles containing special chemicals, called neurotransmitters, to merge with the cell membrane, therefore releasing the neurotransmitter into the synaptic cleft, an area between the end of one cell and the beginning of another neuron. The neurotransmitter released is glutamate, a neurotransmitter whose receptors are often excitatory.

In the cone pathway glutamate:

- *Hyperpolarizes on-centre bipolar cells*. Glutamate that is released from the photoreceptors in the dark binds to metabotropic glutamate receptors (mGluR6), which, through a G-protein coupling mechanism, causes non-specific cation channels in the cells to close, thus hyperpolarizing the bipolar cell.
- *Depolarizes off-centre bipolar cells.* Binding of glutamate to ionotropic glutamate receptors results in an inward cation current that depolarizes the bipolar cell.

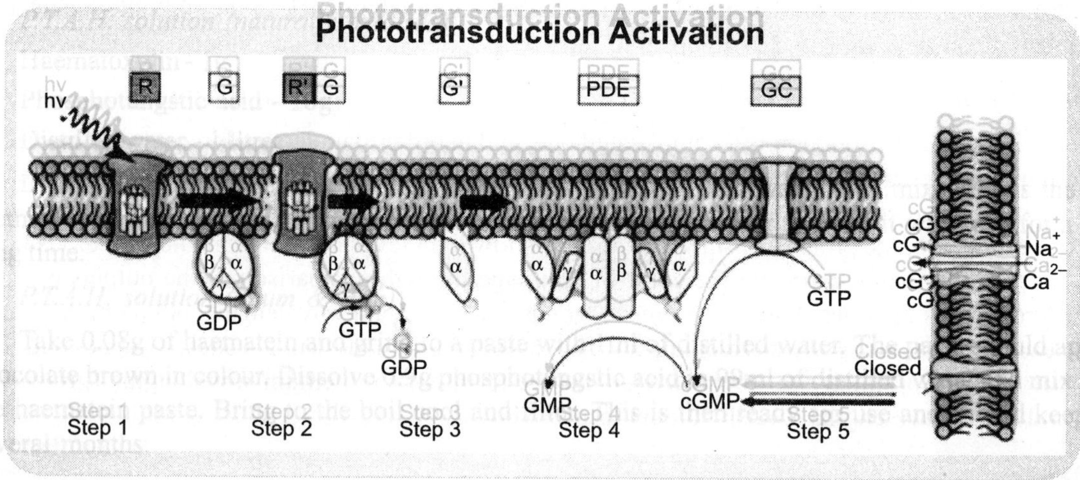

Fig 12.4: The Representation of Molecular Steps in Photoactivation. Depicted is an outer membrane disk in a rod. Step 1: Incident photon (hi) is absorbed and activates a rhodopsin by conformational change in the disk membrane to R*. Step 2: Next, R* makes repeated contacts with transducin molecules, catalyzing its activation to G* by the release of bound GDP in exchange for cytoplasmic GTP. The α and γ subunits Step 3: G* binds inhibitory γ subunits of the phosphodiesterase (PDE) activating its α and β subunits. Step 4: Activated PDE hydrolyzes cGMP. Step 5: Guanylyl cyclase (GC) synthesizes cGMP, the second messenger in the phototransduction cascade. Reduced levels of cytosolic cGMP cause cyclic nucleotide gated channels to close preventing further influx of Na^+ and Ca_2^+.

In the Light

1. A light photon interacts with the retinal in a photoreceptor. The retinal undergoes isomerisation, changing from the 11-*cis* to all-*trans* configuration.
2. Retinal no longer fits into the opsin binding site.
3. Opsin therefore undergoes a conformational change to metarhodopsin II.
4. Metarhodopsin II is unstable and splits, yielding opsin and all-*trans* retinal.
5. The opsin activates the regulatory protein transducin. This causes transducin to dissociate from its bound GDP, and bind GTP, then the alpha subunit of transducin dissociates from the beta and gamma subunits, with the GTP still bound to the alpha subunit.
6. The alpha subunit-GTP complex activates phosphodiesterase.
7. Phosphodiesterase breaks down cGMP to 5'-GMP. This lowers the concentration of cGMP and therefore the sodium channels close.
8. Closure of the sodium channels causes hyperpolarization of the cell due to the ongoing potassium current.
9. Hyperpolarization of the cell causes voltage-gated calcium channels to close.
10. As the calcium level in the photoreceptor cell drops, the amount of the neurotransmitter glutamate that is released by the cell also drops. This is because calcium is required for the glutamate-containing vesicles to fuse with cell membrane and release their contents.
11. A decrease in the amount of glutamate released by the photoreceptors causes depolarization of On center bipolar cells (rod and cone On bipolar cells) and hyperpolarization of cone Off bipolar cells.

Deactivation of the Phototransduction Cascade

GTPase Activating Protein (GAP) interacts with the alpha subunit of transducin, and causes it to hydrolyse its bound GTP to GDP, and thus halts the action of phosphodiesterase, stopping the transformation of cGMP to GMP.

Guanylate Cyclase Activating Protein (GCAP) is a calcium binding protein, and as the calcium levels in the cell have decreased, GCAP dissociates from its bound calcium ions, and interacts with Guanylate Cyclase, activating it. Guanylate Cyclase then proceeds to transform GTP to cGMP, replenishing the cell's cGMP levels and thus reopening the sodium channels that were closed during phototransduction.

Finally, Metarhodopsin II is deactivated. Recoverin, another calcium binding protein, is normally bound to Rhodopsin Kinase when calcium is present. When the calcium levels fall during phototransduction, the calcium dissociates from recoverin, and rhodopsin kinase is released, when it proceeds to phosphorylate metarhodopsin II, which decreases its affinity for transducin. Finally, arrestin, another protein, binds the phosphorylated metarhodopsin II, completely deactivating it. Thus, finally, phototransduction is deactivated, and the dark current and glutamate release is restored. It is this pathway, where Metarhodopsin II is phosphorylated and bound to arrestin and thus deactivated, which is thought to be responsible for the S2 component of dark adaptation. The S2 component represents a linear section of the dark adaptation function present at the beginning of dark adaptation for all bleaching intensities.

All-*trans* retinal is transported to the pigment epithelial cells to be reduced to all-*trans* retinol, the precursor to 11-*cis* retinal. This is then transported back to the rods. All-*trans* retinol cannot be synthesised by humans and must be supplied by vitamin A in the diet. Deficiency of all-*trans* retinol can lead to night blindness. This is part of the bleach and recycle process of retinoids in the photoreceptors and retinal pigment epithelium.

VITAMIN B

The **B vitamins** are eight water-soluble vitamins that play important roles in cell metabolism. The B vitamins were once thought to be a single vitamin, referred to as **vitamin B** (much as people refer to vitamin C or vitamin D). Later research showed that they are chemically distinct vitamins that often coexist in the same foods. In general, supplements containing all eight are referred to as a **vitamin B complex**. Individual B vitamin supplements are referred to by the specific name of each vitamin (e.g., B_1, B_2, B_3 etc.).

LIST OF B VITAMINS

- **Vitamin B_1** (thiamine)
- **Vitamin B_2** (riboflavin)
- **Vitamin B_3** (niacin or niacinamide, sometimes also known as vitamin PP)
- **Vitamin B_5** (pantothenic acid)
- **Vitamin B_6** (pyridoxine, pyridoxal, or pyridoxamine, or pyridoxine hydrochloride)
- **Vitamin B_7** (biotin)
- **Vitamin B_9** (folic acid)
- **Vitamin B_{12}** (various cobalamins; commonly cyanocobalamin in vitamin supplements)

HEALTH BENEFITS

The B vitamins are necessary in order to:

- Support and increase the rate of metabolism
- Maintain healthy skin and muscle tone
- Enhance immune and nervous system function
- Promote cell growth and division, including that of the red blood cells that help prevent anemia
- Reduce the risk of pancreatic cancer, one of the most lethal forms of cancer, when consumed in food, but not when ingested in vitamin tablet form.

All B vitamins are water-soluble, and are dispersed throughout the body. Most of the B vitamins must be replenished regularly, since any excess is excreted in the urine.

B-VITAMIN DEFICIENCY

Several named vitamin deficiency diseases may result from the lack of sufficient B-vitamins. Deficiencies of other B vitamins result in symptoms that are not part of a named deficiency disease.

Vitamin	Name	Deficiency effects
1	2	3
Vitamin B_1	thiamine	Deficiency causes beriberi. Symptoms of this disease of the nervous system include weight loss, emotional disturbances, Wernicke's encephalopathy (impaired sensory perception), weakness and pain in the limbs, periods of irregular heartbeat, and edema (swelling of bodily tissues). Heart failure and death may occur in advanced cases. Chronic thiamine deficiency can also cause Korsakoff's syndrome, an irreversible psychosis characterized by amnesia and confabulation.
Vitamin B_2	riboflavin	Deficiency causes ariboflavinosis. Symptoms may include cheilosis (cracks in the lips), high sensitivity to sunlight, angular cheilitis, glossitis (inflammation of the tongue), seborrheic dermatitis or pseudo-syphilis (particularly affecting the scrotum or labia majora and the mouth), pharyngitis, hyperemia, and edema of the pharyngeal and oral mucosa.
Vitamin B_3	niacin	Deficiency, along with a deficiency of tryptophan causes pellagra. Symptoms include aggression, dermatitis, insomnia, weakness, mental confusion, and diarrhea. In advanced cases, pellagra may lead to dementia and death.
Vitamin B_5	pantothenic acid	Deficiency can result in acne and paresthesia, although it is uncommon.
Vitamin B_6	pyridoxine	Deficiency may lead to microcytic anemia (because pyridoxyl phosphate is the cofactor for heme synthesis), depression, dermatitis, high blood pressure (hypertension), water retention, and elevated levels of homocysteine.
Vitamin B_7	biotin	Deficiency does not typically cause symptoms in adults but may lead to impaired growth and neurological disorders in infants. Multiple carboxylase deficiency, an inborn error of metabolism, can lead to biotin deficiency even when dietary biotin intake is normal.
Vitamin B_9	folic acid	Deficiency results in a macrocytic anemia, and elevated levels of homocysteine. Deficiency in pregnant women can lead to birth defects. Supplementation is often recommended during pregnancy. Researchers have shown that folic acid might also slow the insidious effects of age on the brain.

1	2	3
Vitamin B_{12}	cobalamin	Deficiency results in a macrocytic anemia, elevated homocysteine, peripheral neuropathy, memory loss and other cognitive deficits. It is most likely to occur among elderly people, as absorption through the gut declines with age; the autoimmune disease pernicious anemia is another common cause. It can also cause symptoms of mania and psychosis. In rare extreme cases, paralysis can result.

B-VITAMIN TOXICITY

Although most B vitamins are eliminated regularly in the urine, taking large doses of certain B vitamins may produce harmful effects.

Vitamin	Name	Tolerable Upper Intake Level	Harmful effects
Vitamin B_1	thiamine	None	No known toxicity from oral intake. There are some reports of anaphylaxis caused by high-dose thiamin injections into the vein or muscle. However, the doses were greater than the quantity humans can physically absorb from oral intake.
Vitamin B_2	riboflavin	None	No known toxicity
Vitamin B_3	niacin	35 mg/day from supplements, drugs or fortified food[6]	Flushing (redness of the skin, often accompanied by itching or a mild burning sensation). Intake of 3000 mg/day of nicotinamide and 1500 mg/day of nicotinic acid are associated with nausea, vomiting, and signs and symptoms of liver toxicity.
Vitamin B_5	pantothenic acid	None	No known toxicity
Vitamin B_6	pyridoxine	100 mg/day from supplements, drugs or fortified food	sensory neuropathy and dermatological lesions
Vitamin B_7	biotin	None	No known toxicity
Vitamin B_9	folic acid	1 mg/day	Masks B_{12} deficiency, which can lead to permanent neurological damage[8]
Vitamin B_{12}	cobalamin	None	No known toxicity

B-VITAMIN SOURCES

B-vitamins are found in all whole, unprocessed foods. Processing, as with sugar and white flour, tends to significantly reduce B vitamin content. B vitamins are particularly concentrated in meat and meat products such as liver, turkey, and tuna. Other good sources for B vitamins are potatoes, bananas, lentils, chile peppers, tempeh, beans, nutritional yeast, brewer's yeast, and molasses. Marmite and Vegemite bill themselves as "one of the world's richest known sources of vitamin B". Although the yeast used to make beer results in beer's being a source of B vitamins, their bioavailability ranges from poor to negative given the fact consumption of ethanol is known to inhibit absorption of thiamine (B_1), riboflavin (B_2), niacin (B_3), biotin (B_7), and folic acid (B_9). In addition, each of the preceding studies further emphasizes that elevated consumption of beer and other ethanol-based drinks results in a net deficit of those B vitamins and the health risks associated with such deficiencies.

The B_{12} vitamin is of note because it is not available from plant products, making B_{12} deficiency a concern for vegans. Manufacturers of plant-based foods will sometimes report B_{12} content, leading to confusion about what sources yield B_{12}. The confusion arises because the standard US Pharmacopeia (USP) method for measuring the B_{12} content does not measure the B_{12} directly. Instead, it measures a bacterial response to the food. Chemical variants of the B_{12} vitamin found in plant sources are active for bacteria, but cannot be used by the human body. This same phenomenon can cause significant over-reporting of B_{12} content in other types of foods as well.

Vitamin B may also be delivered by injection to reverse deficiencies.

Another popular means of increasing one's vitamin B intake is through the use of dietary supplements purchased at supermarkets, health centers, or natural food stores. B vitamins are also commonly added to energy drinks. Many energy drinks have been marketed with large amounts of B vitamins (5-Hour Energy contains an astounding 8333% of the recommended dietary allowance of vitamin B_{12} and 2000% of the RDA for vitamin B_6. Many energy drinks offer "360% of the RDA for vitamin B_6, 120% of B_{12}, 140% of niacin (vitamin B_3)") with claims that this will cause the consumer to "sail through your day without feeling jittery or tense." Nutritionists, such as Case Western University Professor Hope Barkoukis, dismiss these claims: "It's brilliant marketing, but it doesn't have any basis [in fact]."

While B vitamins do "help unlock the energy in foods... Just about everyone in America already gets all of the B vitamins they could possibly need in their diets... In general, extra B vitamins are just flushed out of the system, although everyone's limit of absorption is different in regards to B complex vitamins, and no-one knows how much is needed on an individual basis of these vitamins…" The elderly and athletes may need to supplement their intake of B_{12} and other B vitamins due to problems in absorption and increased needs for energy production. Also, Vitamin B_9 (folic acid) deficiency in early embryo development has been linked to neural tube defects. Thus, women planning to become pregnant are usually encouraged to increase daily dietary folic acid intake and/or take a supplement. However, for "most typical consumers of energy supplements or drinks, B vitamins are nothing more than a 'gimmick' when they are making these false claims."

RELATED NUTRIENTS

Many of the following substances have been referred to as vitamins because they were believed to be vitamins at one time, and they are relevant to vitamin nomenclature in that the numbers that were assigned to them form "gaps" in the series of B-vitamin names. Some of them, though not essential to humans, are essential in the diets of other organisms; others have no known nutritional value.

- **Vitamin B_4**: adenine, a nucleobase, is synthesized by the human body.
- **Vitamin B_7**: "vitamin I" of Centanni E. (1935)—also called "Enteral factor"—is a water and alcohol-soluble rice-bran factor that prevents digestive disturbance in pigeons. It governs the anatomical and functional integrity of the intestinal tract. Later found in yeast. Possible candidates for this substance are inositol, niacin (nicotinic acid), and biotin. Carnitine was also claimed to be a candidate but is not soluble in alcohol.
- **Vitamin B_8**: adenosine monophosphate, or alternately myo-inositol, is synthesized by the human body.
- **Vitamin B_{10}**: *para*-aminobenzoic acid (PABA)
- **Vitamin B_{11}**: pteryl-hepta-glutamic acid—chick growth factor, which is a form of folic acid. Later found to be one of five folates necessary for humans; also known as vitamin S or factor S. L-Carnitine is called vitamin B_{11} in France.
- **Vitamin B_{13}**: orotic acid, now known to not be a vitamin.
- **Vitamin B_{14}**: cell proliferant, anti-anemia, rat growth, and antitumor pterin phosphate named by Earl R. Norris. Isolated from human urine at 0.33ppm (later in blood), but later abandoned by him as further evidence did not confirm this. He also claimed this was not xanthopterin.
- **Vitamin B_{15}**: pangamic acid
- **Vitamin B_{16}**: dimethylglycine (DMG)
- **Vitamin B_{17}**: amygdalin, nitrilosides, or laetrile. These substances are found in a number of seeds, sprouts, beans, tubers, and grains. While toxic in large quantities, proponents claim that it is effective in cancer treatment and prevention.
- **Vitamin B_{18}**:
- **Vitamin B_{19}**:
- **Vitamin B_{20}**: carnitine
- **Vitamin B_{21}**:
- **Vitamin B_{22}**: often claimed as an ingredient of Aloe vera extracts but also in many other foods. Claimed by one source to be vitamin B12b-ä.
- **Vitamin B_h**: biotin
- **Vitamin B_m**: "mouse factor": also used to designate inositol
- **Vitamin B_p**: choline Choline is only required for survival of some mutants. Most commonly it is synthesized in vivo de novo May be added as supplement especially when methionine supply is limited.

- **Vitamin B_t**: L-carnitine
- **Vitamin B_v**: a type of B_6 but not pyridoxine
- **Vitamin B_w**: a type of biotin but not d-biotin
- **Vitamin B_x**: *para*-aminobenzoic acid

Note: B_{16}, B_{17}, B_{18}, B_{19}, B_{20}, B_{21} & B_{22} do not appear to be animal factors but are claimed by some naturopaths as human therapeutic factors.

VITAMIN B$_1$ THIAMINE

Thiamine was the first of the water-soluble vitamins to be described, leading to the discovery of more such trace compounds essential for survival and to the notion of vitamin. Chinese medical texts referred to beriberi (a thiamine deficiency disease) as early as 2700 BC. It was not until AD 1884 that Kanehiro Takaki (1849-1920), a surgeon general in the Japanese navy, rejected the previous germ theory and attributed the disease to insufficient nitrogen intake (protein deficiency).

In 1897 Christiaan Eijkman (1858-1930), a military doctor in the Dutch Indies, discovered that fowl fed on a diet of cooked, polished rice developed paralysis, which could be reversed by discontinuing rice polishing. He attributed that to a nerve poison in the endosperm of rice, from which the outer layers of the grain gave protection to the body. Eijkman was awarded the Nobel Prize in Physiology and Medicine in 1929, because his observations led to the discovery of vitamins. An associate, Gerrit Grijns (1865-1944), correctly interpreted the connection between excessive consumption of polished rice and beriberi in 1901: he concluded that rice contained an essential nutrient in the outer layers of the grain that was removed by polishing.

In 1911 Casimir Funk isolated an antineuritic substance from rice bran that he called a "vitamine" (on account of its containing an amino group). Dutch chemists, Barend Coenraad Petrus Jansen (1884-1962) and his closest collaborator Willem Frederik Donath (1889-1957), went on to isolate and crystallize the active agent in 1926, whose structure was determined by Robert Runnels Williams (1886-1965), a US chemist, in 1934. Thiamine ("sulphur-containing vitamin") was synthesized in 1936 by the same group.

It was first named "aneurin" (for anti-neuritic vitamin). Sir Rudolph Peters, in Oxford, introduced thiamine-deprived pigeons as a model for understanding how thiamine deficiency can lead to the pathological-physiological symptoms of beriberi. Indeed, feeding the pigeons upon polished rice leads to an easily recognizable behaviour of head retraction, a condition called opisthotonos. If not treated, the animal will die after a few days. Administration of thiamine at the stage of opithotonos will lead to a complete.cure of the animal within 30 min. As no morphological modifications were observed in the brain of the pigeons before and after treatment with thiamine,. Peeters introduced the concept of biochemical lesion

When Lohman and Schuster (1937) showed that the diphosphorylated thiamine derivative (thiamine diphosphate, ThDP) was a cofactor required for the oxydative decarboxylation of pyruvate, (a reaction now known to be catalyzed by pyruvate dehydrogenase), the mechanism of action of thiamine in the cellular metabolism seemed to be elucidated. Presently, this view seems to be oversimplified: pyruvate dehydrogenase is only one of several enzymes requiring thiamine diphosphate as a cofactor, moreover other thiamine phosphate derivatives have been discovered since then, and they may also contribute to the symptoms observed during thiamine deficiency.

Finally, the mechanism by which the thiamine moiety of ThDP exerts its coenzyme function by proton substitution on position 2 of the thiazolium ring was elucidated by Ronald Breslow in 1958.

THE CHEMISTRY OF THIAMINE

The crystalline vitamin thiamine hydrochloride is named by IUPAC as 2, 5 dimethyl – 6 amionopyrimidine-4 methyl – 5 hydroxyethyl thiozole

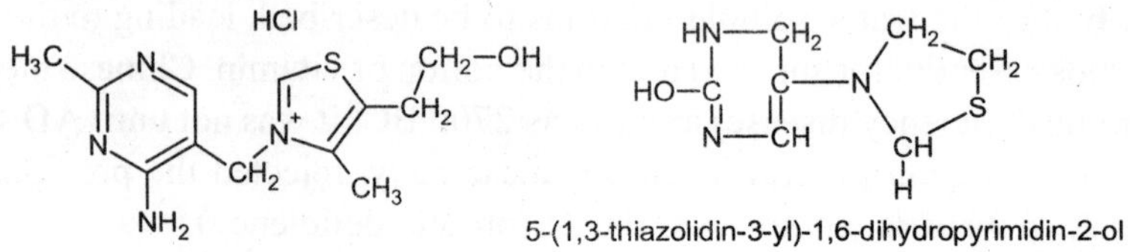

5-(1,3-thiazolidin-3-yl)-1,6-dihydropyrimidin-2-ol

3-[(4-amino-2-methylpyrimidin-5-yl)methyl]-5-(2-hydroxyethyl)-4-methyl-1,3-thiazol-3-ium

Thiamine hydrochloride

Pyrimidine thiazole

Vitamin B_1 was synthesized by Williams has his co workers in 1936 and the process is utilized for large scale commercial production of Thiamin hydrochloride.

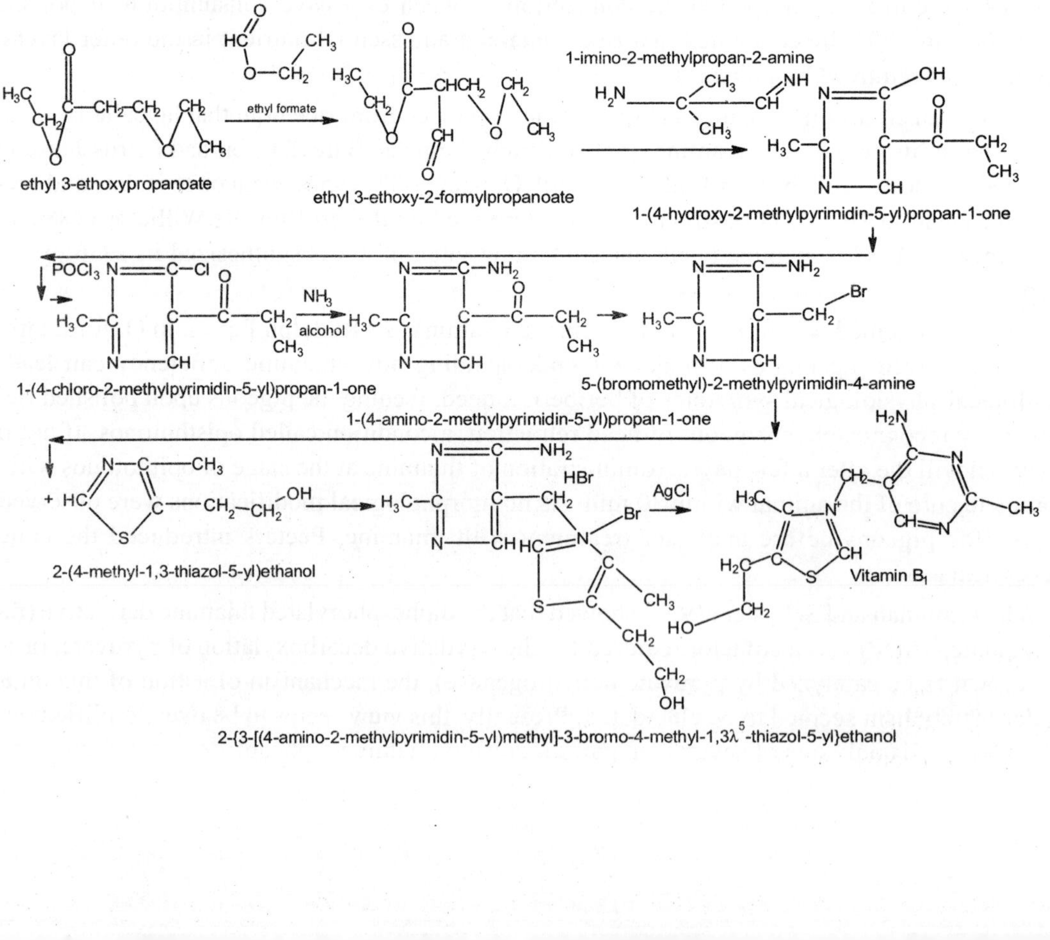

ethyl 3-ethoxypropanoate → (ethyl formate) → ethyl 3-ethoxy-2-formylpropanoate → (1-imino-2-methylpropan-2-amine) → 1-(4-hydroxy-2-methylpyrimidin-5-yl)propan-1-one → ($POCl_3$) → 1-(4-chloro-2-methylpyrimidin-5-yl)propan-1-one → (NH_3, alcohol) → 1-(4-amino-2-methylpyrimidin-5-yl)propan-1-one → 5-(bromomethyl)-2-methylpyrimidin-4-amine → (+ 2-(4-methyl-1,3-thiazol-5-yl)ethanol) → 2-{3-[(4-amino-2-methylpyrimidin-5-yl)methyl]-3-bromo-4-methyl-1,3λ^5-thiazol-5-yl}ethanol → (AgCl) → Vitamin B_1

Thiamin hydrochloride is a white crystalline solid with a molecular weight of 301.81. The crystals are stable at 100°C, but decomposes at 249°C. The vitamin is hydroscopic dissolves in water to form an acid solution which is optically inactive. One gram of vitamin is dissolved in the 1 ml of water; it is insoluble in ether, benzene and chloroform. This vitamin tastes and smell like yeast. The vitamin is stable in acid medium but in alkaline medium it is unstable and slowly gets destroyed due to hydrolysis.

Thiamine itself is a basic substance and reacts with picric acid to form insoluble compounds, it also reacts with tannic acid and phosphotungstic acid, it also reacts with the salts of noble metals, AS the pH increases towards neutrality the thiamine becomes unstable. Thiamine in food is stable, but at high temperature considerable amount of thiamine extracted and often discarded in the cooking water (rice making).

The thiamine pyrophosphate also known as diphosphothiamine is the main ester of thiamine. Thiamine pyrophosphate works as a co-enzyme in the decarboxylation of pyruvic acid.

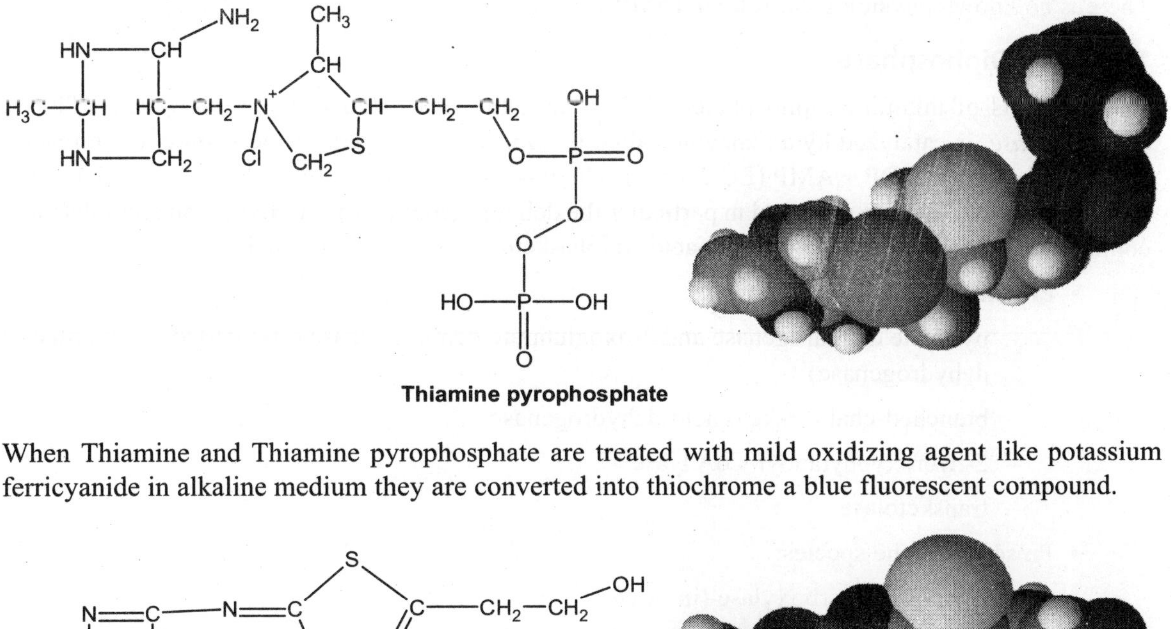

Thiamine pyrophosphate

When Thiamine and Thiamine pyrophosphate are treated with mild oxidizing agent like potassium ferricyanide in alkaline medium they are converted into thiochrome a blue fluorescent compound.

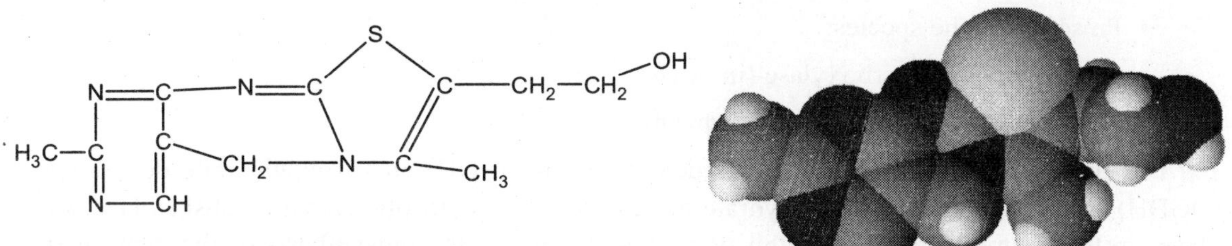

thiochrome

Thiamine or thiamine pyrophosphate diazotized p-amino acetophenone to produced dyes. The reaction products obtained from the pyrophosphate are insoluble in organic solvents and this serves as a means to distinguish one from the other. The thiochrome or the diazotization reactions are utilized for their quantitative estimations.

In the thiochrome preparation the vitamin is oxidized by an alkaline solution of potassium ferricyanide. The compounds is extracted with isobutanol in which the fluorescence is determined in photocolourimeter against standard solutions of the vitamin treated similarly using a blank test where all the reactions except ferricyanide is added.

Thiamine phosphate Derivatives and Function

Thiamine is mainly the transport form of the vitamin, while the active forms are phosphorylated thiamine derivatives. There are four known natural thiamine phosphate derivatives: thiamine monophosphate (ThMP), thiamine diphosphate (ThDP), also sometimes called thiamine pyrophosphate (TPP), thiamine triphosphate (ThTP), and the recently discovered adenosine thiamine triphosphate (AThTP) and adenosine thiamine diphosphate (AThDP).

Thiamine Monophosphate

There is no known physiological role of ThMP.

Thiamine Diphosphate

The synthesis of thiamine diphosphate (ThDP), also known as *thiamine pyrophosphate* (TPP) or *cocarboxylase*, is catalyzed by an enzyme called thiamine diphosphokinase according to the reaction thiamine + ATP $\rightarrow$ ThDP + AMP (EC 2.7.6.2). ThDP is a coenzyme for several enzymes that catalyze the transfer of two-carbon units and in particular the dehydrogenation (decarboxylation and subsequent conjugation with coenzyme A) of 2-oxoacids (alpha-keto acids). Examples include:

- Present in most species
  - pyruvate dehydrogenase and 2-oxoglutarate dehydrogenase (also called α-ketoglutarate dehydrogenase)
  - branched-chain α-keto acid dehydrogenase
  - 2-hydroxyphytanoyl-CoA lyase
  - transketolase
- Present in some species:
  - pyruvate decarboxylase (in yeast)
  - several additional bacterial enzymes

The enzymes transketolase, pyruvate dehydrogenase (PDH) and 2-oxoglutarate dehydrogenase (OGDH) are all important in carbohydrate metabolism. The cytosolic enzyme transketolase is a key player in the pentose phosphate pathway, a major route for the biosynthesis of the pentose sugars deoxyribose and ribose. The mitochondrial PDH and OGDH are part of biochemical pathways that result in the generation of adenosine triphosphate (ATP), which is a major form of energy for the cell. PDH links glycolysis to the citric acid cycle, while the reaction catalyzed by OGDH is a rate-limting step in the citric acid cycle. In the nervous system, PDH is also involved in the production of acetylcholine, a neurotransmitter, and for myelin synthesis.

Thiamine Triphosphate

Thiamine triphosphate (ThTP) was long considered a specific neuroactive form of thiamine. However, recently it was shown that ThTP exists in bacteria, fungi, plants and animals suggesting a much more general cellular role. In particular in *E. coli*, it seems to play a role in response to amino acid starvation.

Adenosine Thiamine Triphosphate

Adenosine thiamine triphosphate (AThTP) or thiaminylated adenosine triphosphate has recently been discovered in *Escherichia coli* where it accumulates as a result of carbon starvation. In *E. coli*, AThTP may account for up to 20 % of total thiamine. It also exists in lesser amounts in yeast, roots of higher plants and animal tissue.

Adenosine Thiamine Diphosphate

Adenosine thiamine diphosphate (AThDP) or thiaminylated adenosine diphosphate exists in small amounts in vertebrate liver, but its role remains unknown.

DEFICIENCY

Thiamine derivatives and thiamine-dependent enzymes are present in all cells of the body, thus, a thiamine deficiency would seem to adversely affect all of the organ systems. However, the nervous system and the heart are particularly sensitive to thiamine deficiency, because of their high oxidative metabolism.

Thiamine deficiency can lead to severe fatigue of eyes and myriad problems including neurodegeneration, wasting and death. A lack of thiamine can be caused by malnutrition, a diet high in thiaminase-rich foods (raw freshwater fish, raw shellfish, ferns) and/or foods high in anti-thiamine factors (tea, coffee, betel nuts) and by grossly impaired nutritional status associated with chronic diseases, such as alcoholism, gastrointestinal diseases, HIV-AIDS, and persistent vomiting. It is thought that many people with diabetes have a deficiency of thiamine and that this may be linked to some of the complications that can occur.

Well-known syndromes caused by thiamine deficiency include beriberi and Wernicke-Korsakoff syndrome, diseases also common with chronic alcoholism.

Beriberi

Beriberi is a neurological and cardiovascular disease. The three major forms of the disorder are dry beriberi, wet beriberi, and infantile beriberi.

- *Dry beriberi* is characterized principally by peripheral neuropathy consisting of symmetric impairment of sensory, motor, and reflex functions affecting distal more than proximal limb segments and causing calf muscle tenderness.
- *Wet beriberi* is associated with mental confusion, muscular wasting, edema, tachycardia, cardiomegaly, and congestive heart failure in addition to peripheral neuropathy.

- *Infantile beriberi* occurs in infants breast-fed by thiamin-deficient mothers (who may show no sign of thiamine deficiency). Infants may manifest cardiac, aphonic, or pseudomeningitic forms of the disorder. Infants with cardiac beriberi frequently exhibit a loud piercing cry, vomiting, and tachycardia. Convulsions are not uncommon, and death may ensue if thiamine is not administered promptly.

Following thiamine treatment, rapid improvement occurs generally within 24 hours. Improvements of peripheral neuropathy may require several months of thiamine treatment.

Alcoholic Brain Disease

Nerve cells and other supporting cells (such as glial cells) of the nervous system require thiamine. Examples of neurologic disorders that are linked to alcohol abuse include Wernicke's encephalopathy (WE, Wernicke-Korsakoff syndrome) and Korsakoff's psychosis (alcohol amnestic disorder) as well as varying degrees of cognitive impairment.

Wernicke's encephalopathy is the most frequently encountered manifestation of thiamine deficiency in Western society, though it may also occur in patients with impaired nutrition from other causes, such as gastrointestinal disease, those with HIV-AIDS, and with the injudicious administration of parenteral glucose or hyperalimentation without adequate B-vitamin supplementation.[36] This is a striking neuro-psychiatric disorder characterized by paralysis of eye movements, abnormal stance and gait, and markedly deranged mental function.

Alcoholics may have thiamine deficiency because of the following:

- inadequate nutritional intake: alcoholics tend to intake less than the recommended amount of thiamine.
- decreased uptake of thiamine from the GI tract: active transport of thiamine into enterocytes is disturbed during acute alcohol exposure.
- liver thiamine stores are reduced due to hepatic steatosis or fibrosis.
- impaired thiamine utilization: magnesium, which is required for the binding of thiamine to thiamine-using enzymes within the cell, is also deficient due to chronic alcohol consumption. The inefficient utilization of any thiamine that does reach the cells will further exacerbate the thiamine deficiency.
- Ethanol *per se* inhibits thiamine transport in the gastrointestinal system and blocks phosphorylation of thiamine to its cofactor form (ThDP).

Korsakoff Psychosis is generally considered to occur with deterioration of brain function in patients initially diagnosed with WE.. This is an amnestic-confabulatory syndrome characterized by retrograde and anterograde amnesia, impairment of conceptual functions, and decreased spontaneity and initiative.

Following improved nutrition and the removal of alcohol consumption, some impairments linked with thiamine deficiency are reversed; particularly poor brain functionality, although in more severe cases, Wernicke-Korsakoff syndrome leaves permanent damage.

VITAMIN B_2 RIBOFLAVIN

Vitamin B-2 (Riboflavin) is a water soluble, stable vitamin. It plays an important role in several enzymes and coenzymes as a hydrogen atom acceptor in various metabolic roles, including protein metabolism. It is a constituent of coenzymes involved in the release of energy from glucose and fatty acids in the mitochondria. It is important in DNA synthesis, and in the formation of red blood cells. Riboflavin also takes part in the activation of vitamin B-6 and conversion of folacin to its coenzymes.

METHODS OF ACTION

Riboflavin is absorbed through the walls of the small intestine. Food in the gastrointestinal tract increases the absorption of riboflavin; when ingested without food, only about 15% of riboflavin is absorbed. Within the intestinal cells, much of the absorbed riboflavin is phosphorylated to flavin mononucleotide (FMN) and, subsequently, is carried by the blood to the tissues.

Free riboflavin is transported by blood albumin to the liver, where it is converted to flavin-adenine dinucleotide (FAD). Both FAD and FMN play important roles in catalyzing oxidation-reduction reactions in the cells, and are essential coenzymes in the oxidative phosphorylation which occurs in cellular mitochondria.

Riboflavin is a component of the enzymes L- and D-amino acid oxidase, which catalyze the oxidation of amino acids, thereby involving the vitamin in protein metabolism. Riboflavin plays a primary role in the conversion of folacin to its coenzymes, some of which are needed for DNA synthesis and cell proliferation. It also activates vitamin B-6 and may participate in red blood cell formation.

Loss of protein from the body is accompanied by a loss in riboflavin. Riboflavin is not stored in the body and must be supplied in the diet, as any excess is excreted in the urine.

THE CHEMISTRY OF FLAVIN AND RIBOFLAVIN

Riboflavin has been used effectively in treating the following symptoms of deficiencies: soreness and burning of the lips, mouth, and tongue; fissures and cracks in the lips, a purple, swollen tongue; and possible anemia.

Riboflavin needs increase during growth, pregnancy, wound healing, and lactation. This phenomenon is related to the fact that riboflavin plays a role in protein metabolism, which is increased during these periods.

The yellow-coloured compounds with the basic structure of 7,8-dimethyl-10-alkylisoalloxazine are generally termed as flavins. Flavins are ubiquitous innature, and they take part in many biochemical reactions as coenzymes andphotoreceptors. Riboflavin, the precursor of all the biologically important flavins, was first reported as lactochrome, a bright yellow pigment isolated from cow milk in 1879.1 Later, in the late 1920s and early 1930s, yellow pigments with bright greenish fluorescence were isolated from different sources, and they were named as lactoflavin, ovoflavin, etc., indicating the source from which they had been isolated. Concomitantly, it was recognized that the yellow pigment was a constituent of the vitamin B complex. Two important groups determined the structure and proved it by chemical synthesis. The name riboflavin (RF) was given to the compound; it derives from the ribityl side chain and from the yellow-conjugated ring system (see figure 1 below). There is a broad distribution of flavins in tissues, but little is present as free RF. The majority is found in flavocoenzymes, mainly as flavin adenine dinucleotide (FAD), and in lesser amounts as flavin mononucleotide (FMN), the common name of riboflavin-50-phosphate, despite the known fact that RF is

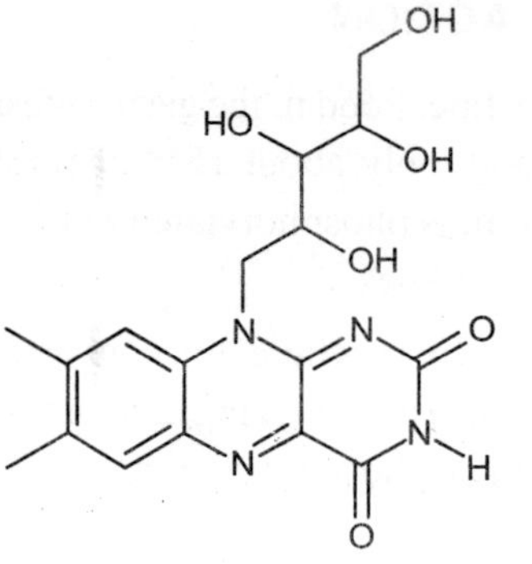

RF
(1)

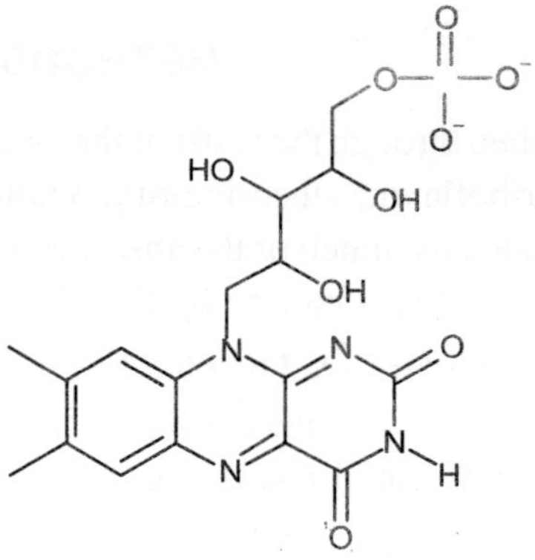

FMN
(2)

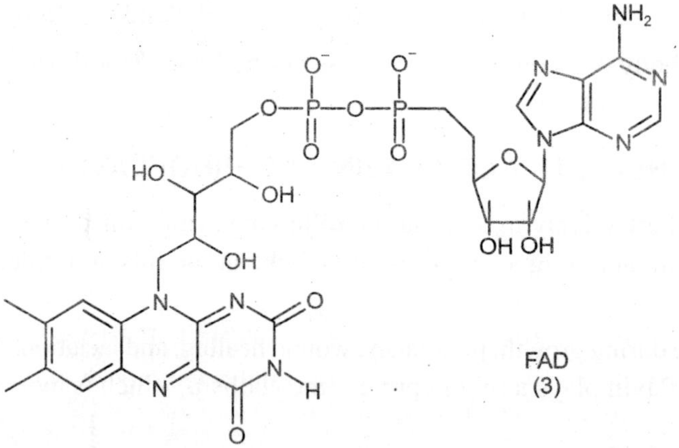

FAD
(3)

Chemical structures of riboflavin (RF), flavin momonucleotide (FMN) and flavin adenine dinucleotide (FAD) not a real nucleoside because the linkage between the ribityl chain and the N10 of the flavin is not glycosidic; therefore, FMN and FAD are not real nucleotides. The structures are shown in Figure above

Since the pioneer study of Theorell, who demonstrated in 1935 that the biochemical basis for the necessity of RF as a vitamin is its role as precursor of the FMN cofactor in enzyme catalysis (coenzyme), and those by Krebs and Warburg who showed its role as precursor of FAD cofactors, hundreds of flavoprotein enzymes have been known, and new ones are reported every year. Most of them contain non-covalent-bound FAD or FMN, and are specific for binding either of the two flavin forms as nature initially provided them with. Nowadays, there are many flavoproteins crystal structures known, which reveals that the majority of the flavin-protein interactions are with the N-10 ribityl side chain of FMN or FAD. A recent study on the sequence-structure relationship in 32 families of FAD-containing proteins, showed that in every case the pyrophosphate moiety binds to the most strongly conserved sequence motif, suggesting that pyrophosphate binding is a significant component of molecular recognition.

Properties of Flavins

The redox potential for the two electron reduction of the flavin is about –200mV. However, this value can greatly vary in flavoproteins, due to the crucial role of the protein environment in the properties of flavins, spanning a range from approximately –400 mV to +60 mV. In general, the proximity of a positive charge is believed to increase the redox potential and a negative charge or a hydrophobic environment are expected to lower it. A few flavoenzymes have a covalent-bound FAD molecule, and site-directed mutagenesis studies suggest that the covalent interaction could increase the oxidative power of the flavin.

Since the discovery and characterization of RF and its derivatives FMN and FAD, they have been recognized by their ability of participate in both one- and two-electron transfer processes. This means that flavin molecules can exist in three different redox states: oxidized, one-electron reduced (semiquinone) and two-electron reduced states. Therefore, they can participate in redox reactions as either one- and two-electron mediator making the flavoenzymes very versatile in terms of substrate and type of catalysed reactions. This is a major reason for the ubiquity of flavin-dependent enzymes in biological systems. Flavins have the potential for transfer of single electrons, of hydrogen atoms and of hydride ions. Therefore, they can participate in redox reactions as either one- and two-electron mediator making the flavoenzymes very versatile in terms of substrate and type of reactions, which is a major reason for the ubiquity of flavin-dependent enzymes in biological systems. In addition, the oxidized flavin molecule is susceptible to nucleophilic attack, especially at N-5 and C-4α.

In free solution (when not bound to an enzyme), the equilibrium of the different flavin species is pH dependent, as shown in Scheme 1 (proposed by Heelis, 1982).

This Scheme presents the different redox states: oxidized, one-electron reduced (semiquinone) and two-electron reduced (fully reduced) and also the different protonation states for each of them. From the nine forms in Scheme 1, at least six are physiologically possible on the basis of their pK_a values.

At pH 7, only about a 5% radical is stabilized in an equimolar mixture of oxidized and reduced flavin. The semiquinone can exist in a neutral or anionic form, with a pKa of ~8.5.

On binding to a specific protein, this equilibrium can change dramatically: some enzymes show no stabilization of semiquinone, while others give almost 100% stabilization. In some cases, if the protein can stabilize the neutral radical species over the whole range of pH values at which the enzyme is stable, the pK_a is shifted up significantly from 8.5. In other cases if the semiquinone anion is stabilized, the pKa is decreased significantly. There are some enzymes, of which glucose oxidase was the first example, which show such a pK_a that the identification of both forms is possible. In addition to these redox/ionic forms (each of them with different canonical forms), there are other electronic states, known as charge-transfer states. They are electronic states that do not belong to any of the three redox

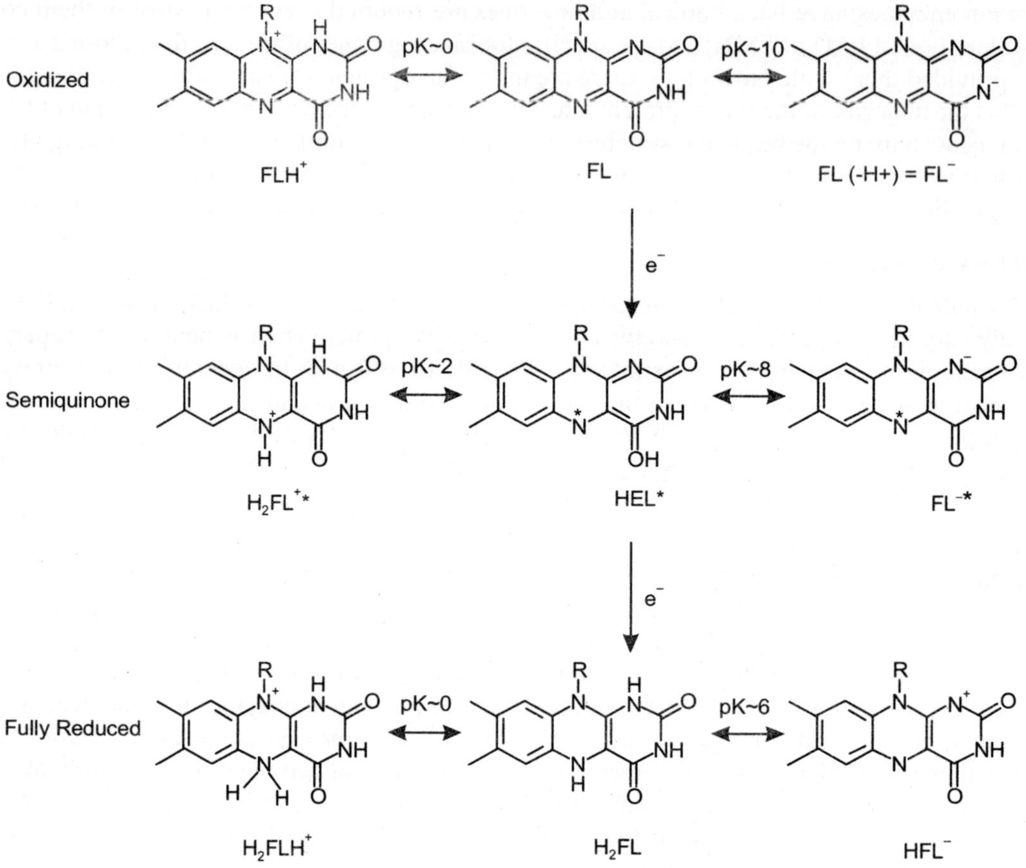

Oxidized

FLH$^+$ pK~0 FL pK~10 FL (-H+) = FL$^-$

e^-

Semiquinone

H_2FL^{+*} pK~2 HEL* pK~8 FL^{-*}

e^-

Fully Reduced

H_2FLH^+ pK~0 H_2FL pK~6 HFL^-

Scheme 1. Redox and acid-base equilibria of flavins

states, but are those in which partial charge is transferred to or from one of the three redox states. All these redox, ionic and charge-transfer states are the origin of the different colours of flavins and flavoproteins. The large spectral differences between the various flavin redox-ionic-electronic states make it possible to monitor the events occurring in catalysis using the flavin itself as a reporter.

CLASSIFICATION OF FLAVOPROTEINS

Flavins show an extremely high chemical versatility, which is reflected in the remarkable versatility of flavoenzymes; however, each of the enzymes is also characterized by a strict specificity. Many attempts have been made to achieve a rational classification of the many different types of flavoproteins, depending on their different properties.

In 1970, researchers proposed four classes of flavoproteins according to their behaviour at half reduction in the absence of substrates:

(A) Flavoproteins exhibiting stoichiometric amounts of blue semiquinone

(B) Flavoproteins exhibiting stoichiometric amounts of red semiquinone

(C) Flavoproteins exhibiting no paramagnetic intermediates

(D) Flavoproteins exhibiting non-stoichiometric amounts of semiquinone.

The red radicals of class B have been identi.ed as .avosemiquinone anions (Fl^-), and the blue radicals of class A as neutral semiquinones (FlH*d*). According to this classi.cation, glucose oxidase belongs to class A and B, depending on pH. The authors proposed that in class A and B .avopropteins N-5 is protected by forming a strong hydrogen bond in the radical state, so that no "in plane" transfer of electrons can occur. These radicals are stable as long as the hydrogen bond at N-5 is not removed. They identi.ed class D as metalloflavoproteins, where the radical yields at half reduction never exceed 50% of total flavin. Class C includes those flavoproteins that exhibit a cyst(e)ine residue participating in the catalysis. Both internal redox systems (flavin and disulfide) are in covalent contact at 2e-reduction of the enzyme. The point of attachment of the sulfur component to the flavin has been proposed to be C-4α.

Despite the fact that several hundred flavoenzymes have been described, proposed that a large number of flavoproteins may be classified into two classes, based on distinct spectroscopic properties: regiospecific reactivities of the flavin skeleton and hydrogen bonding patterns with the apoprotein. The results reported by researchers support this proposal. The first group (dehydrogenases/oxidases) is characterized by a "red" semiquinone radical, a bent structure of the reduced form, and hydrogen bonding to the N-1 nitrogen atom, activating the N-5 position.

The second group (electron transferases) is characterized by a blue semiquinone radical, a planar structure of the reduced form, and hydrogen bonding to N-5, activating of the C-4α position.

With the aim of addressing some characteristics of the two classes of flavoenzymes described above, the electronic properties of the three different protonated forms HFlox, H_2Flox^+ (N-5) and H_2Flox^+ (N-1) using lumiflavin as model compound. Lumiflavin is a photoproduct of RF, in which the ribityl side chain is replaced by a methyl group. In order to describe the electrophilic/nucleophilic properties, they investigated the geometry, charge distribution, and HOMO-LUMO topologies of the three forms of lumiflavin adopting the RHF/3-21G level. The electronic spectra were evaluated at the CNDO/CI level with a particular attention to the blue or red shifts respect to the oxidized form. The authors concluded that a larger electron delocalization, demonstrated by a smaller bond length alternation upon protonation, as well as a larger basicity associated with a more negative charge can explain the favourable protonation of N-1 with respect to N-5, and the unusual electrophilic affinity of N-5 when lumiflavin is protonated at N-1. The calculations correctly place the π-π^* transition as the lowest energy transition for neutral oxidized lumiflavin and predict a blue shift of the low-lying electronic transition upon monoprotonation (formation of the stable N-1 protonated form). They found in their calculation on radical intermediates of lumiflavin, a theoretical rationalization for the experimental classi.cation of flavoenzymes in two distinct classes. Hydrogen bonding at N-5 and the fact that the semiquinone radical is red appear consistent with protonation at N-1 and larger wavelength absorption. Hydrogen bonding at N-1 and the fact that the semiquinone radical is blue are consistent with protonation of N-5 and a smaller wavelength absorption.

Among the known redox coenzymes, flavocoenzymes are unique in that they can participate in both one-electron and two-electron processes. The other redox cofactors usually catalyse exclusively either one- or two-electron processes. Active redox metalloenzymes catalyse only one-electron process, and nicotinamide nucleotides, with wide distribution in biological systems, are involved in only two electron redox reactions, because the radical forms of the pyridine ring are not sufficiently stable as to be involved in enzymaticreactions. For these reasons, flavoenzymes mediate two-electron and one electron processes, as is the case of the well-known mitochondrial and chloroplast electron transport chains.

The reactions catalysed by a flavoenzyme always involve two separate halfreactions: reductive and oxidative half-reactions, both of which are necessary for the turn over of the enzyme. The former is the process in which a substrate or an electron donor is oxidized with the concomitant flavin reduction. In the latter process, the reduced flavin is oxidized by another substrate or an electron donor. Researchers proposed a scheme consisting of five broad classes of flavoenzymes based on the nature of the substrate involved in the twoseparate half-reactions:

(*i*) *Transhydrogenase*, where two-electron equivalents are transferred, along with the appropriate hydrogen ions, from one organic substrate to another.

(*ii*) *Dehydrogenase–oxidase*, where two-electron equivalents are transferred to the flavin from an organic substrate, where molecular oxygen is the oxidizing substrate, being reduced to H_2O_2.

(*iii*) *Dehydrogenase–monooxygenase*, where the flavin is reduced generally by a reduced pyridine nucleotide, and where on oxidation with O_2 in the presence of a co substrate one atom of oxygen is inserted into the co substrate, while the other is reduced to H_2O.

(*iv*) *Dehydrogenase*–electron transferase, where the flavin is reduced by two electron transfer from a reduced substrate and then reoxidized in sequential single electron transfers to acceptors, such as cytochromes and iron–sulphur proteins. This class might be further subdivided to distinguish those enzymes which function in the reverse sense, i.e., those which receive electrons one at a time and then transmit them in a two electron step in the reduction of a pyridine nucleotide.

(v) *Electron transferase,* where the flavin is reduced and reoxidized in one electron steps.

This scheme was adopted by the International Union of Biochemistry.

In the last decade, several interesting studies about flavins properties have been published. Recently, Miura has proposed an alternative classification of flavoenzymes based on the number of electrons involved in the catalytic cycle. Because these half-reactions can be either one-electron or two-electron processes, there are four possible combinations of the two half-reactions; therefore, the author proposed that flavoenzymes can be classified according to the combination of one- and two-electron processes for reductive and oxidative half-reactions. Although four categories are theoretically possible; 1/1, 1/2, 2/1 and 2/2, the author points that flavoenzymes of the 1/2 category are not known at present. The author proposed that specificity of each flavoenzyme is understood in terms of the regulatory mechanism of the broad reactive potential of flavin. The regulatory mechanisms include hydrogen-bonding networks, electrostatic effects, charge transfer interactions, positioning between substrate and flavin, and modulation of resonance hybridization. The reaction between the fully reduced flavin H_2Fl and oxygen is essentially irreversible in free solution, according to the redox potentials, but it is also slow because of the spin inversion associated with the reaction of the singlet-reduced flavin and the triplet molecular oxygen. In flavoproteins, the reaction with oxygen may be orders of magnitude faster or slower, depending on the specific flavoprotein.

They have been classified in four well-de.ned groups, according to the rate of their reactions with oxygen and the nature of the products formed, such as electron transferases, dehydrogenases or transhydrogenases, oxidases and monooxygenases. In 1994, Massey published an interesting review about the reactions of flavin and flavoproteins with oxygen. The author discussed the common characteristics of the flavoproteins that are classified according their reactivity with oxygen, and proposed that they have many characteristic properties of a particular group, not shared with other groups of flavoproteins. Considering that the ratios of the rates of reaction with oxygen are at least 106 among flavoproteins, and that enormous differences in the decay rates of the hydroperoxides are found, the author also pointed out that there are important questions that remain to be answered What are the flavin–protein interactions that control the rate of reaction with oxygen? What factors determine whether the C-4a hydroperoxide should be formed? What are the factors that determine the stability of this hydroperoxide for different flavoenzymes?

In 1997, a study of the electronic properties of lumiflavin published as model compound. The authors found in their calculation a theoretical rationalization for the behaviour of flavoproteins and its classification in two groups, as described above. They also gave an explanation, in terms of energy level differences, for the red shift observed for the first absorption band of the isoalloxazine ring upon complex formation with appropriate ligands. The favourable stacking observed in the crystal structures of lumiflavin–hydroquinone complexes was also rationalized in terms of the complementarity of the molecular electrostatic potential generated by the two molecules, and/or by the superposition of the

HOMO of hydroquinone and the

LUMO of the oxidizedflavin.18

In 2000, researcher published an interesting review, where the catalytic versatility of flavoproteins is explained by the catalytic mechanisms of selected flavoenzymes. The author studied the catalytic mechanisms of the different enzymes by replacing the native flavin with artificial flavins, such as 8-chloro-flavins, 8-mercaptoflavins, and the photoreactive 6-azido-FAD. He also replaced the substrate with derivatives such as β-chloro amino acids or β-chloro a-hydroxyacids, and used the information available from crystal structures. He proposed possible reaction mechanisms for each of the groups of flavoenzymes: that catalyse the oxidation of α-hidroxyacids and a-amino acids, disulphide reductases, monooxygenases, and reductases, dehydrogenases and electron transferases (including the old yellow enzyme). Flavoproteins are now known to have a variety of folding topologies.

In 2000, researchers examined the three-dimensional structures of flavoenzymes that were available in the protein data bank (PDB), searching for recurrent features in their catalytic apparatus. They focused their analysis to the dehydrogenases, a group of enzymes that catalyse a reaction involving the rupture of a kinetically stable C-H bond, coupled with the transfer of two electrons to the flavin. The authors found that these enzymes share a few invariant features in the hydrogen-bond interactions between the protein and the flavin and that the positioning of the reactive part of the substrate with respect to the flavin is generally conserved. The authors recognized that the mechanistic problems cannot be solved solely on the basis of structural data; however, they pointed out that all the different mechanistic proposals (hydride transfer, radical mechanism and carbanion mechanisms) require, at least to some extent, juxtaposition between the flavin N-5-C4a locus and the reactive C-H group of the substrate. Although many of the reviewed enzymes are proposed to function via hydride transfer, there is no general consensus about the exact mechanisms for some of them. Nonetheless, the stereochemical principles underlying the mutual interactions between the substrate C–H group and the flavin are surprisingly well conserved, and they represent a validation test for the plausibility of any proposed mechanism, which must be compatible with the observed stereochemistry of substrate binding. The authors proposed that in the future the stereochemical requirements of each of the proposed mechanisms would be de.ned, to evaluate their compatibility with the three dimensional structures.

Researchers published an interesting review, where they analyzed structure–sequences relationships in 32 families of FAD-binding proteins. Their work illustrates how similar cofactors are utilized by nature in a wide variety of proteins families. The availability of the three-dimensional structures of a large collection of FAD-proteins (150 flavoprotein X-ray structures solved in the presence of FAD were retrieved from PDB) allow to identify four different FAD-family folds, each with distinctive conserved sequence motifs. Furthermore, the FAD pocket shape can be distinguished from one fold to another.

In contrast to the diversity of the FAD fold and its sequence motif, the pyrophosphate moiety makes hydrogen bonds with residues from the mostconserved sequence motifs of each fold family. The authors proposed that the pyrophosphate moiety is crucial for molecular recognition, whereas the isoalloxazine ring is involved in catalytic function and the adenine ring stabilizes cofactor binding, both rings interacting with protein residues in partially conserved sequence motifs. As a result, the isoalloxazine and adenine rings interact with different residues in the different proteins belonging to the same FAD-family fold. This is consistent with the variability in bond type (covalent or non-covalent) and cofactor conformation (elongated or bent) observed in the FAD-family folds. However, in each FAD-family fold there is a clear correlation between the fold, the shape and position of the pocket and the cofactor conformation. This is in agreement with the finding that the cofactor exhibits directionality, which is highly conserved within a family fold.

The same year, a review about recent progress on the elucidation of the mechanisms of oxidation of organic substrates of flavoenzymes reported that catalyse dehydrogenation of bonds between carbon and either nitrogen or oxygen. The work is focused on the oxidation of alcohols, amino and hydroxy acids, amines and nitroalkanes. The author combined information obtained from site-directed mutations, effect of inhibitors and substrate analogues, kinetic isotope effects and structural information, with an initial goal to develop uni.ed mechanisms for each of the different families of flavoproteins oxidases. In most, if not all the cases, multiple mechanisms still remain viable despite intensive study.

This uncertainty can be attributed to the wide range of reaction mechanisms potentially available to the flavin cofactor. The author concluded that radical and nucleophilic mechanisms must be considered in addition to the more accepted hydride transfer mechanism. The possibility that the reaction observed with substrate analogues may diverge from the normal catalytic mechanisms must also be considered. Our understanding of flavin chemistry has increased in the last years by the wealth of information obtained by catalysis and biomimetics as well as from structural studies of flavoproteins.

However, as discussed in this work, versatility and specificity are essential concepts to describe flavoproteins flavoenzymes. The wide range of possible reactions give such a high versatility to these proteins as to make very complex any attempt to describe their behaviour by general patterns. However, each flavoprotein has a strict specificity, thusimplying that one of the most critical roles of the protein component is to limit the whole range of possible flavin–protein interactions to those beneficial to the reaction to be catalysed. Despite the significant progress that has been made in recent years, the factors that determine the specificity of flavoproteins are still poorly understood, and many research groups are contributing to this area. In this context, advances in the research on the photochemical and photobiological properties of flavins and flavoproteins are necessary for a complete understanding of these important compounds.

CONSEQUENCE OF DEFICIENCY

Early signs of riboflavin deficiency are: soreness and burning of the lips, mouth, and tongue; fissures and cracks in the lips; a purple, swollen tongue; burning and itching of the eyes; photophobia; a loss of visual acuity; and possible anemia. Individuals who ingest a diet devoid of animal protein sources and green leafy vegetables are prone to ariboflavinosis, a severe riboflavin deficiency. If the intake of riboflavin is low for several months, the following symptoms will appear: cheilosis; angular stomatitis (cracks in the skin at the corners of the mouth); a greasy eruption on the skin of the nasolabial folds and scrotum; capillary overgrowth around the cornea of the eye; and a purplish, swollen tongue.

Riboflavin deficiency rarely appears alone; most often it appears in conjunction with multiple nutritional deficiencies. This makes diagnosis quite difficult, especially as niacin, iron, and pyridoxine deficiencies mimic the symptoms of riboflavin deficiency. Ocular manifestations such as photophobia and itching, burning, and redness of the eyes, are believed to be the earliest signs of riboflavin deficiency. Urinary excretion of riboflavin can confirm diagnosis.

VITAMIN B_3 NIACIN

Vitamin B_3 is one of 8 B vitamins. It is also known as niacin (nicotinic acid) and has 2 other forms, niacinamide (nicotinamide) and inositol hexanicotinate, which have different effects from niacin. All B vitamins help the body to convert food (carbohydrates) into fuel (glucose), which is "burned" to produce energy. These B vitamins, often referred to as B complex vitamins, also help the body metabolize fats and protein. B complex vitamins are necessary for healthy skin, hair, eyes, and liver. They also help the nervous system function properly. Niacin also helps the body make various sex and stress-related hormones in the adrenal glands and other parts of the body. Niacin is effective in improving circulation and reducing cholesterol levels in the blood.

All the B vitamins are water-soluble, meaning that the body does not store them. You can meet all of your body's needs for B_3 through diet; it is rare for anyone in the developed world to have a B_3 deficiency. In the United States, alcoholism is the prime cause of vitamin B_3 deficiency. Symptoms of mild deficiency include indigestion, fatigue, canker sores, vomiting, and depression. Severe deficiency can cause a condition known as pellagra. Pellagra is characterized by cracked, scaly skin, dementia, and diarrhea. It is generally treated with a nutritionally balanced diet and niacin supplements. Niacin deficiency also results in burning in the mouth and a swollen, bright red tongue. Very high doses of B_3 (available by prescription) have been shown to prevent or improve symptoms of the following conditions. However, taken at high doses niacin can be toxic, so you should take doses higher than the Recommended Daily Allowance only under your doctor's supervision. Researchers are trying to determine if inositol hexanicotinate has similar benefits without serious side effects, but so far results are preliminary.

THE CHEMISTRY OF NIACIN

Nicotinic acid or niacin is obtained by the oxidation of the alkaloid nicotine with nitric acid or potassium permanganate.

Nicotine → (oxidation; pot permanganate, nitric acid) → Niacin (Vitamin B_3) → (NH_3) → Nicotinamide

Nicotinic acid is a crystalline solid with a molecular weight of 123.2 and melts at 236°C. It sublimes without destruction. The vitamin is stable to boiling in neutral acidic or alkaline solution. Chemically niacin is a simple carboxyl derivative of pyridine (pyridine-3-carboxylic acid)

The amide and the sodium salt are as active biologically as the free acid. Benzene helps to crystallize the nicotinamide in the form of light needles, which melts at 129°C. Nicotinamide is more soluble in organic solvents than the free acid and may be extracted from water with ether. Nicotinamide is readily hydrolyzed to free niacin by heating with acid or alkali, Niacin or niacinamide reacts with cyanogens bromide in presence of a primary or secondary amine to produce a yellow-green compound. Coloured products are also produced when niacin or niacinamide reacts with trogonelline, nicotinuric acid, nicotine and certain other pyridine compounds. This reaction is used in quantitative estimation of the vitamin.

The amide of the nicotinic acid was isolated by Knhn and Vetter in 1935 from heart muscle. The amide is a component of the pyridine nucleotide co –enzymes – dyphosphopyridine nucleotide DPN co-enzyme II and triphosphopyridine or TPN co –enzyme II. Resistant to heat light and alkalies nicotinic acid is one of the stable vitamins

EFFECT IN THE METABOLISM

Nicotinic acid reacts with hemoglobin and myoglobin in meat to form a brightly coloured complex, and thus has been used as a food additive, typically to improve the colour of minced (ground) meat. However, sometimes excess niacin is added to the meat during processing. Though still licensed as a food colouring agent in some countries, it is not licensed as such in Europe.

The body can synthesize niacin from the essential amino acid tryptophan, but the synthesis is extremely slow; 60 mg of tryptophan are required to make one milligram of niacin. For this reason, eating lots of tryptophan is not an adequate substitute for consuming niacin. As serotonin synthesis is reliant on tryptophan availability, inadequate dietary intake of vitamin B_3 may also therefore lead to depression.

Because niacin in large quantities is a vasodilator, large doses of niacin (either from vitamin B_3 tablets or from treated meats) may cause harmless and short-lived but unpleasant symptoms such as extreme skin flushing resembling a sunburn, itching, gastric disturbances, and lowering of blood pressure. The amide form (strictly speaking a provitamin) does not cause these side effects, but is also not as easily assimilated by the body.

Large doses of niacin are sometimes prescribed to combat high blood pressure, and also to lower blood cholesterol levels. Pharmacologic doses of niacin (1.5 to 6 grams/day) reduces LDL cholesterol levels by 10 to 25 percent and triglyceride levels by 20 to 50 percent. HDL cholesterol levels are also increased by 15 to 35 per cent. Vitamin B_3 has also been used to treat schizophrenia and other mental illnesses by orthomolecular practitioners. Usually the nicotinamide form is used, as it is considered to be more effective. Unfortunately there is little scientific evidence that this treatment is effective.

Niacin opens blood vessels wider. Niacin, when used properly, offers benefits in terms of cholesterol reduction, as well as for detoxification. People with high cholesterol levels should see improvements once their follow the proper diet of raw foods, preferably through a good juicing program, along with the proper healing fats in their diet. However for people who don't respond, it would seem that niacin would be a far safer and less expensive alternative to the statin drugs,* which have their serious complications. Niacin is a water-soluble B vitamin – vitamin B_3 - and the common name for 2 very different compounds: "nicotinic acid" and "niacinamide". High doses of niacin (as nicotinic acid) can lower cholesterol levels (although the exact mechanism of action is still not known). The other form of niacin (nicotinamide or niacinamide) does not open blood vessels wider nor provide a cholesterol-lowering effect.

THE DEFICIENCY

At the present time, niacin deficiency is rarely seen in developed countries and is usually apparent in conditions of poverty and malnutrition and chronic alcoholism. Alcoholic patients typically experience increased intestinal permeability leading to negative health outcomes. Studies have indicated that in patients with alcoholic pellagra, niacin deficiency may be an important factor influencing both the onset and severity of this condition. Severe deficiency of niacin in the diet causes the disease pellagra. Pellagra is characterized by diarrhea, dermatitis and dementia as well as "necklace" lesions on the lower neck, hyperpigmentation, thickening of the skin, inflammation of the mouth and tongue, digestive disturbances, amnesia, delirium, and eventually death, if left untreated. Common psychiatric symptoms of niacin deficiency include irritability, poor concentration, anxiety, fatigue, restlessness, apathy, and depression. Mild niacin deficiency has been shown to slow metabolism, causing decreased tolerance to the cold. Dietary niacin deficiency tends to occur in areas where people eat maize ("corn") as a staple food. Maize is the only grain low in niacin, and nixtamalization is needed to increase the bioavaiability of niacin during meal/flour production. Nixtamalization refers to the process of cooking maize with alkaline lime. This is the primary processing step during the manufacture of maize products, including chips,

*The statins (or HMG-CoA reductase inhibitors) are a class of drug used to lower plasma cholesterol level.

tortillas, and taco shells. The basic pre-Columbian technique involves cooking whole maize in water for 12-16 hours in large tanks. The steeped maize is referred to as nixtamal, and the cooked liquid is nejayote. This process functions to soften the pericarp of the maize, and allows the endosperm to absorb water, enabling its milling. The nixtamal is washed and then stone-ground to produce masa, which is used to produce a variety of products with improved bioavailability of niacin.

The recommended daily allowance of niacin is 2-12 mg/day for children, 14 mg/day for women, 16 mg/day for men, and 18 mg/day for pregnant or breast-feeding women. The upper limit for adult men and women is 35 mg/day, which is based on flushing as the critical adverse effect, this dose-dependent flushing effect consists of a single episode 10 to 20 minutes after niacin is taken. Skin flushing is not technically considered a side-effect as such; this flushing is caused by increased blood flow caused by the niacin and, in general, ceases after several days of taking it. Hartnup's disease is a hereditary nutritional disorder resulting in niacin deficiency. This condition was first identified in the 1950s by the Hartnup family in London. It is due to a deficit in the intestines and kidneys, making it difficult for the body to break down and absorb dietary tryptophan. The resulting condition is similar to pellagra, including symptoms of red, scaly rash, and sensitivity to sunlight. Oral niacin is given as a treatment for this condition in doses ranging from 40-200 mg with a good prognosis if identified and treated early. Niacin synthesis is also deficient in carcinoid syndrome, because of metabolic diversion of its precursor, tryptophan, to form serotonin.

In general, niacin status is tested through urinary biomarkers, which are believed to be more reliable than plasma levels.

VITAMIN B_4 ADENINE

Adenine is a nucleobase (a purine derivative) with a variety of roles in biochemistry including cellular respiration, in the form of both the energy-rich adenosine triphosphate (ATP) and the cofactors nicotinamide adenine dinucleotide (NAD) and flavin adenine dinucleotide (FAD), and protein synthesis, as a chemical component of DNA and RNA.[1] The shape of adenine is complementary to either thymine in DNA or uracil in RNA. In older literature, adenine was sometimes called **Vitamin B_4**. It is no longer considered a true vitamin or part of the Vitamin B complex. However, two B vitamins, niacin and riboflavin, bind with adenine to form the essential cofactors nicotinamide adenine dinucleotide (NAD) and flavin adenine dinucleotide (FAD), respectively. Hermann Emil Fischer was one of the early scientists to study adenine. Some think that, at the origin of life on Earth, the first adenine was formed by the polymerization of five hydrogen cyanide (HCN) molecules. However, this has been criticized by some chemists.

VITAMIN B_5 PANTOTHENIC ACID

INTRODUCTION

Pantothenic acid was discovered in 1933 and belongs to the group of water-soluble B vitamins. Its name originates from the Greek word "pantoo", meaning "every where", as it can be found throughout all living cells.

THE CHEMISTRY OF PANTOTHENIC ACID

Pantothenic acid is isolated from the liver extract. The chemical analysis yields that a butyroacetone and β - alanine which were condensation in the resultant vitamin

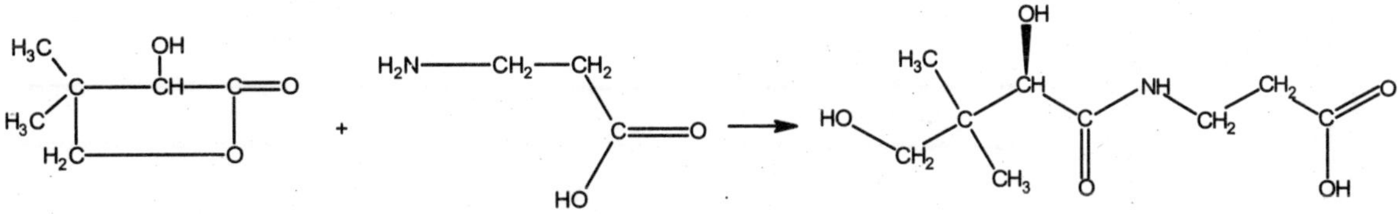

3-hydroxy-4,4-dimethyl dihydrofuran-2(3H)-one + 3-aminopropanoic acid → 3-{[(2R)-2,4-dihydroxy-3,3-dimethylbutanoyl] amino}propanoic acid

pantothenic acid

Pantothenic acid is pale yellow viscous oil, with a molecular weight of 219. The synthesized variety prepared commercially is white salt of Pantothenic acid calcium salt $(C_9H_{16}O_5N)_2Ca$ The structural formula shown below:

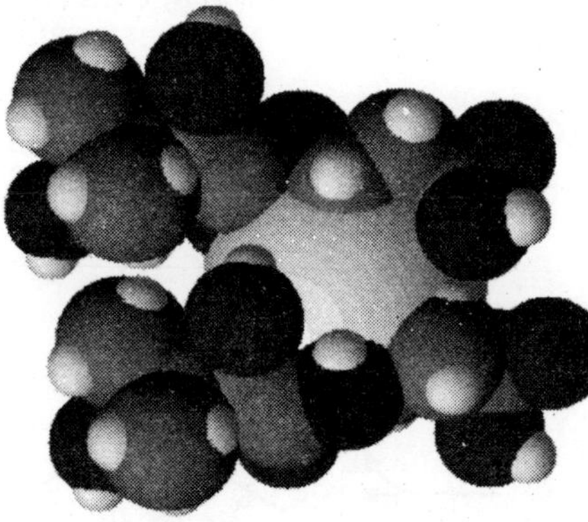

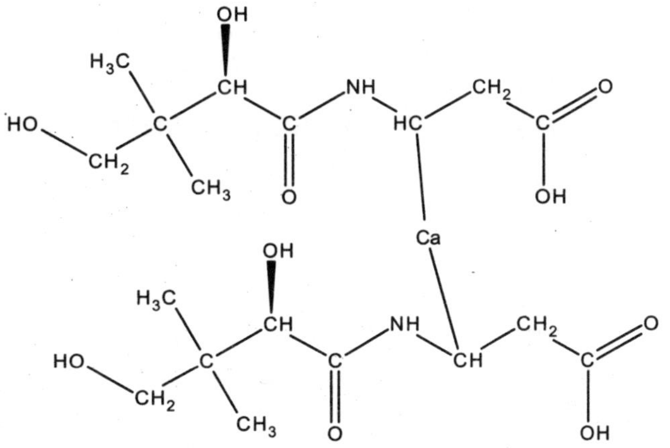

Pantothenic acid is absorbed by Activated charcoal or activated aluminum (previously active with hydrochloric acid) and may be extracted with ammonia or pyridine and methanol. It is soluble in water ethyl acetate and acetic acid. It is slightly soluble in ether and amyl alcohol. It does not dissolve in chloroform and benzene. It is unstable to heat; in the dry state or in acidic or in alkaline medium. The dexo-rotator form of the calcium salt is generally used since the calcium salt is generally used since the other (leavo form) is of little biological value. In animal tissue the vitamin exits in the free state (acids) as well as in the bound state. An example of the bound state is co – enzyme A in its active form, Pantothenic acid is a constituent of coenzyme A which is also called co – acetylase – the co enzyme involved in acetylation relations CoA is a complex of adenine ribose phosphoric acid, Pantothenic acid and mercaptoethanol amine.

Pantothenic acid in the living cells combine with mercaptoethanolamine to form pantothein, which in turn is built up into the important substance co enzyme A which plays an important role in the fat and the carbohydrate metabolism.

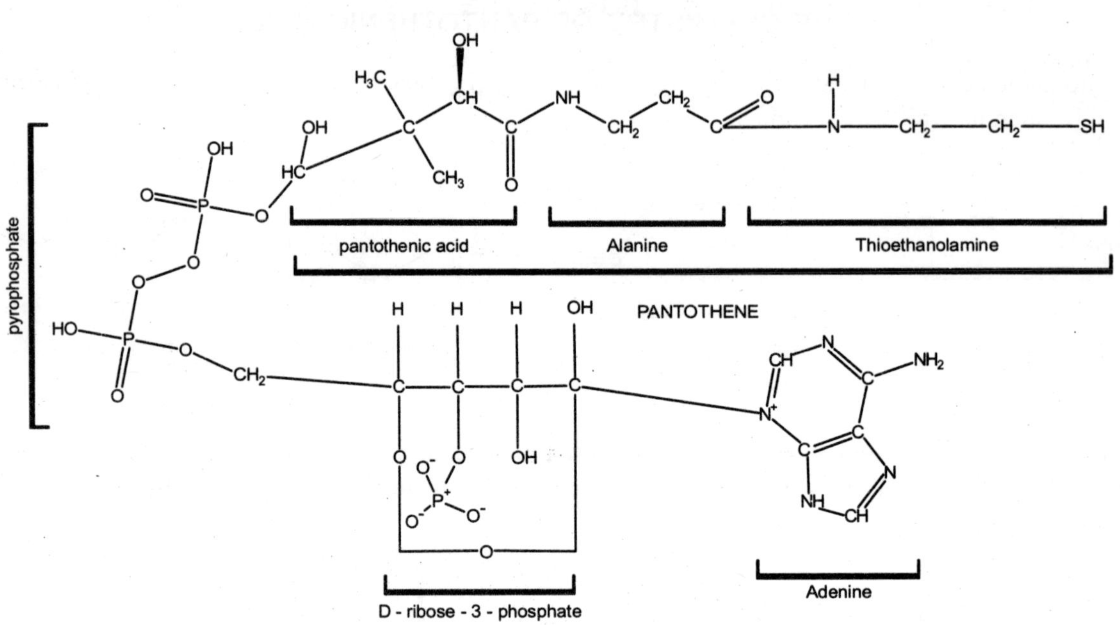

pyrophosphate
pantothenic acid
Alanine
Thioethanolamine
PANTOTHENE
D - ribose - 3 - phosphate
Adenine

THE BIOSYNTHESIS OF PANTOTHENIC ACID AND CO ENZYME A

The biosynthesis of Pantothenic acid and co enzyme A take place according to the pathway shown below.

THE SYNTHESIS OF PANTOTHENIC ACID IN OTHER LIFE FORMS

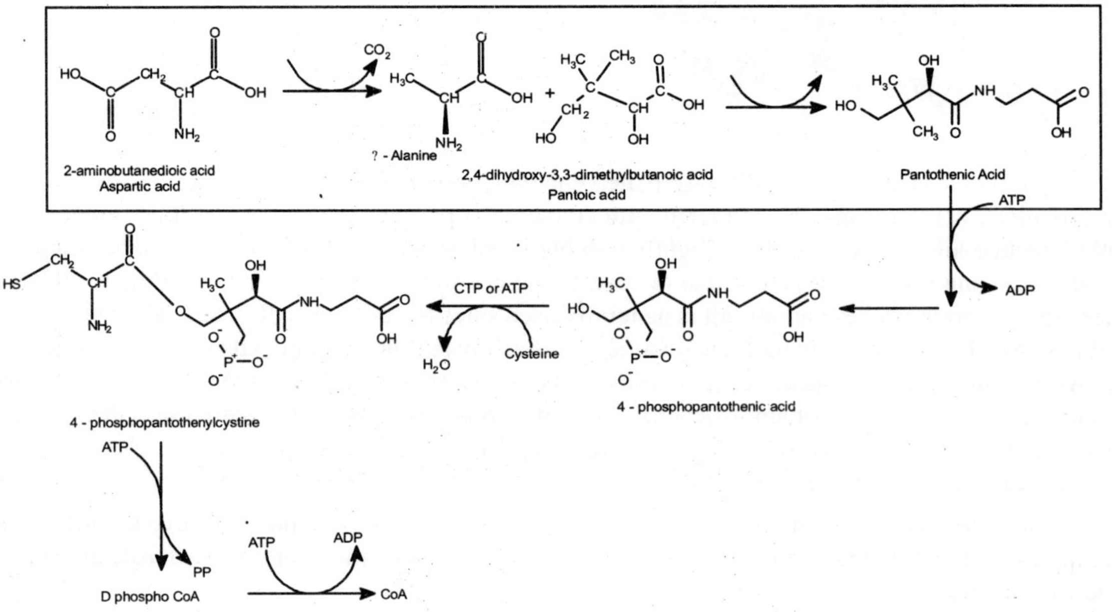

2-aminobutanedioic acid
Aspartic acid
CO_2
? - Alanine
2,4-dihydroxy-3,3-dimethylbutanoic acid
Pantoic acid
Pantothenic Acid
ATP
ADP
4 - phosphopantothenic acid
CTP or ATP
Cysteine
H_2O
4 - phosphopantothenylcystine
ATP
PP
D phospho CoA
ATP
ADP
CoA

FUNCTIONS

Pantothenic acid, as a constituent of coenzyme A (a coenzyme of acetylation), plays a key role in the metabolism of carbohydrates, proteins and fats, and is therefore important for the maintenance and repair of all cells and tissues. Coenzyme A is involved in reactions that supply energy, in the synthesis of essential lipids (e.g. sphingolipids, phospholipids), sterols (e.g. cholesterol), hormones (e.g. growth, stress and sex hormones), neurotransmitters (e.g. acetylcholine), porphyrin (a component of haemoglobin, the oxygen-carrying red blood cell pigment) and antibodies, and in the metabolism of drugs (e.g. sulphonamides) and in alcohol detoxification. Another essential role of pantothenic acid concerns acyl carrier protein, an enzyme involved in the synthesis of fatty acids. In the process of fat burning, pantothenic acid works in concert with coenyzme Q10 and L-carnitine. Most of the pantothenic acid in food exists in the form of coenzyme A, and pantothenic acid is released by a series of enzyme reactions in the small intestine. It is then absorbed by passive diffusion into the portal circulation and transported to the tissues, where re-synthesis of the coenzyme occurs. About half of the pantothenic acid in the diet is actually absorbed. If calcium pantothenate or pantothenic acid are ingested as nutritional supplements, they must first be converted to pantetheine by intestinal enzymes before being absorbed. Topical and orally applied D-panthenol (the alcoholic form of pantothenic acid that can, e.g., be found in many cosmetic products) is also absorbed by passive diffusion and transformed to pantothenic acid by enzymatic oxidation. The highest concentrations in the body are in the liver, adrenal glands, kidneys, brain, heart and testes. Total pantothenic acid levels in whole blood are at least 1 mg/L in healthy adults; most of it exists as coenzyme in the red blood cells. Urinary excretion in the form of pantothenic acid generally correlates with dietary intake, but variation is large (2-7 mg daily). During lactation, a large proportion of the intake reaches the milk (1-5 mg daily).

> The reactions responsible for the synthesis of pantothenic acid are indicated as they occur in the plant and bacteria so human metabolism requires this vitamin to be introduced through diet.

THE DEFICIENCY

Pantothenic acid deficiency is exceptionally rare and has not been thoroughly studied. In the few cases where deficiency has been seen (victims of starvation and limited volunteer trials), nearly all symptoms can be reversed with the return of pantothenic acid. Symptoms of deficiency are similar to other vitamin B deficiencies. There is impaired energy production, due to low CoA levels, which could cause symptoms of irritability, fatigue, and apathy. Acetylcholine synthesis is also impaired, therefore, neurological symptoms can also appear in deficiency. They include numbness, paresthesia, and muscle cramps. Deficiency in pantothenic acid can also cause hypoglycemia, or an increased sensitivity to insulin. Insulin receptors are acylated with palmitic acid when they do not want to bind with insulin. Therefore, more insulin will bind to receptors when acylation decreases, causing hypoglycemia. Additional symptoms could include: restlessness, malaise, sleep disturbances, nausea, vomiting, and abdominal cramps. In a few rare circumstances more serious (but reversible) conditions have been seen, such as adrenal insufficiency and encephalopathy. It has been noted that painful burning sensations of the feet were reported in tests conducted on volunteers. Deficiency of pantothenic acid may explain similar sensations reported in malnourished prisoners of war. Deficiency symptoms in other non-ruminant animals include disorders of the nervous, gastrointestinal, and immune systems, reduced growth rate, decreased food intake, skin lesions and changes in hair coat, alterations in lipid and carbohydrate metabolism.

VITAMIN B_6 PYRIDOXINE

INTRODUCTION

Vitamin B_6 is a water-soluble compound that was discovered in the 1930s during nutrition studies on rats. In 1934, a Hungarian physician, Paul Gyorgy discovered a substance that was able to cure a skin disease in rats (dermititis acrodynia), this substance he named vitamin B_6. In 1938, Lepkovsky isolated vitamin B_6 from rice bran. Harris and Folkers in 1939 determined the structure of pyridoxine, and, in 1945, Snell was able to show that there are two forms of vitamin B_6, pyridoxal and pyridoxamine. Vitamin B_6 was named pyridoxine to indicate its structural homology to pyridine. All three forms of vitamin B_6 are precursors of an activated compound known as pyridoxal 5'-phosphate (PLP), which plays a vital role as the cofactor of a large number of essential enzymes in the human body. Enzymes dependent on PLP focus a wide variety of chemical reactions mainly involving amino acids. The reactions carried out by the PLP-dependent enzymes that act on amino acids include transfer of the amino group, decarboxylation, racemization, and beta- or gamma-elimination or replacement. Such versatility arises from the ability of PLP to covalently bind the substrate, and then to act as an electrophilic catalyst, thereby stabilizing different types of carbanionic reaction intermediates. Overall, the Enzyme Commission has catalogued more than 140 PLP-dependent activities, corresponding to 4% of all classified activities.

FORMS OF PYRIDOXINE

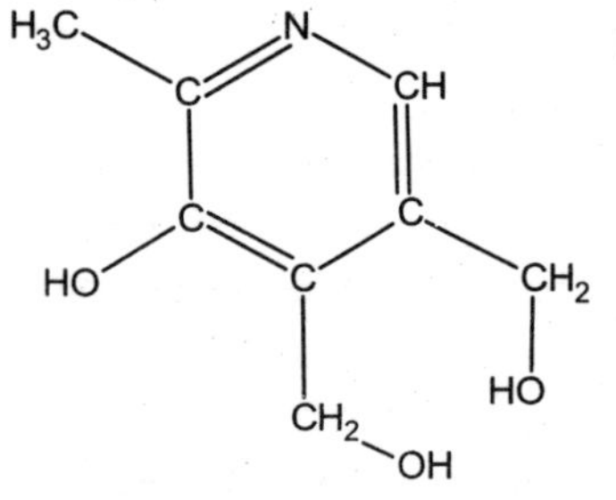

Pyridoxine

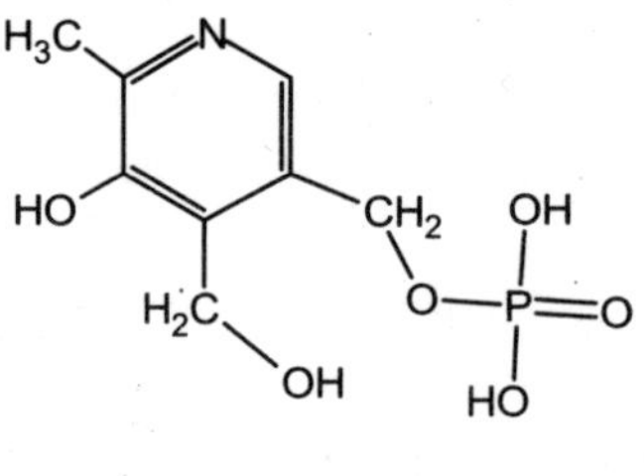

Pyridoxal Phosphate

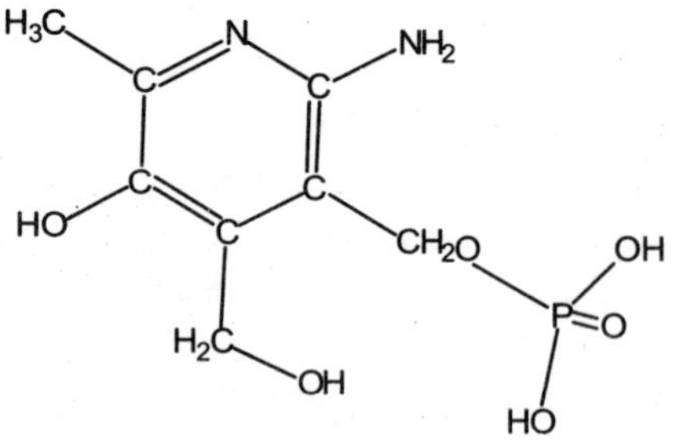

Pyridoxamine Phosphate

Seven forms of this vitamin are known:

- Pyridoxine (PN), the form that is given as vitamin B_6 supplement
- Pyridoxine 5'-phosphate (PNP)
- Pyridoxal (PL)
- Pyridoxal 5'-phosphate (PLP), the metabolically active form
- Pyridoxamine (PM)
- Pyridoxamine 5'-phosphate (PMP)
- 4-Pyridoxic acid (PA), the catabolite which is excreted in the urine

All forms except PA can be interconverted.

THE CHEMISTRY OF PYRIDOXINE

Pyridoxine is not the most active element form of the vitamin available in nature and is convertible to other derivatives which are more active in their function in the metabolic processes. They are pyridoxal and pyridoxamine phosphate, in nature the Vitamin B_6 occurs in the all three compounds. The main constitution of the Vitamin B_6 is 4-pyridoxic acid. (3-hydroxy-2,5-dimethylpyridine-4-carboxylic acid) it is extracted from urine.

FUNCTIONS IN THE METABOLISM

Pyridoxal phosphate, the metabolically active form of vitamin B_6, is involved in many aspects of macronutrient metabolism, neurotransmitter synthesis, histamine synthesis, hemoglobin synthesis and function and gene expression. Pyridoxal phosphate generally serves as a coenzyme for many reactions and can help facilitate decarboxylation, transamination, racemization, elimination, replacement and beta-group interconversion reactions. The liver is the site for vitamin B_6 metabolism.

Amino Acid Metabolism

Pyridoxal phosphate (PLP) is a cofactor in transaminases that can catabolize amino acids. PLP is also an essential component of two enzymes that converts methionine to cysteine via two reactions. Low vitamin B_6 status will result in decreased activity of these enzymes. PLP is also an essential cofactor for enzymes involved in the metabolism of selenomethionine to selenohomocysteine and then from selenohomocysteine to hydrogen selenide. Vitamin B_6 is also required for the conversion of tryptophan to niacin and low vitamin B_6 status will impair this conversion. PLP is also used to create physiologically active amines by decarboxylation of amino acids. Some notable examples of this include: histidine to histamine, tryptophan to serotonin, glutamate to GABA (gamma-aminobutyric acid), and dihydroxyphenylalanine to dopamine.

Gluconeogenesis

Vitamin B_6 also plays a role in gluconeogenesis. Pyridoxal phosphate can catalyze transamination reactions that are essential for the providing amino acids as a substrate for gluconeogenesis. Also, vitamin B_6 is a required coenzyme of glycogen phosphorylase, the enzyme that is necessary for glycogenolysis to occur.

Lipid Metabolism

Vitamin B_6 is an essential component of enzymes that facilitate the biosynthesis of sphingolipids. Particularly, the synthesis of ceramide requires PLP. In this reaction serine is decarboxylated and combined with palmitoyl-CoA to form sphinganine which is combined with a fatty acyl CoA to form dihydroceramide. Dihydroceramide is then further desaturated to form ceramide. In addition, the breakdown of sphingolipids is also dependent on vitamin B_6 since S1P lyase, the enzyme responsible for breaking down sphingosine-1-phosphate, is also PLP dependent.

Metabolic Functions

The primary role of vitamin B_6 is to act as a coenzyme to many other enzymes in the body that are involved predominantly in metabolism. This role is performed by the active form, pyridoxal phosphate. This active form is converted from the two other natural forms founds in food: pyridoxal, pyridoxine and pyridoxamine.

Vitamin B_6 is involved in the following metabolic processes:

- amino acid, glucose and lipid metabolism
- neurotransmitter synthesis
- histamine synthesis
- hemoglobin synthesis and function
- gene expression

Amino Acid Metabolism

Pyridoxal phosphate is involved in almost all amino acid metabolism, from synthesis to breakdown.

1. **Transamination:** transaminase enzymes needed to break down amino acids are dependent on the presence of pyridoxal phosphate. The proper activity of these enzymes are crucial for the process of moving amine groups from one amino acid to another.
2. **Transsulfuration:** Pyridoxal phosphate is a coenzyme needed for the proper function of the enzymes cystathionine synthase and cystathionase. These enzymes work to transform methionine into cysteine.
3. **Selenoamino acid metabolism:** Selenomethionine is the primary dietary form of selenium. Pyridoxal phosphate is needed as a cofactor for the enzymes that allow selenium to be used from the dietary form. Pyridoxal phosphate also plays a cofactor role in releasing selenium from selenohomocysteine to produce hydrogen selenide. This hydrogen selenide can then be used to incorporate selenium into selenoproteins.
4. **Vitamin B_6** is also required for the conversion of tryptophan to niacin and low vitamin B_6 status will impair this conversion.

Neurotransmitter Synthesis

Pyridoxal phosphate-dependent enzymes play a role in the biosynthesis of four important neurotransmitters: serotonin, epinephrine, norepinephrine and gamma-aminobutyric acid. Serine racemase, which synthesizes the neuromodulator D-serine, is also a pyridoxal phosphate-dependent enzyme.

Histamine Synthesis

Pyridoxal phosphate is involved in the metabolism of histamine.

Hemoglobin Synthesis and Function

Pyridoxal phosphate aids in the synthesis of heme and can also bind to two sites on hemoglobin to enhance the oxygen binding of hemoglobin.

Gene Expression

It transforms homocysteine into cistation then into cysteine. Pyridoxal phosphate has been implicated in increasing or decreasing the expression of certain genes. Increased intracellular levels of the vitamin will lead to a decrease in the transcription of glucocorticoid hormones. Also, vitamin B_6 deficiency will lead to the increased expression of albumin mRNA. Also, pyridoxal phosphate will influence gene expression of glycoprotein IIb by interacting with various transcription factors. The result is inhibition of platelet aggregation.

DEFICIENCIES

The classic clinical syndrome for B_6 deficiency is a seborrhoeic dermatitis-like eruption, atrophic glossitis with ulceration, angular cheilitis, conjunctivitis, intertrigo, and neurologic symptoms of somnolence, confusion, and neuropathy. While severe vitamin B_6 deficiency results in dermatologic and neurologic changes, less severe cases present with metabolic lesions associated with insufficient activities of the coenzyme pyridoxal phosphate. The most prominent of the lesions is due to impaired tryptophan-niacin conversion. This can be detected based on urinary excretion of xanthurenic acid after an oral tryptophan load. Vitamin B_6 deficiency can also result from impaired transsulphuration of methionine to cysteine. The pyridoxal phosphate-dependent transaminases and glycogen phosphorylase provide the vitamin with its role in gluconeogenesis, so deprivation of vitamin B_6 results in impaired glucose tolerance. A deficiency of vitamin B_6 alone is relatively uncommon and often occurs in association with other vitamins of the B complex. The elderly and alcoholics have an increased risk of vitamin B_6 deficiency, as well as other micronutrient deficiencies. Renal patients undergoing dialysis may experience vitamin B_6 deficiency. The availability of vitamin B_6 to the body can be affected by certain drugs such as anticonvulsants and corticosteriods.

VITAMIN B_7 BIOTIN (VITAMIN H)

Biotin is a water-soluble B-complex vitamin (Vitamin B_7) that is composed of an ureido (tetrahydroimidizalone) ring fused with a tetrahydrothiophene ring. A valeric acid substituent is attached to one of the carbon atoms of the tetrahydrothiophene ring. Biotin is a coenzyme in the metabolism of fatty acids and leucine, and it plays a role in gluconeogenesis. Biotin is necessary for cell growth, the production of fatty acids, and the metabolism of fats and amino acids. It plays a role in the citric acid cycle, which is the process by which biochemical energy is generated during aerobic respiration. Biotin not only assists in various metabolic reactions but also helps to transfer carbon dioxide. Biotin may also be helpful in maintaining a steady blood sugar level. Biotin is often recommended for strengthening hair and nails. Consequently, it is found in many cosmetics and health products for the hair and skin, though it cannot be absorbed through the hair or skin itself. Biotin deficiency is rare, as, in general, intestinal bacteria produce an excess of the body's recommended daily requirement. For that reason, statutory agencies in many countries, for example the USA and Australia, do not prescribe a recommended daily intake of biotin.

THE BIOCHEMISTRY OF BIOTIN

Biotin D(+) is a cofactor responsible for carbon dioxide transfer in several carboxylase enzymes:

- Acetyl-CoA carboxylase alpha
- Acetyl-CoA carboxylase beta
- Methylcrotonyl-CoA carboxylase
- Phosphoenolpyruvate carboxylase
- Propionyl-CoA carboxylase
- Pyruvate carboxylase

and, so, is important in fatty acid synthesis, branched-chain amino acid catabolism, and gluconeogenesis. Biotin covalently attaches to the epsilon-amino group of specific lysine residues in these carboxylases. This biotinylation reaction requires ATP and is catalyzed by holocarboxylase synthetase. The attachment of biotin to various chemical sites can be used as an important laboratory technique to study various processes including protein localization, protein interactions, DNA transcription and replication. Biotinidase itself is known to be able to biotinylate histones, but little biotin is found naturally attached to chromatin. Biotin binds very tightly to the tetrameric protein avidin (also streptavidin and neutravidin), with a dissociation constant K_d in the order of 10^{-15}, which is one of the strongest known protein-ligand interactions, approaching the covalent bond in strength. This is often used in different biotechnological applications. Until 2005, very harsh conditions were required to break the biotin-streptavidin bond.

5-(5-oxo-1,4,6-thiadiazocan-2-yl)pentanoic acid

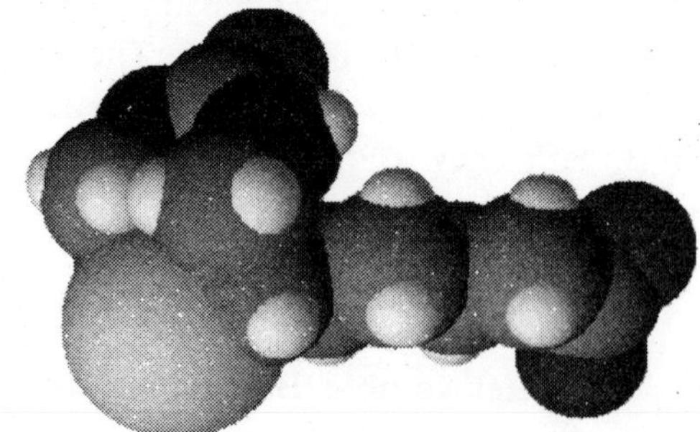

Biotin is a monocarboxylic acid with two fused rings.

The acid is valeric acid $CH_3 - CH_2 - CH_2 - CH_2 - COOH$. One of the fused ring is of the mimidiazole variety and the other is thiophene derivative. It is structurally represented below:

Biotin is a crystalline compound melting at 230°C-232°C, it is a water soluble compound also soluble in alcohol. Biotin has a property of resting autoclaving with strong inorganic acids. The biotin and thiamine are only two vitamins with sulphur as an element. Biotin is obtained from liver is identical with that from milk. Biotin as obtained from egg yolk (so called α - biotin) is different from liver (β - biotin) in chemical nature of the side chain. This difference has not been conclusively proved.

Desthiobiotin can be used by an organism for the synthesis of biotin. Imidiazoleonecarporic acid is closely related to destiobiotin and prevents the later conversion to biotin thus acting violently to it.

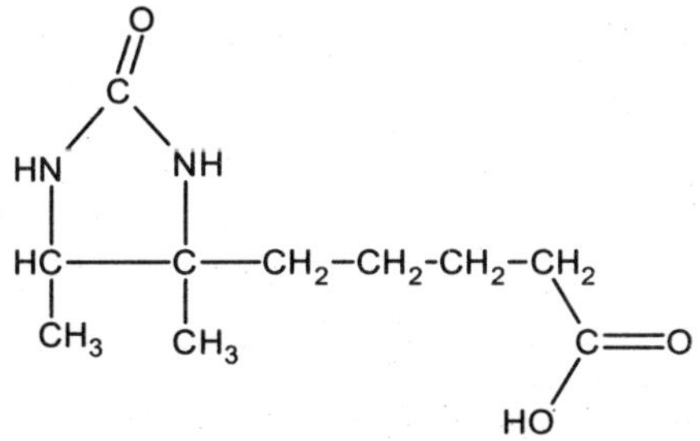

5-(4,5-dimethyl-2-oxoimidazolidin-4-yl)pentanoic acid
desthiobiotin

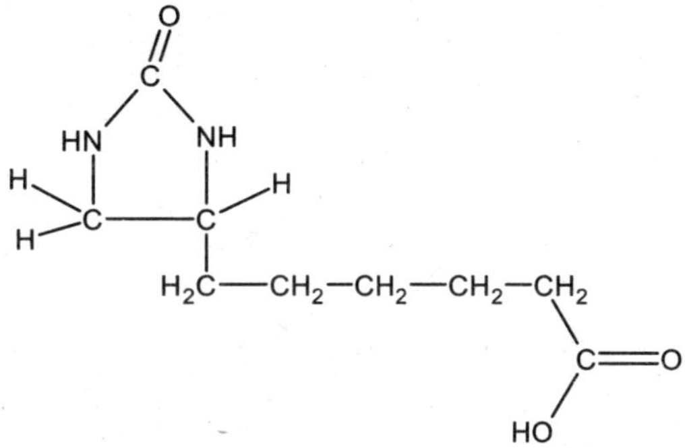

6-(2-oxoimidazolidin-4-yl)hexanoic acid
imidazolin 1 carporic acid

Biotin bound to a specific enzyme protein is intimately associated with carboxylation reactions, biotin participates in three well known carboxylation catalyzed by different enzymes, all the three reactions utilize biotin as the actual CO_2 carrier. In the carboxylation (CO_2 fixation) reactions, Biotin co –enzyme complex is attached to the CO_2 which is transferred to other substances. The CO_2 on the

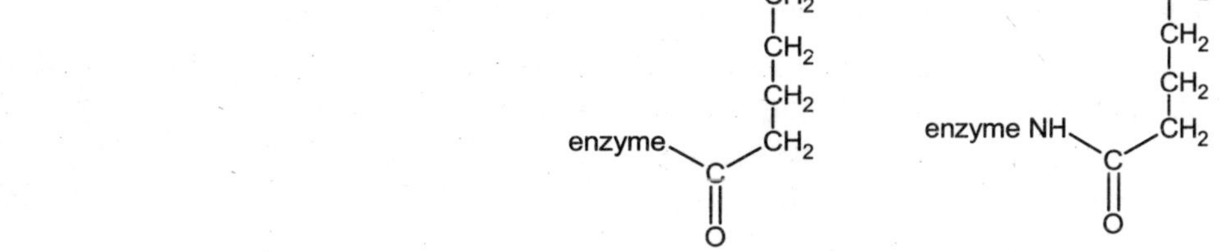

CO_2 biotine enzyme

complex is transferred in a reaction catalyzed by acetyl CoA carboxylase to form malonyl CoA. This constitutes an important step in the entire mitochondrial biosynthesis of fatty acids.

Malonyl CoA is formed from acetyl CoA and bicarbonate with biotin as co –enzyme and ATP to Supply the energy, Biotin is first converted into carboxy biotin by reaction with bicarbonate and ATP ($ATP \rightarrow ADP + P_i$). The carboxy – biotin then reacts with acetyl CoA to give biotin and malonyl – CoA

$$CH_3COs - CoA \xrightarrow[\text{acetylCoA}]{\text{BiotinHCO}_3^-\text{Mn}^{++}} HOOC - CH_2 - CO - S - CoA$$

The malonyl CoA reacts with fatty acid synthase which binds the malonyl group and another acetyl CoA with loss of CO_2 and formation of acetylaceto – co enzyme

$HOOC - CH_2COS -$ enzyme $+ CH_3COSCoA +$ ACETYL CoA $\rightarrow CoASH + CO_2 + CH_3COCH_2CoS$-enzyme

Acetyl CoA is reduced by NADPH

$$CH_3COH_2CoS\text{ - enzyme} \longrightarrow CH_3CHOHCH_2CO\text{ - S - enzyme}$$

NADPH + H^+ NADP$^+$

A water molecule is lost

$$H_3C—CH{=}CH—\overset{O}{\overset{\|}{C}}—S—\text{enzyme} \longrightarrow H_3C—CH_2—CH_2—\overset{O}{\overset{\|}{C}}—S—\text{enzyme}$$

NADPH + H^+ NADP +

Butyryl – enzyme is finally converted to butyryl CoA

$CH_3 - CH_2 - CH_2\ CoS -$ enzyme $+ CoASH \rightarrow CH_3CH_2CoSCoA +$ enzyme
Butyryl CoA

Biotin CoA reacts with malonyl enzyme and the cycle is repeated until the chain length reaches up to 16 or 18 carbons.

The fatty acid is released from the enzyme CoA – SH complex or converted to fats. CO_2 biotin complex is the source of the active CO_2 required for the formation of carbamyl phosphate, which in turn provides the active CO_2 and ammonia to ornithine.

THE FUNCTION OF BIOTIN

Biotin participates in the formation of an essential co–enzyme for a series of enzymes catalyzed reactions involved in the urea cycle and the synthesis of arginine as well as well as for the functioning of biosynthetic pathway for pyrimidines and the extra mitochondrial pathway of fatty acid biosynthesis.

Biotin may also involved in the conversion of pyruvate to oxaloacetate carboxylation reaction Amino group for Glutamate is transferred to form aspartic acid by transamination reaction. Addition of aspartic acid and oxaloacetic acid to the nutrient solution reduces the requirement of the vitamin may be involved in the series of transamination reactions. he primary role of biotin is in the metabolism of fats, proteins and carbohydrates.

Biotin functions as a critical component of several enzymes (where it functions as a coenzyme) involved in energy metabolism (such as pyruvate carboxylase). Biotin helps in the synthesis of fatty acids, in energy metabolism, and in the synthesis of amino acids and glucose. Biotin is also involved in making glucose, some amino acids and in energy production. Vitamin H plays a special role in enabling the body to use blood sugar (glucose), a major source of energy for body fluids. Biotin supplements may improve thin, splitting, or brittle toe and fingernails as well as hair health. Biotin has also been used to combat alopecia (partial or complete loss of hair) in both children and adults. Biotin is typically included in most multi-vitamin supplements.

Biotin is used to treat the biotin-responsive inborn errors of metabolism holocarboxylase synthetase deficiency and biotinidase deficiency. Large doses of biotin may be given to babies with a condition called infantile seborrhea or to patients with genetic abnormalities in biotin metabolism. Biotin supplements are sometimes given to help reduce blood sugar in diabetic patients. People with type 2 diabetes often have low levels of biotin. The major benefit of biotin as a dietary supplement is in strengthening hair and nails. Biotin supplements may improve thin or splitting toenails or fingernails and improve hair health. Some skin disorders, such as "cradle cap," improve with biotin supplements. Biotin has also been used to combat premature graying of hair, though it is likely to be useful only for those with a low biotin level. Biotin has been used for people in weight loss programs to help them metabolize fat more efficiently.

THE BIOTIN DEFICIENCY

Biotin deficiency is a rare nutritional disorder which can become serious, even fatal, if allowed to progress untreated. It can occur in people of any age, ancestry, or gender. Biotin deficiency rarely occurs among healthy people, since the daily requirement of biotin is low, many foods provide adequate amounts of it, intestinal bacteria synthesize small amounts of it, and the body effectively scavenges and recycles it from bodily waste. However, deficiencies can be caused by consuming raw egg whites over a period of months to years. Egg whites contain high levels of avidin, a protein that binds biotin strongly. When cooked, avidin is denatured and becomes entirely non-toxic.

VITAMIN B_8 INOSITOL

Inositol or **cyclohexane-1, 2, 3, 4, 5, 6-hexol** is a chemical compound with formula $C_6H_{12}O_6$ or $(\text{-CHOH-})_6$, a sixfold alcohol (polyol) of cyclohexane. It exists in nine possible stereoisomers, of which the most prominent form, widely occurring in nature, is ***cis*-1, 2, 3, 5-*trans*-4, 6-cyclohexanehexol**, or

***myo*-inositol**. Inositol is a carbohydrate, though not a classical sugar. It is almost tasteless, with a small amount of sweetness.*Myo*-inositol plays an important role as the structural basis for a number of secondary messengers in eukaryotic cells, including inositol phosphates, phosphatidylinositol (PI) and phosphatidylinositol phosphate (PIP) lipids. In plants, the hexaphosphate of inositol is found as phytic acid. Inositol or its phosphates and associated lipids are found in many foods, in particular, in cereals with high bran content, nuts, beans, and fruit, especially cantaloupe melons and oranges. *Myo*-Inositol was once classified as a member of the vitamin B complex (B_8). However, because it is produced by the human body from glucose, it is not an essential nutrient. Some substances such as niacin can also be synthesized in the body, but are not made in amounts considered adequate for good health, and thus are still classified as essential nutrients. However, there is no convincing evidence that this is the case for *Myo*-inositol.

STRUCTURES AND ISOMERS OF INOSITOL

Besides *myo*-inositol, the other naturally occurring isomers (though in minimal quantities) are ***scyllo*-**, ***muco*-**, **D-*chiro*-**, and ***neo*-inositol**. The other possible isomers are **L-*chiro*-**, ***allo*-**, ***epi*-**, and ***cis*-inositol**.

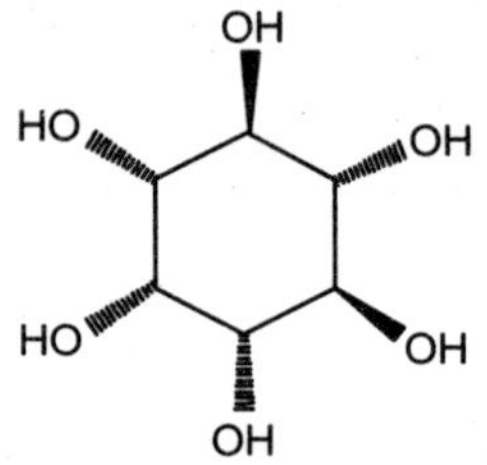

myo- 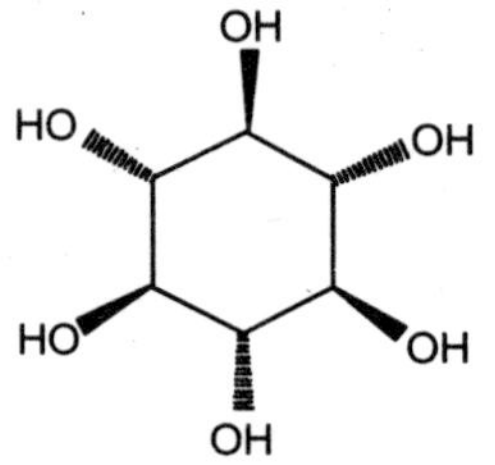 *scyllo-* 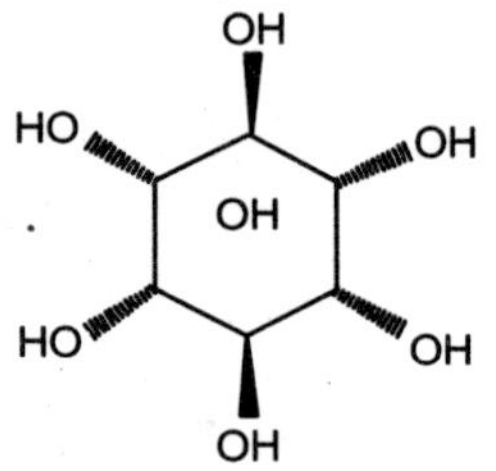 *muco-*

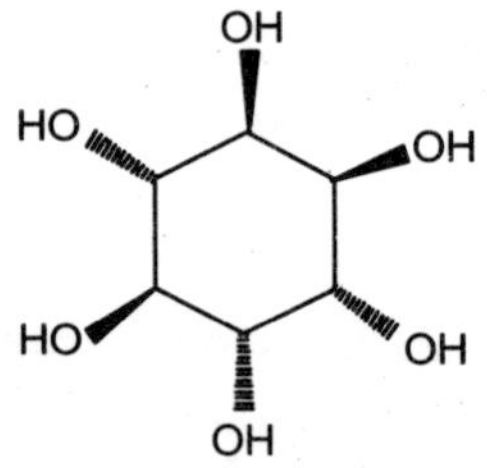

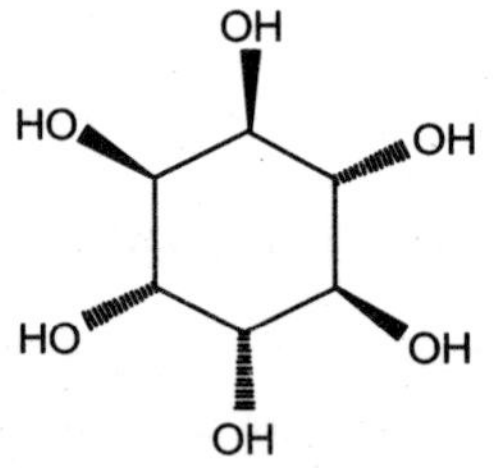

Chiro- 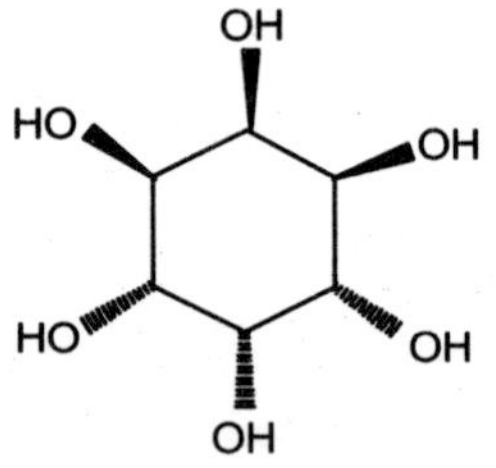 *Neo-*

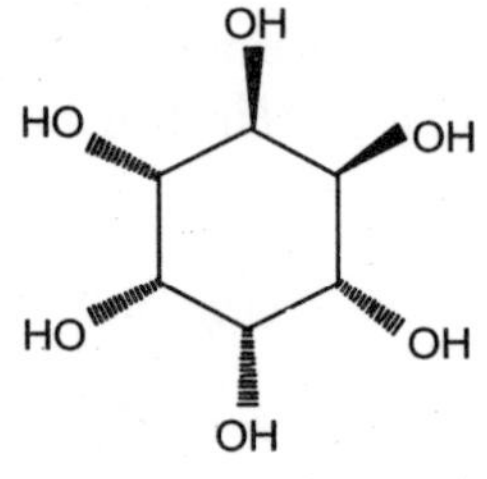

allo- 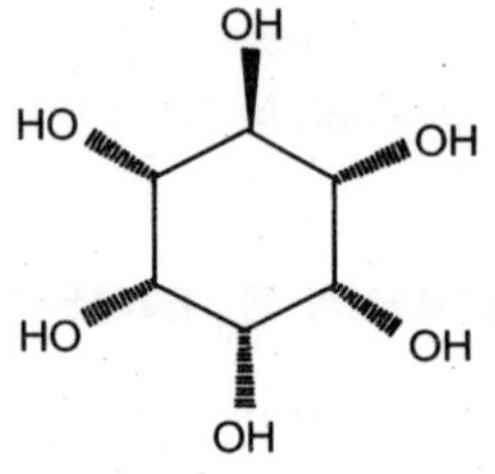 *epi-* 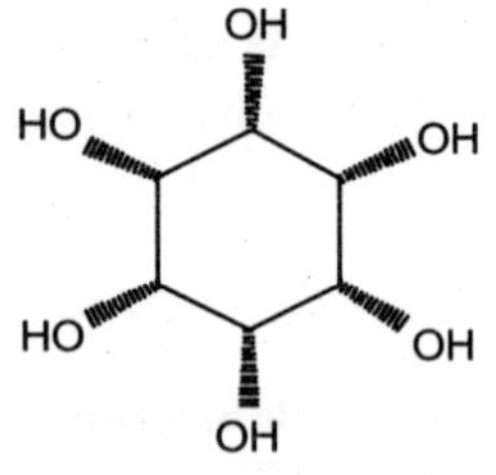 *cis-*

In its most stable conformational geometry, the *myo*-inositol isomer assumes the chair conformation, which puts the maximum number of hydroxyls to the equitoral position, where they are farthest apart from each other. In this conformation the natural *myo* isomer has a structure in which five of the six hydroxyls (the 1st, 3rd, 4th, 5th, and 6th are equatorial, whereas the 2nd hydroxyl group is axial.

THE FUNCTION OF INOSITOL

Inositol and a number of its mono and polyphosphates function as the basis for a number of signaling and secondary messenger molecules. They are involved in a number of biological processes, including:

- Insulin signal transduction
- Cytoskeleton assembly
- Nerve guidance (Epsin)
- Intracellular calcium (Ca^{2+}) concentration control
- Cell membrane potential maintenance
- Serotonin activity modulation
- Breakdown of fats and reducing blood cholesterol
- Gene expression

The **cytoskeleton** (also CSK) is a cellular "scaffolding" or "skeleton" contained within the cytoplasm and is made out of protein. The cytoskeleton is present in both plant and animal cells. The cytoskeleton was once thought to be unique to eukaryotes, but recent research has identified the prokaryotic cytoskeleton. It is a dynamic structure that maintains cell shape, protects the cell, enables cellular motion (using structures such as flagella, cilia and lamellipodia), and plays important roles in both intracellular transport (the movement of vesicles and organelles, for example) and cellular division. The concept and the term (*cytosquelette*, in French) was first introduced by French embryologist Paul Wintrebert in 1931.

DAILY INTAKE AND DEFICIENCY

An **inositol deficiency** could be a contributing cause of abnormal platelet aggregation, alcoholism, heart disease (along with B-vitamin deficiencies), liver disease, memory loss, panic/anxiety attacks, and tardive dyskinesia. Pharmaceutical drugs that can cause an inositol deficiency include aminoglycosides, cephalosporins, chlortetracycline, demeclocycline, doxycycline, fluoroquinolones, macrolides, minocycline, oxytetracycline, penicillins, sulphonamides, tetracyclines and trimethoprim. If the inositol intake is not sufficient, symptoms such as eczema, hair loss, constipation, and abnormalities of the eyes and raised cholesterol may be observed. People with diabetes have increased excretion and may benefit from inositol supplementation. People with diabetes experience much greater losses of inositol than other people, which may be caused by an increase in the amount of glucose that appears in the urine in diabetes.

Most people do not need to take inositol. In addition, the small amounts commonly found in multivitamin supplements are probably unnecessary and ineffective. Doctors sometimes suggest 500 mg twice per day. For depression, anxiety, and obsessive-compulsive disorder, 12-18 grams per day has been shown to be effective in double-blind trials. There is no recommended daily allowance for inositol, but the normal human dietary intake is about 1 gram per day. Dosages in the range of 12-18 grams daily have been used for certain conditions such as panic disorder. For the management of depression and panic attacks, 12 grams of myo-inositol daily, in divided doses, were used in clinical studies. In the clinical studies performed with myo-inositol, effects, if any, were seen in about one month. Compliance with such doses may be a problem.

WHAT IS THE SCIENTIFIC EVIDENCE FOR INOSITOL?

Depression

Small double-blind studies have found inositol helpful for depression. In one such trial, 28 depressed individuals were given a daily dose of 12 g of inositol for 4 weeks. By the fourth week, the group receiving inositol showed significant improvement compared to the placebo group.

However, a double-blind study of 42 people with severe depression that was not responding to standard antidepressant treatment found no improvement when inositol was added.

Panic Disorder

People with panic disorder frequently develop panic attacks, often with no warning. The racing heartbeat, chest pressure, sweating, and other physical symptoms can be so intense that they are mistaken for a heart attack. A small double-blind study (21 participants) found that people given 12 g of inositol daily had fewer and less severe panic attacks as compared to the placebo group. A double-blind, crossover study of 20 individuals compared inositol to the antidepressant drug fluvoxamine (Luvox), a medication related to Prozac. The results over 4 weeks of treatment showed that the supplement was at least as effective as the drug.

Bipolar Disorder

In a 6-week, double-blind study, 24 individuals with bipolar disorder received either placebo or inositol (2 g three times daily for a week, then increased to 4 g three times daily) in addition to their regular medical treatment.[5] The results of this small study failed to show statistically significant benefits; however, promising trends were seen that suggest a larger study is warranted.

Polycystic Ovary Syndrome

Polycystic ovary syndrome (PCOS) is a chronic endocrine disorder in women that leads to infertility, weight gain, and many other problems. In a double-blind, placebo-controlled trial, 136 women with PCOS were given inositol at a dose of 100 mg twice daily, while 147 were given placebo. Over the study period of 14 weeks, participants given inositol showed improvement in ovulation frequency as compared to those given placebo. Benefits were also seen in terms of weight loss and levels of HDL ("good") cholesterol. A subsequent study of 94 people found similar results. However, both of the studies were performed by the same research group. Independent confirmation will be necessary before inositol could be considered an effective treatment for PCOS.

VITAMIN B_9 FOLIC ACID

Folic acid (also known as vitamin B_9 or folacin) and folate (the naturally occurring form), as well as pteroyl-L-glutamic acid and pteroyl-L-glutamate, are forms of the water-soluble vitamin B_9. Folic acid is itself not biologically active, but its biological importance is due to tetrahydrofolate and other derivatives after its conversion to dihydrofolic acid in the liver. Vitamin B_9 (folic acid and folate inclusive) is essential to numerous bodily functions ranging from nucleotide biosynthesis to the remethylation of homocysteine. The human body needs folate to synthesize DNA, repair DNA, and methylate DNA as well as to act as a cofactor in biological reactions involving folate. It is especially important during periods of rapid cell division and growth. Both children and adults require folic acid to produce healthy red blood cells and prevent anemia. Folate and folic acid derive their names from the Latin word *folium* (which means "leaf"). Leafy vegetables are a principal source, although in Western diets fortified cereals and bread may be a larger dietary source. A lack of dietary folic acid leads to folate deficiency (FD). This can result in many health problems, most notably neural tube defects in developing embryos. Low folate can also lead to homocysteine accumulation as a result of one carbon metabolism mechanism methylation being impaired. DNA synthesis and repair are impaired and this could lead to cancer development. Supplementation in patients with ischaemic heart disease may however lead to increased rates of cancer and all-cause mortality. A 2003 opinion article in the *New York Times* named micronutrients, especially folic acid, the "world's most luscious food" since absence of folic acid and a handful of other micronutrients causes otherwise-preventable deformities and diseases, especially in fetal development. Adding folic acid and micronutrients to the food supply of developing countries would have a greater impact than any other single action in improving world health.

Over the years, folic acid has been called by many names, but *folacin* is the name now officially used. The term "folate" remains popular and is used to encompass the entire group of folate vitamin forms: the natural "folypolyglutamates" found in food, and "folic acid" (pteroylglutamic acid), the synthetic form added to dietary supplements and fortified foods. Other names include: Vitamin Bc, folate, Wills' factor, pteroylmonoglutamic acid (PGA), antianemia factor PGA, Vitamin M, SLR factor, factor R, factor U, factor CF, CF, factor LC, leucovorin, Vitamin U, Liver lactobacillus casei factor, citrovoram factor, yeast norit eluate factor, folsaure, Cytofol, Foldine, Folacin, Folaemin, Foliamin, Folicet, Folipac, Folettes, Folsan, Folvite, Incafolic, Millafol, Vitamin B9.

THE BIOCHEMISTRY OF FOLIC ACID AND ITS DERIVATIVES

The vitamin has been identified in several forms containing one three and seven molecules of glutamic acid. It is synthesized by the combination of pteridine nucleus, *p* – aminobenzoic acid and 1 (+) glutamic acid.

Folinic acid (INN) or leucovorin (USAN), generally administered as calcium or sodium folinate (or leucovorin calcium/sodium), is an adjuvant used in cancer chemotherapy involving the drug methotrexate. It is also used in synergistic combination with the chemotherapy agent 5-fluorouracil.

2-aminopentanedioic acid + 4-aminobenzoic acid + 2-amino-6-methylpteridin-4-ol →

Folinic Acid

Levofolinic acid and its salts are the enantiopure drugs. They are configurated *S* at the remaining asymmetric carbon atom (see below). Folinic acid should be distinguished from folic acid. Folinic acid is a 5-formyl derivative of tetrahydrofolic acid. It is readily converted to other reduced folic acid derivatives (e.g. tetrahydrofolate), and thus has vitamin activity which is equivalent to folic acid. However, since it does not require the action of dihydrofolate reductase for its conversion, its function as a vitamin is unaffected by inhibition of this enzyme by drugs such as methotrexate.

Folinic acid, therefore, allows for some purine/pyrimidine synthesis to occur in the presence of dihydrofolate reductase inhibition, so that some normal DNA replication and RNA transcription processes can proceed.

Rhizopterin

Rhizopterin or Streptococcus Lactic R (SLR factor) is a naturally occurring compound which has a formyl group at position 10, it is a formyl derivative of pteroic acid. The pterdine group is chemically related to xanthoprotein, a yellow pigment present in certain natural materials including urine. Pteroylglutamic acid (PGA) has a low toxicity level.

A second formation of the vitamin fermentation L-casei factor* has been isolated from an aerobic fermentation of an organism of the genus corynebacterium. It has three molecules of glutamic acid condensed with the pteroyl group.

THE FUNCTION OF FOLIC ACID

DNA and Cell Division

Folate is necessary for the production and maintenance of new cells, for DNA synthesis and RNA synthesis, and for preventing changes to DNA and thus preventing cancer. It is especially important during periods of rapid cell division and growth such as infancy and pregnancy. Folate is needed to carry one carbon groups for methylation reactions and nucleic acid synthesis (most notably thymine, but also purine bases). Thus, folate deficiency hinders DNA synthesis and cell division, affecting hematopoietic cells and neoplasms the most because of rapid cell division. RNA transcription, and subsequent protein synthesis, are less affected by folate deficiency, as the mRNA can be recycled and used again (as opposed to DNA synthesis where a new genomic copy must be created). Since folate deficiency limits cell division, erythropoiesis, production of red blood cells is hindered and leads to megaloblastic anemia which is characterized by large immature red blood cells. This pathology results from persistently thwarted attempts at normal DNA replication, DNA repair, and cell division, and produces abnormally large red cells called megaloblasts (and hypersegmented neutrophils) with abundant cytoplasm capable of RNA and protein synthesis, but with clumping and fragmentation of nuclear chromatin. Some of these large cells, although immature (reticulocytes), are released early from the marrow in an attempt to compensate for the anemia. Both adults and children need folate to make normal red and white blood cells and prevent anemia. Deficiency of folate in pregnant women has been implicated in neural tube defects (NTD); therefore, many developed countries have implemented mandatory folic acid fortification in cereals, etc. It must be noted that NTD's occur early in pregnancy (first month) therefore women must have abundant folate upon conception. Folate is required to make red blood cells and white blood cells and folate deficiency may lead to anemia which further leads to fatigue and weakness and inability to concentrate.

*Lactobacillus casei factor (LCF), the folic acid conjugate

Biochemistry of DNA base and Amino Acid Production

Metabolism of folic acid to produce methyl-vitamin B_{12}.

In the form of a series of tetrahydrofolate (THF) compounds, folate derivatives are substrates in a number of single-carbon-transfer reactions, and also are involved in the synthesis of dTMP (22 -deoxythymidine-52 -phosphate) from dUMP (22 -deoxyuridine-52 -phosphate). It is a substrate for an important reaction that involves vitamin B_{12} and it is necessary for the synthesis of DNA, and so required for all dividing cells.

The pathway leading to the formation of tetrahydrofolate (FH_4) begins when folate (F) is reduced to dihydrofolate (DHF) (FH_2), which is then reduced to THF. Dihydrofolate reductase catalyses the last step. Vitamin B_3 in the form of NADPH is a necessary cofactor for both steps of the synthesis.

Methylene-THF (CH_2FH_4) is formed from THF by the addition of methylene groups from one of three carbon donors: formaldehyde, serine, or glycine. Methyl tetrahydrofolate (CH_3-THF) can be made from methylene-THF by reduction of the methylene group with NADPH. It is important to note that Vitamin B_{12} is the only acceptor of methyl-THF. There is also only one acceptor for methyl-B_{12} which is homocysteine in a reaction catalyzed by homocysteine methyltransferase. This is important because a defect in homocysteine methyltransferase or a deficiency of B_{12} can lead to a methyl-trap of THF and a subsequent deficiency. Thus, a deficiency in B_{12} can generate a large pool of methyl-THF that is unable to undergo reactions and will mimic folate deficiency. Another form of THF, formyl-THF or folinic acid) results from oxidation of methylene-THF or is formed from formate donating formyl group to THF. Finally, histidine can donate a single carbon to THF to form methenyl-THF.

In other words:

folate $\rightarrow$ dihydrofolate $\rightarrow$ tetrahydrofolate $\leftrightarrow$ methylene-THF $\rightarrow$ methyl-THF

Overview of Drugs that interfere with Folate Reactions

A number of drugs interfere with the biosynthesis of folic acid and THF. Among them are the dihydrofolate reductase inhibitors such as trimethoprim, pyrimethamine and methotrexate; the sulfonamides (competitive inhibitors of 4-aminobenzoic acid in the reactions of dihydropteroate synthetase).

The National Health and Nutrition Examination Survey (NHANES III 1988–91) and the Continuing Survey of Food Intakes by Individuals (1994–96 CSFII) indicated that most adults did not consume adequate folate. However, the folic acid fortification program in the United States has increased folic acid content of commonly eaten foods such as cereals and grains, and as a result diets of most adults now provide recommended amounts of folate equivalents.

Human Reproduction

Folic acid is an important nutrient for women who may become pregnant, because a woman's blood levels of folate fall during pregnancy due to an increased maternal RBC synthesis in the first half of the pregnancy and fetal demands in the second half. The first four weeks of pregnancy (when most women do not even realize they are pregnant) require folic acid for proper development of the brain, skull, and spinal cord. Serious birth defects like neural tube defects are less likely to occur when women take 0.4 mg of folic acid daily. Adequate folate intake during the periconceptional period, the time right before and just after a woman becomes pregnant, helps protect against a number of congenital malformations including neural tube defects (which are the most notable birth defects that occur from folate deficiency). Neural tube defects (NTDs) result in malformations of the spine (spina bifida), skull, and brain (anencephaly). The risk of neural tube defects is significantly reduced when supplemental folic acid is consumed in addition to a healthy diet prior to and during the first month following conception. The protective effect of folate during pregnancy goes beyond NTDs. Supplementation with folic acid has been shown to reduce the risk of congenital heart defects, cleft lip, limb defects, and urinary tract anomalies. Women who could become pregnant are advised to eat foods fortified with folic acid or take supplements in addition to eating folate-rich foods to reduce the risk of some serious birth defects. Having enough folic acid supplements in the months before pregnancy is very important to prevent neural tube defects. Taking 400 micrograms of synthetic folic acid daily from fortified foods and/or supplements has been suggested. The RDA for folate equivalents for pregnant women is 600-800 micrograms, twice the normal RDA of 400 micrograms for women who are not pregnant.

A study published by Milunski et al. has indicated that women who took folic acid supplements during the course of pregnancy can dramatically reduce the prevalence of infant neural tube defects by 3.9 times. The prevalence had dropped from 3.5 to 0.9 defects per 1000 births.

Although the recommended folic acid intake for women planning for pregnancy is 400 micrograms per day, the mechanisms and reasons why folic acid prevents birth defects is unknown. It is hypothesized that the insulin-like growth factor 2 gene is differentially methylated and these changes in IGF2 result in improved intrauterine growth and development.

Folate deficiency during pregnancy can increase the risk of preterm delivery, infant low birth weight, and fetal growth retardation. Folate deficiency in the mother increases homocysteine level in the blood which may lead to spontaneous abortion and pregnancy complications such as placental abruption and preeclampsia.

Recently studies have been conducted to test the hypothesis that folic acid supplementation reduces the risk of childhood acute lymphoblastic leukemia, but evidence so far has been weak.

Folic acid may also reduce chromosomal defects in sperm to some extent, which may be relevant for men considering to father a child. A benefit is indicated even for more than 700 mcg folate per day, which though below the tolerable upper intake levels of 1,000 μg/day was 1.8 times the recommended dietary allowance.

It is estimated that approximately 85% of women use folic acid supplements before they become pregnant but only 18% use enough folic acid supplements to meet the current folic acid requirements due to socio-economic challenges facing some women.

Folic acid supplements may even protect the fetus against disease when the mother is battling a disease or taking medications or smoking during pregnancy.

Heart Disease

An estimated 13,500 deaths occur annually due to folate deficiency's effect on coronary artery disease and the risk of ischemic heart disease and stroke has been reduced by 15% since folate fortification regulations were enforced. Adequate concentrations of folate, vitamin B_{12}, or vitamin B_6 may decrease the circulating level of homocysteine, an amino acid normally found in blood. There is evidence that an elevated homocysteine level is an independent risk factor for heart disease and stroke. The evidence suggests that high levels of homocysteine may damage coronary arteries or make it easier for blood clotting cells called platelets to clump together and form a clot. However, there is currently no evidence available to suggest that lowering homocysteine with vitamins will reduce risk of heart disease. The NORVIT trial suggests that folic acid supplementation may do more harm than good.

As of 2006, studies have shown that giving folic acid to reduce levels of homocysteine does not result in clinical benefit. One of these studies suggests that folic acid in combination with B_{12} may even increase some cardiovascular risks.

However a 2005 study found that 5 mg of folate daily over a three-week period reduced pulse pressure by 4.7 mmHg compared with a placebo, and concluded that Folic acid is a safe and effective supplement that targets large artery stiffness and may prevent isolated systolic hypertension.

Also, as a result of new research, "heart experts" at Johns Hopkins Medical Center reported in March 2008 in favour of therapeutic folate, although they cautioned that it is premature for people to begin to self-medicate by taking high doses of folic acid.

Hyperhomocysteinemia is a predictor of cardiovascular disease and hypertension among children and folic acid is a safe and effective supplement because it reduces serum homocysteine levels as well as systolic and diastolic blood pressure, thus preventing cardiovascular disease in children.

Folic acid supplements may improve the integrity of the vascular endothelium. Folic acid supplements consumed before and during pregnancy may reduce the risk of heart defects in infants. Folic acid supplementation may reduce the risk of children developing metabolic syndrome. Folic acid supplements may worsen the outcomes in patients with cardiovascular disease such as angina and myocardial infarction.

Stroke

Folic acid appears to reduce the risk of stroke. The reviews indicate only that in some individuals the risk of stroke appears to be reduced, but a definite recommendation regarding supplementation beyond the current recommended daily allowance has not been established for stroke prevention. Observed stroke reduction is consistent with the reduction in pulse pressure produced by folate supplementation of 5 mg per day, since hypertension is a key risk factor for stroke. Folic supplements are inexpensive and relatively safe to use and that is why stroke or hyperhomocysteinemia patients are encouraged to consume daily B vitamins including folic acid.

Cancer

Folate deficiency decreases intracellular S-adenosylmethionine (SAM) which inhibits cytosine methylation in DNA, activates proto-oncogenes, induces malignant transformations, causes DNA precursor imbalances, misincorporates uracil into DNA, and promotes chromosome breakage; all of these mechanisms increase the risk of prostate cancer development.

The association between folate and cancer appears to be complex. Even though theoretically it has been suggested that folate may help prevent cancer actual trials have found that supplementation increases rates of cancer.

Some investigations have proposed that good levels of folic acid may be related to lower risk of esophageal, stomach, and ovarian cancer, but benefices of folic acid against cancer may depend on when it is taken and on individual conditions. In addition folic acid may not be helpful, and could even be damaging, in people who already are suffering from cancer or from a precancerous condition. Conversely, it has been suggested that excess folate may promote tumor initiation. Folate has shown to play a dual role in cancer development; low folate intake protects against early carcinogenesis but high folate intake promotes advanced carcinogenesis. Therefore public health recommendations should be careful not to encourage too much folate intake.

Diets great in folate are associated with decreased risk of colorectal cancer; some studies show an association which is stronger for folate from foods alone than for folate from foods and supplements, while other studies find that folate from supplements is more effective due to greater bioavailability. A 2007 randomized clinical trial found that folate supplements did not reduce the risk of colorectal adenomas, but do in fact increase the presence of advanced lesions and adenoma multiplicity.[66] Colorectal cancer is the most studied type of cancer in relation to folate and one carbon metabolism and most research studies associate high folate intake with a reduced risk of prostate cancer. However folic acid supplement intake increased advanced colorectal cancer development by 67% in a 14 year European research study involving 520,000 men.

A 2006 prospective study of 81,922 Swedish adults found that diets great in folate from foods, but not from supplements, were associated with a reduced risk of pancreatic cancer.

Most epidemiologic studies suggest that diets high in folate are associated with decreased risk of breast cancer, but results are not uniformly consistent: one broad cancer screening trial reported a potential harmful effect of much folate intake on breast cancer risk, suggesting that routine folate supplementation should not be recommended as a breast cancer preventive, but a 2007 Swedish prospective study found that much folate intake was associated with a lower incidence of postmenopausal breast cancer. A 2008 study has shown no significant effect of folic acid on overall risk of total invasive cancer or breast cancer among women. Folate intake may not have any effect on the risk of breast cancer but may have an effect for women who consume at least 15 g/d of alcohol. Folate intake of more than 300 μg/d may reduce the risk of breast cancer in women who consume alcohol.

In men, folic acid supplementation appears to double the risk of prostate cancer. Recently a clinical trial showed that daily supplementation of 1 mg of folic acid increased the risk of prostate cancer while dietary and plasma folate levels among non vitamin users actually decreased the risk of prostate cancer. The reasons why high levels of folic acid may increase cancer is because of its role in nucleotide synthesis (proliferating neoplastic cells need this and folate receptors are increased in cancers). Folate's role in DNA methylation is important in prostate cancer. Unmetabolized folic acid is associated with a reduction in natural killer cell cytotoxicity which reduces the immune system's ability to defend against malignant cells. However, the study also showed that dietary baseline intake of folate may have inverse effects of prostate cancer.

The cancer drug methotrexate is designed to inhibit the metabolism of folic acid. Folic acid may interact unexpectedly with the cancer drug fluorouracil. The exact mechanism of interaction is unknown.

The low dihydrofolate reductase activity in the liver of humans compared to other animals and so the low conversion of folic acid into its active derivatives might be due to the control of this enzyme by transcription factors such as E2F-1 involved in cell proliferation. It has been suggested that "the low level of DHFR, and the other proteins under the control of E2F-1, in humans may have evolved to hinder the development of cancer. If this is the case, other animals with slow tissue turnover rates, possibly related to long life span, might also have low DHFR activity.

Although the relationship between folate and prostate cancer is not yet clear, there has been suicide gene studies that show a target vector for folate to prostate and nasopharyngeal cancer cells. Growth of tumor cells are significantly inhibited when a folate-linked nanoparticle is injected intratumorally. The mechanism might be due to the interference of transfection and communication failures of intracellular gap junctions.

A Finnish study consisting of 29,133 older male smokers resulted in the observation that prostate cancer risk had no relationship with serum folate levels. Folic acid supplements prevent mistakes from occurring during DNA replication and repair, for example the mistake of inserting uracils into the DNA. This is a proposed mechanism for folic acid's protection against colorectal cancer. Folic acid supplements stimulate the PI3k/Akt signaling cascade which leads to improved cell survival but this could be beneficial or harmful for the body because cancer cells may use this pathway to survive. Folic acid may also reduce the levels of PTEN (a tumour suppressor gene), making this relationship even more controversial.

Antifolates

Folate is important for cells and tissues that rapidly divide. Cancer cells divide rapidly, and drugs that interfere with folate metabolism are used to treat cancer. The antifolate methotrexate is a drug often used to treat cancer because it inhibits the production of the active form of THF from the inactive dihydrofolate (DHF). Unfortunately, methotrexate can be toxic, producing side effects such as inflammation in the digestive tract that make it difficult to eat normally. Also, bone marrow depression (inducing leukopenia and thrombocytopenia), acute renal and hepatic failure have been reported.

Folinic acid, under the drug name leucovorin, is a form of folate (formyl-THF) that can help "rescue" or reverse the toxic effects of methotrexate. Folinic acid is *not* the same as folic acid. Folic acid supplements have little established role in cancer chemotherapy. There have been cases of severe adverse effects of accidental substitution of folic acid for folinic acid in patients receiving methotrexate cancer chemotherapy. It is important for anyone receiving methotrexate to follow medical advice on the use of folic or folinic acid supplements. The supplement of folinic acid in patients undergoing methotrexate treatment is to give non rapidly dividing cells enough folate to maintain normal cell functions. The amount of folate given will be depleted by rapidly dividing cells (cancer) very fast and so will not negate the effects of methotrexate. Low dose methotrexate is used to treat a wide variety of non-cancerous diseases such as rheumatoid arthritis, lupus, scleroderma, psoriasis, asthma, sarcoidosis, primary biliary cirrhosis, polymyositis, and inflammatory bowel disease. Low doses of methotrexate can deplete folate stores and cause side effects that are similar to folate deficiency. Both high folate diets and supplemental folic acid may help reduce the toxic side effects of low dose methotrexate without decreasing its effectiveness. Anyone taking low dose methotrexate for the health problems listed above should consult with a physician about the need for a folic acid supplement.

While the role in folate as a cancer treatment is well established its long term effectiveness is diminished by cellular response. In response to decreased THF the cell begins to transcribe more DHF reductase, the enzyme that reduces DHF to THF. Because methotrexate is a competitive inhibitor of DHF reductase increased concentrations of DHF reductase can overcome the drugs inhibition.

Obesity

Folic acid increases lipolysis in adipocytes and may have a role in the prevention of obesity and type 2 diabetes. This mechanism involves the beta adrenoceptors in the adbdominal adipocytes. Folic acid supplements may reduce the accumulation of cholesterol in the liver and in the blood; this may be due to folic acid's role in incorporating cholesterol into bile acid. In fact folic acid supplements have been shown to increase bile acid production and flow.

Depression

Some evidence links a shortage of folate with depression. There is some limited evidence from randomised controlled trials that using folic acid in addition to antidepressants, specifically SSRIs, may have benefits. Research at the University of York and Hull York Medical School has found a link between depression and low levels of folate. One study by the same team involved 15,315 subjects. However, the evidence is probably too limited at present for this to be a routine treatment recommendation. Folic acid supplementation affects noradrenaline and serotonin receptors within the brain and this could be the cause of folic acid's possible ability to act as an anti-depressant.

Memory and Mental Agility

In a 3-year trial on 818 people over the age of 50, short-term memory, mental agility, and verbal fluency were all found to be better among people who took 800 micrograms of folic acid daily, twice the current RDA, than those who took placebo. The study was reported in *The Lancet* on 20 January 2007.

Schizophrenia

Folate deficiency may increase the risk of schizophrenia because by increasing homocysteine levels folate also increases interleukin 6 and tumor necrosis factor alpha levels and these two cytokines are involved in the development of schizophrenia. The exact mechanisms involved in the development of schizophrenia are not entirely clear but may have something to do with DNA methylation and one carbon metabolism and these are the precise roles of folate in the body and that is why folate deficiency has been linked to schizophrenia.

Allergic Diseases

There is a relationship between folic acid and allergic diseases. In one study that examined the relationship between serum folate levels and markers of atopy, wheeze, and asthma in 8083 subjects serum folate levels were found to be inversely related to IgE level, atopy, and wheeze in a dose-response relationship. Increased folate levels were also associated with decreased risk of doctor-diagnosed asthma. Folic acid supplementation during late pregnancy is associated with an increased risk of childhood asthma, increased risk of persistent asthma, and poorer respiratory function in young children.

Rheumatoid arthritis

Folic acid supplementation of 5-27 mg per week has shown to have a protective effect against rheumatoid arthritis.

Fertility

Folate is necessary for fertility in both men and women. In men, it contributes to spermatogenesis. In women, on the other hand, it contributes to oocyte maturation, implantation, placentation, in addition to the general effects of folic acid and pregnancy. Therefore, it is necessary to receive sufficient amounts through the diet, in order to avoid subfertility.

Renal Disease

Folic acid supplements may reduce the risk of children developing renal diseases or injuries such as microalbuminuria.

Type 1 diabetes mellitus

Type 1 diabetes mellitus patients have lower plasma levels of folic acid and may benefit from folic acid supplements or folic acid fortified food products.

Macular Degeneration

A substudy of the Women's Antioxidant and Folic Acid Cardiovascular Study published in 2009 reports that use of a nutritional supplement that contains folic acid, pyridoxine, and cyanocobalamin decreased the risk of developing age-related macular degeneration by 34 per cent.

Bone health

It has been hypothesized that folate deficiency can lead to elevated homocysteine levels which in turn lead to an increased risk of bone fractures, osteoporosis, and reduction in BMD but research studies so far show controversial results.

Menopause

Folic acid supplements help relieve hot flushes in postmenopausal women. Just like in estrogen hormone replacement therapy, folic acid interacts with neurotransmitters (norepinephrine, serotonin) in the brain to reduce hot flushes.

Infectious Disease

Folate deficiency is linked to anemia causing Plasmodium falciparum malaria in areas such as Colombia where malaria has reached endemic proportions.

Bone loss in Parkinson's Disease (PD)

Folate lowers homocysteine (Hcy) levels which in turn prevents bone loss in Parkinson's disease (PD) patients taking levodopa (a psychoactive drug taken to treat Parkinson's disease). Improvements in bone health include increased BMD at the lumbar spine, total femur, and femur shaft.

Folic acid Supplements and Masking of B_{12} deficiency

There has been concern about the interaction between vitamin B_{12} and folic acid. Folic acid supplements can correct the anemia associated with vitamin B_{12} deficiency. Unfortunately, folic acid will not correct changes in the nervous system that result from vitamin B_{12} deficiency. Permanent nerve damage could theoretically occur if vitamin B_{12} deficiency is not treated. Therefore, intake of supplemental folic acid should not exceed 1000 micrograms (1000 μg = 1 mg) per day to prevent folic acid from masking symptoms of vitamin B_{12} deficiency. In fact, to date the evidence that such masking actually occurs is scarce, and there is no evidence that folic acid fortification in Canada or the U.S. has increased the prevalence of vitamin B_{12} deficiency or its consequences.

However, one recent study has demonstrated that high folic or folate levels, when combined with low B_{12} levels, are associated with significant cognitive impairment among the elderly.

In any case, it is important for older adults to be aware of the relationship between folic acid and vitamin B_{12}, because they are at greater risk of having a vitamin B_{12} deficiency. Patients 50 years of age or older should ask their physicians to check their vitamin B_{12} status before taking a supplement that contains folic acid.

Health Risk of too much Folic Acid

The risk of toxicity from folic acid is low because folate is a water soluble vitamin and is regularly removed from the body through urine. The Institute of Medicine has established a tolerable upper intake level (UL) for folate of 1 mg for adult men and women, and a UL of 800 μg for pregnant and lactating (breast-feeding) women less than 18 years of age. Supplemental folic acid should not exceed the UL to prevent folic acid from masking symptoms of vitamin B_{12} deficiency.

Research suggests high levels of folic acid can interfere with some antimalarial treatments.

A 10,000-patient study at Tufts University in 2007 concluded that excess folic acid worsens the effects of B_{12} deficiency and in fact may affect the absorption of B_{12}.

A study at the University of Adelaide concluded that the intake of folic acid supplements during late pregnancy increases the risk of babies developing childhood asthma by 30%, although researchers emphasized that their finding did not contradict recommendations to supplement folic acid in first trimester, when no additional risk was found.

Elderly Population

There are benefits and risks of food folic acid fortification for elderly populations.[126] Elevated exposure to folic acid due to fortification can improve folate and homocysteine levels but can also mask symptoms of vitamin B_{12} deficiency. A study where 747 subjects aged 67 to 96 years were measured for B vitamin and homocysteine status showed that diets with folic acid fortification of 140 μg/100 g of grain product decreased homocysteine level and heart disease risk. However, Canada's food supply is fortified with 150 μg/100 g of grain and much of the elderly population also take a supplement which includes a folic acid component of 400 μg. Therefore it is important not to consume quantities over the recommended DRI.

THE DEFICIENCY

Folate deficiency may lead to glossitis, diarrhea, depression, confusion, anemia, and fetal neural tube defects and brain defects (during pregnancy). Folate deficiency is diagnosed by analyzing CBC and plasma vitamin B_{12} and folate levels. CBC may indicate megaloblastic anemia but this could also be a sign of vitamin B_{12} deficiency. A serum folate of 3 μg/L or lower indicates deficiency. Serum folate level reflects folate status but erythrocyte folate level better reflects tissue stores after intake. An erythrocyte folate level of 140 μg/L or lower indicates inadequate folate status. Increased homocysteine level suggests tissue folate deficiency but homocysteine is also affected by vitamin B_{12} and vitamin B_6, renal function, and genetics. One way to differentiate between folate deficiency from vitamin B_{12} deficiency is by testing for methylmalonic acid levels. Normal MMA levels indicate folate deficiency and elevated MMA levels indicate vitamin B_{12} deficiency. Folate deficiency is treated with supplemental oral folate of 400 to 1000 μg per day. This treatment is very successful in replenishing tissues even if deficiency was caused by malabsorption. Patients with megaloblastic anemia need to be tested for vitamin B_{12} deficiency before folate treatment because if the patient has vitamin B_{12} deficiency, folate supplementation can remove the anemia but can also worsen neurologic problems. Morbidly obese patients with BMIs of greater than 50 are more likely to develop folate deficiency. Patients with celiac disease have a higher chance of developing folate deficiency. Cobalamin deficiency may lead to folate deficiency which in turn increases homocysteine levels and finally may result in the development of cardiovascular disease or birth defects

VITAMIN B_{10}

4-Aminobenzoic acid (also known as para-aminobenzoic acid or **PABA**) is an organic compound with the formula $H_2NC_6H_4CO_2H$. PABA is a white crystalline substance that is only slightly soluble in water. It consists of a benzene ring substituted with an amino group and a carboxyl group.

PRODUCTION AND OCCURRENCE

Industrially, PABA is mainly prepared by two routes: reduction of 4-nitrobenzoic acid and the Hoffman degradation of the monoamide derived from terephthalic acid. Food sources of PABA include liver, brewer's yeast (and unfiltered beer), kidney, molasses, and whole grains.

BIOCHEMISTRY

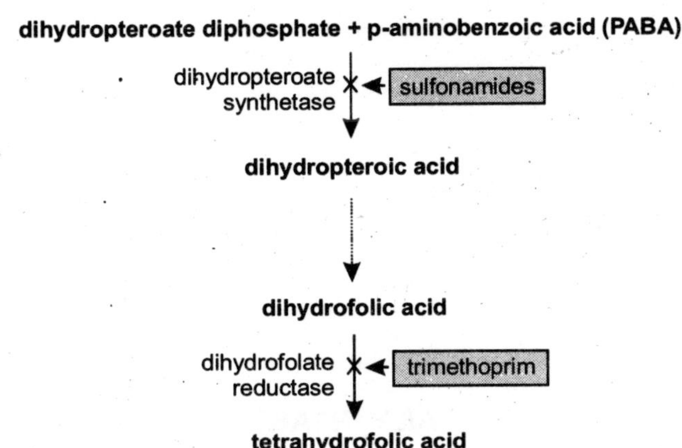

Tetrahydrofolate Synthesis Pathway

PABA is an intermediate in the bacterial synthesis of folate. It has been referred to as Vitamin B_{10} Some bacteria in the human intestinal tract such as *E. coli* require PABA. Humans require folate since we lack the enzymes to convert PABA to folate. Therefore in humans PABA is not a vitamin and is considered nutritionally unimportant. Sulphonamide drugs are structurally similar to PABA, and their antibacterial activity is due to their ability to interfere with the conversion of PABA to folate by the enzyme dihydropteroate synthetase. Thus, bacterial growth is limited through folate deficiency without effect on human cells.

FUNCTION S IN METABOLISM

The potassium salt is used as a drug against fibrotic skin disorders, such as Peyronie's disease, under the trade name Potaba. PABA is also occasionally used in pill form by sufferers of irritable bowel syndrome to treat its associated gastrointestinal symptoms, and in nutritional epidemiological studies to assess the completeness of 24-hour urine collection for the determination of urinary sodium, potassium, or nitrogen levels.

NUTRITIONAL SUPPLEMENT

Despite the lack of any recognized syndromes of PABA deficiency in humans, many claims of benefit are made by commercial suppliers of PABA as a nutritional supplement. Benefit is claimed for fatigue, irritability, depression, weeping eczema (moist eczema), scleroderma (premature hardening of skin), patchy pigment loss in skin (vitiligo), and premature grey hair.

Sources:

- blackstrap molasses
- bran

- brewer's yeast
- eggs
- kidney
- leafy green vegetables such as spinach
- liver
- molasses
- mushrooms
- wheat germ
- whole grains
- yogurt

DAILY INTAKE

- RDA for adults - 400 mcg
- RDA for children 200 - 300 mcg
- RDA for infants - 50 mcg
- RDA for pregnancy - 800 mcg
- RDA for nursing women - 600 mcg

VITAMIN B10 DEFICIENCY

- irritability and depression;
- constipation;
- nervousness;
- skin problems and eczema.

VITAMIN B_{11} PTERYL HEPTA GLUTAMIC ACID

Vitamin B11 (Pteryl-hepta-glutamic acid) is a form of Folic acid (one of five folates necessary for humans) and is now known as 'chick growth factor'.

WHY DO YOU NEED VITAMIN B11?

Pteryl-hepta-glutamic acid functions in:

- DNA and RNA syntheses
- Cell division
- Development of the fetus nervous system

Little is currently known about vitamin B_{11}. It is however also known as factor S and having similar properties of vitamin B_{10} (PABA). Because of this, it was determined that vitamin B11 was in fact a mixture of substances.

FUNCTIONS OF VITAMIN B_{11}

Vitamin B_{11} functions in the DNA and RNA syntheses, essential for the body. It is necessary for cell division. By this way, it helps growth. It is necessary for the development of the fetus nervous system. Deficiency may result in lack of appetite, weight loss, nausea, vomiting, diarrhea, headache, fatigue, forgetfulness, susceptibility to infections and some cardiac problems, like tachycardia.

Vitamin B11 is found in liver, heart, kidney, meat, poultry, fish, milk, cheese and eggs.

VITAMIN B_{12} CYANOCOBALMIN

Vitamin B_{12}, vitamin B12 or vitamin B-12, also called cobalamin, is a water soluble vitamin with a key role in the normal functioning of the brain and nervous system, and for the formation of blood. It is one of the eight B vitamins. It is normally involved in the metabolism of every cell of the body, especially affecting DNA synthesis and regulation, but also fatty acid synthesis and energy production. As the largest and most structurally complicated vitamin, it can be produced industrially only through bacterial fermentation-synthesis.Vitamin B_{12} consists of a class of chemically-related compounds (vitamers), all of which have vitamin activity. It contains the biochemically rare element cobalt. Biosynthesis of the basic structure of the vitamin in nature is only accomplished by simple organisms such as some bacteria and algae, but conversion between different forms of the vitamin can be accomplished in the human body. A common synthetic form of the vitamin, cyanocobalamin, does not occur in nature, but is used in many pharmaceuticals and supplements, and as a food additive, due to its stability and lower cost. In the body it is converted to the physiological forms, methylcobalamin and adenosylcobalamin, leaving behind the cyanide, albeit in minimal concentration. More recently, hydroxocobalamin (a form produced by bacteria), methylcobalamin, and adenosylcobalamin can also be found in more expensive pharmacological products and food supplements. The utility of these is presently debated.Historically, vitamin B_{12} was discovered from its relationship to the disease pernicious anemia, which is an autoimmune disease that destroys parietal cells in the stomach that secrete intrinsic factor. Intrinsic factor is crucial for the normal absorption of B_{12}, so a lack of intrinsic factor, as seen in pernicious anemia, causes a vitamin B12 deficiency. Many other subtler kinds of vitamin B_{12} deficiency, and their biochemical effects, have since been elucidated. The names vitamin B_{12} or vitamin B12 or vitamin B-12, which are sometimes shortened to B_{12} or B12, and the alternative name cyanocobalamin generally refer to all forms of the vitamin. Some medical practitioners have suggested that its use be split into two different categories, however:

- In a broad sense, B_{12} refers to a group of cobalt-containing vitamer compounds known as cobalamins: these include cyanocobalamin (an artifact formed as a result of the use of cyanide in the purification procedures), hydroxocobalamin (another medicinal form, produced by bacteria), and finally, the two naturally occurring cofactor forms of B_{12} in the human body: 5'-deoxyadenosylcobalamin (adenosylcobalamin—AdoB12), the cofactor of Methylmalonyl Coenzyme A mutase (MUT), and methylcobalamin (MeB_{12}), the cofactor of 5-methyltetrahydrofolate-homocysteine methyltransferase (MTR).

- The term B_{12} may be properly used to refer to cyanocobalamin, the principal B_{12} form used for foods and in nutritional supplements. This ordinarily creates no problem, except perhaps in rare cases of eye nerve damage, where the body is only marginally able to use this form due to high cyanide levels in the blood due to cigarette smoking, and thus requires cessation of smoking, or else B_{12} given in another form, for the optic symptoms to abate. However, tobacco amblyopia is a rare enough condition that debate continues about whether or not it represents a peculiar B_{12} deficiency which is resistant to treatment with cyanocobalamin.

Finally, so-called pseudo-B_{12} refers to B_{12}-like substances which are found in certain organisms, including *Spirulina* (a cyanobacterium) and some algae. These substances are active in tests of B_{12} activity by highly sensitive antibody-binding serum assay tests, which measure levels of B_{12} and B_{12}-like compounds in blood. However, these substances do not have B_{12} biological activity for humans, a fact which may pose a danger to vegans and others on limited diets who do not ingest B_{12} producing bacteria, but who nevertheless may show normal "B_{12}" levels in the standard immunoassay which has become the normal medical method for testing for B_{12} deficiency.

THE STRUCTURE OF VITAMIN B_{12}

Vitamin B_{12} Vitamin B_{12} is the only known biomolecule with a stable carbon-metal bond - it is an organometallic compound. The core of the molecule is a corrin ring with various attached sidegroups. The ring consists of 4 pyrrole subunits, joined on opposite sides by a C-CH_3 methylene link, on one side by a C-H methylene link, and with the two of the pyrroles joined directly. It is thus like a porphyrin, but with one of the bridging methylene groups removed. The nitrogen of each pyrolle is coordinated to the central cobalt atom. The sixth ligand below the ring is a nitrogen of a 5,6-dimethylbenzimidazole. The other nitrogen is linked to a five-carbon sugar, which in turn connects to a phosphate group, and thence back onto the corrin ring via one of the seven amide groups attached to the periphery of the corrin ring. The base ligand thus forms a 'strap' back onto the corrin ring.

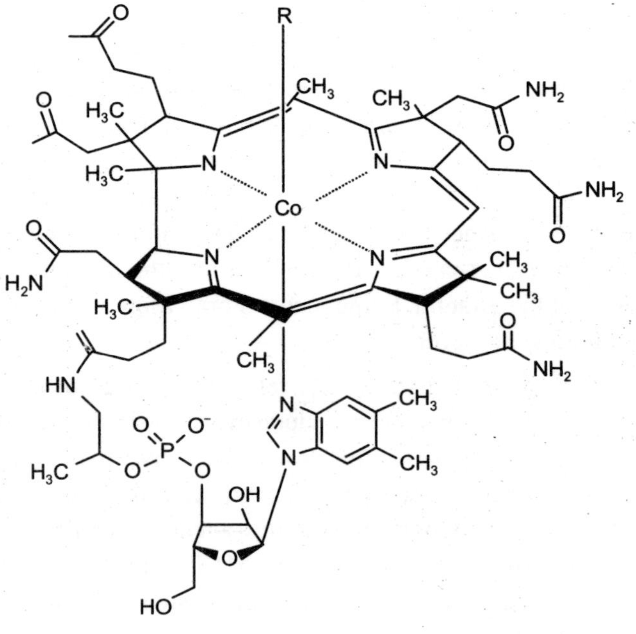

An important aspect of the corrin ring, when compared to the porphyrin, is the relative flexibility of the corrin system, the corrin ring is also less flat when viewed from the side than is a porphyrin ring. This adds up to some considerable differences between the chemistry of a cobalt porphyrin and a cobalt corrin.

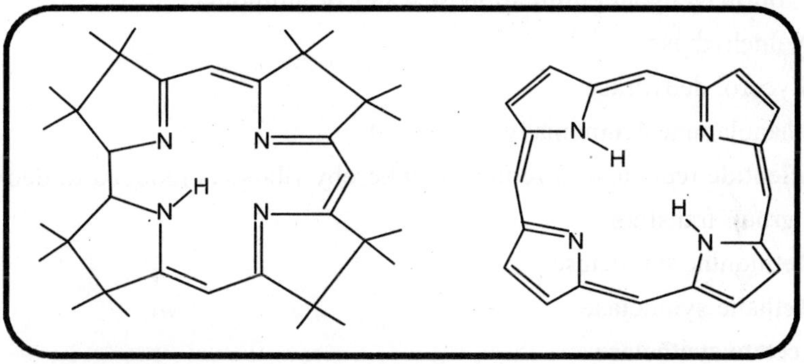

In addition, the corrin only has a conjugated chain around part of the ring system, whereas a porphyrin is delocalised around the whole four pyrolle rings.

BIOCHEMISTRY OF VITAMIN B_{12}

The first type of Vitamin B_{12} to be isolated had a cyanide group attached to the cobalt, this was picked up during the purification of the vitamin, and is the form which is still referred to as "Vitamin B_{12}". The commercial form of the vitamin is usually obtained as the cyanide, it metabolises easily to the coenzyme. Smokers should note that the cyanide absorbed from the smoke into the blood causes the replacement of the 5'-deoxyadenosyl group by a cyanide.

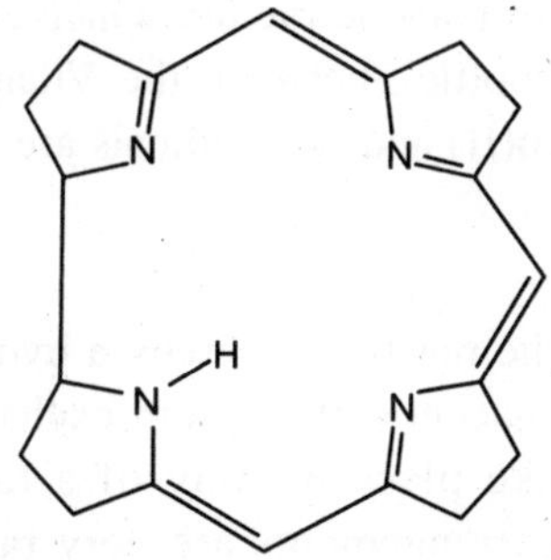

The principal coenzymatic form of Vitamin B_{12} is 5'-deoxyadenosylcobalamin. The notable feature of the molecule is the cobalt-carbon bond between the 5' carbon of the 5'-deoxyadenosyl moiety (the sugar part) and the cobalt of cobalamin. In Vitamin B_{12} as it is extracted, a cyanide replaces this sugar link, this occurs during the final purification with active charcoal. Aquocobalamin and hydroxycobalamin, with water and hydroxide are also known, as is the methylated form, methylcobalamin. A number of other cobalt-carbon bonded have been prepared synthetically, but are not known to occur *in vivo*

Vitamin B_{12} in its various forms, and in cooperation with its coenzymes and various different substrates is involved in three different forms of reaction:

1. several mutases in which a hydrogen and some group on an adjacent carbon exchange places,
   - Glutamate mutase
   - Methylmalonyl CoA mutase
   - Ornithine mutase
   - L-β-lysine mutase
   - α-methyleneglutarate mutase

This can be followed by an elimination of water or ammonia.

- Dioldehydrase
- Glycerol dehydrase
- Ethanolamine Ammonia lyase

2. Ribonucleotide reductase, a reductase whereby ribose is reduced to deoxyribose.
3. Methyl group transfers
   - Methionine synthetase
   - Methane synthetase
   - Acetate synthetase

The first two reactions involve a Co(II) oxidation state intermediate, the third probably involves a Co(I), and in both in its 'resting state' the cobalt is a Co(III). Central to the catalytic rôle of is the relative weakness of the cobalt-carbon bond, with a dissociation energy of around 120 $kJmol^{-1}$. In several of the rearrangements, and in the reductase, EPR signals have been observed for the deoxyadenosyl radical and for a cobalt(II) corrin.In the hydrogen migrations in all of the Vitamin B_{12} coenzyme-dependent rearrangement reactions, there is no exchange of the hydrogen with water protons, and the 1,2 intramolecular shifts are stereospecific for both the hydrogen atom and the exchanging group. Different reactions also proceed with either retention or inversion of the configuration. The mechanistic picture which emerges from this is that the rearrangement involves a kind of 'reaction in a bottle', between the Vitamin B_{12} and the substrate, with the large coenzyme acting as the bottle. Co(II) and free radicals are very oxygen sensitive, so that the whole process must be kept anaerobic.

MECHANISTIC PICTURE

The reaction whereby a hydrogen and a group on an adjacent carbon atom exchange places is believed to take place by way of a radical mechanism, such rearrangements are very rare in organic chemistry, indeed the only known instance is the Kharash-Urey reaction. It is especially surprising that the reactions involve specific configuration changes of the substrates. The evidence for a radical mechanism comes principally from observations of EMR spectra for several of the active enzyme systems. Thus if the dioldehydrase (holoenzyme) is treated with the substrate analogue 1,2-propoanediol, a radical signal believed to be due to the 5'-deoxyadenosyl radical, together with a low spin Co(II) signal is formed within milliseconds, and persists whilst there is substrate present. Similar observations have been made for dioldehydrase-substrate, glycerol-dehydrase, ethanol-ammonia lyase, and also the ribonucleotide reductase system. Since it is necessary to rigorously exclude dioxygen from any reaction involving free radicals, a plausible reaction mechanism {proposed in da Silva and Williams seminal text}.

H X X H
C — C ⇌ C — C

Step 1, and the substrate, 1,2-dihydroxyethane, approaches the holoenzyme, and is about to plug the reaction 'bottle'. Then (**Step 2**) the substrate binds and the cobalt-carbon bond breaks homolytically to give the 5'-adenosyl radical and a Co(II) corrin, anchored to the large coenzyme (are the amide groups involved?) It has been suggested that the cobalt-carbon bond, already a rather weak bond and readily subject to photolysis if the holoenzyme is subjected to visible light, is further weakened by binding of the metalloenzyme to the large coenzyme.

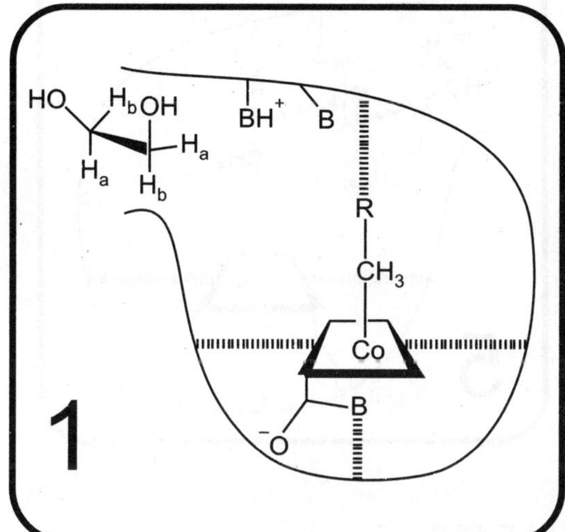

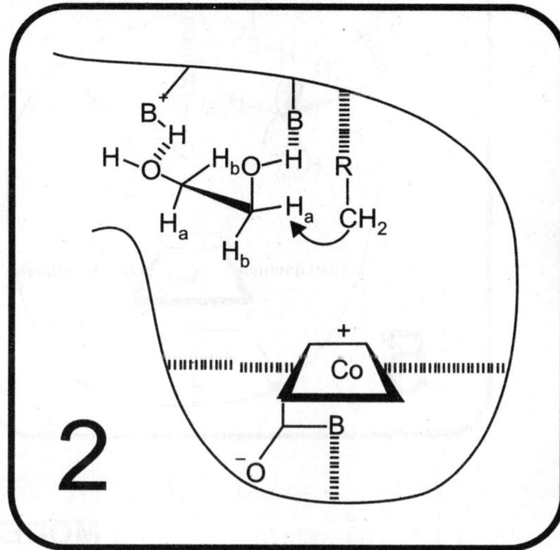

In **Step 3**,a hydrogen atom is abstracted from the substrate by the 5'-deoxyadenosyl radical, converting the dangling CH_2 into a CH_3 Then, **Step 4**, the enzyme base (B) and the acid (BH^+) cause assisted beta-hydroxy fragmentation, that is loss of a water molecule.

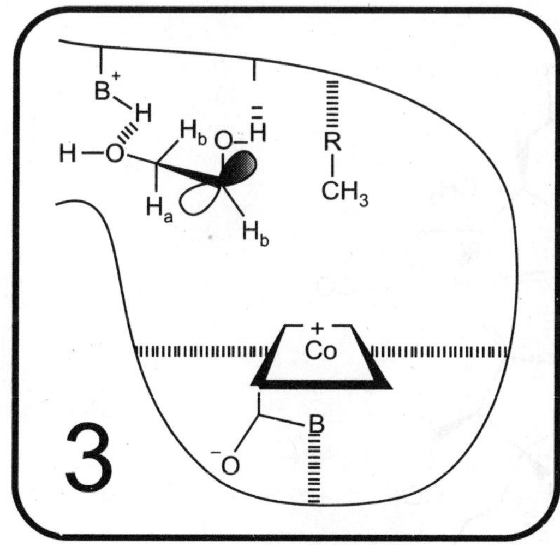

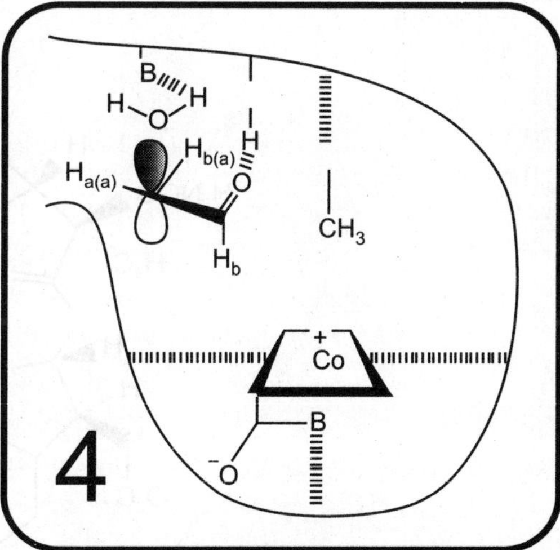

With **Step 5**, comes the re-addition of a water molecule, resulting in a substrate reorientation of C_2 towards the 5'deoxyadenosyl methyl. There must be competing C_1 and C_2 racemisation. A hydrogen atom abstraction occurs, with competing C_2 racemization. The cycle is completed (**Step 6**), by substrate dehydration and release of ethanal and a water molecule, so that the system can pick up another substrate molecule.

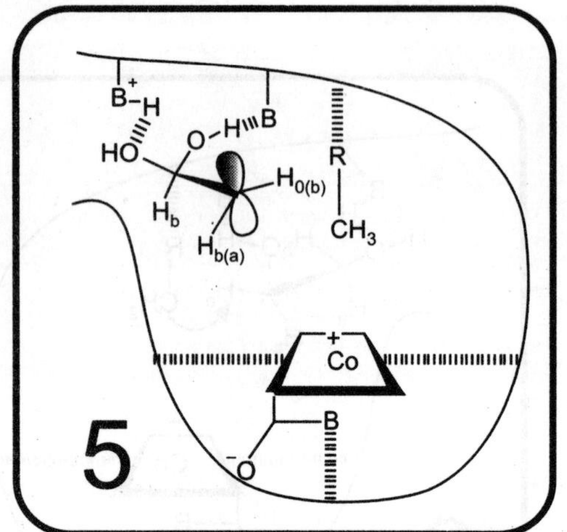

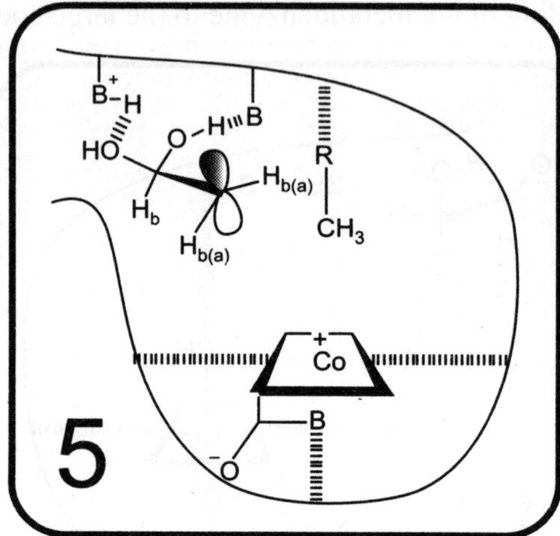

MODEL COMPLEXES

The simplist model for Vitamin B_{12} is the $[MeCo(CN)_5]^{3-}$ ion, and an early observation was that this complex will transfer its methyl group to an Hg(II), generating the highly poisonous $[MeHg^+]$ moiety. A mercury methylation forming this highly toxic ion, and also involving methylcobalamin, and the related.

F430 metalloenzyme is believed to be behind the Minamata tragedy in Japan.

The principle models used for Vitamin B_{12} are the alkylcobalt (III) dimethyl-glyoxinates, so called alkylcobaloximes, which have the advantages of easy preparation and simplicity. The methyl complex photolyses readily to form ethane, and the methyl radicals can be trapped with spin-traps such as PBN. reaction with Hg(II) forms the highly toxic MeHg(II) ion, modeling what happens when Hg(II) attacks methyl Vitamin B_{12}. Of course none of these will ever provide substitutes for Vitamin B_{12}, but they do provide platforms on which various aspects of the mechanisms of the Vitamin B_{12} catalysed reactions can be studied.

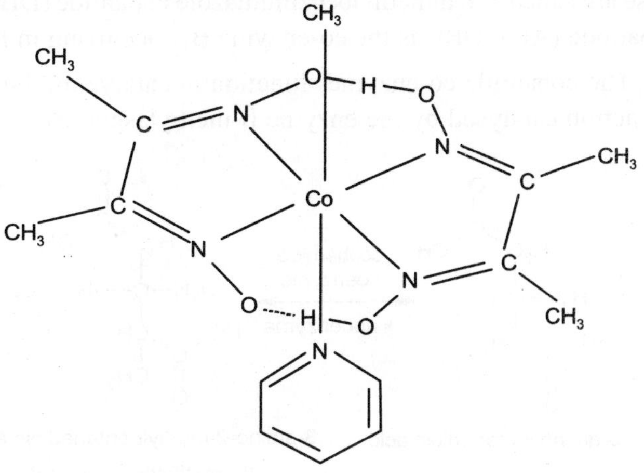

THE METABOLIC FUNCTION

The vitamin B_{12} has been shown by H.A Barter to occur as a part of a Vitamin B_{12} co –enzyme. In this

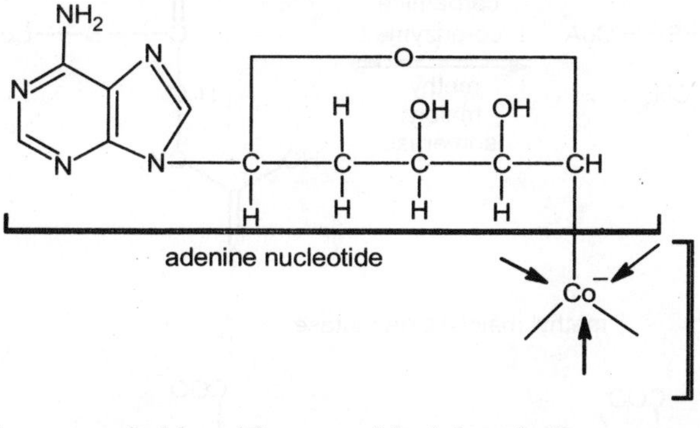

adenine nucleotide

Asin B_{12}

Co-enzyme the position occupied by either cyanide (vitamin B_{12} cyanocobalmin) or a hydroxyl group (vitamin B_{12b} hydroxycobalmin) in the vitamin is bonded directly to the 5'-carbon atom of the ribose of adenosine. The co enzyme is relatively unstable and in the presence of light or cyanide is decomposed respectively to hydroxy cobalmin or cyano-cobalmin from known as the vitamin. A distinct possibility is that the vitamin B_{12} occurs in nature as co-enzyme (cobamide)

This extremely important vitamin plays a significant role as coenzyme B_{12}. The 5 deoxyadenosine that replaces the cobalt bound cyanide is derived from adenosine triphosphide – which then releases the entire three phosphate group as inorganic triphosphate, in the process of formation of co –enzyme B_{12} the cobalt undergoes successive reduction in reactions catalyzes by B_{12} reductase, which is red – c and the cobalt present as Co^{+++}, this progress to B_{12r} (orange) with Co^{++}. The latter form reacts with ATP to form adenosyl co-enzyme. All the three co-enzymes, all the three co-enzymes possess adenine deoxynulceoside differing from one another in the benzimidazole portion of the vitamin B_{12} molecule, these are called 5, 6 dimethylbenzimidiazole cobamide (DBC) benzimidiazole cobamide (BC) and adenyl cobamide (AC). DBC is the co-enzyme B_{12} occurring in largest quantities in materials.

The cobamide co-enzymes function in catalyzing the conversion glutamic acid to mesaconic acid a reaction catalysed by the enzyme β-methylaspertase.

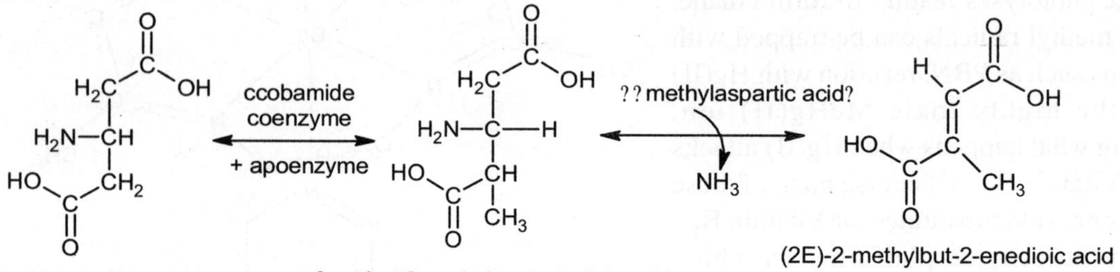

ccobamide coenzyme + apoenzyme

?? methylaspartic acid?

NH_3

3-aminopentanedioic acid

3-amino-2-methylpentanedioic acid

β - methylaspartic acid

(2E)-2-methylbut-2-enedioic acid

Mesaconic acid

The cobamide co-enzyme participates in two reactions, (a) isomerization of dicarboxylic acid and (b) the conversion of vicinal deoxymonohydroxyl grouping.

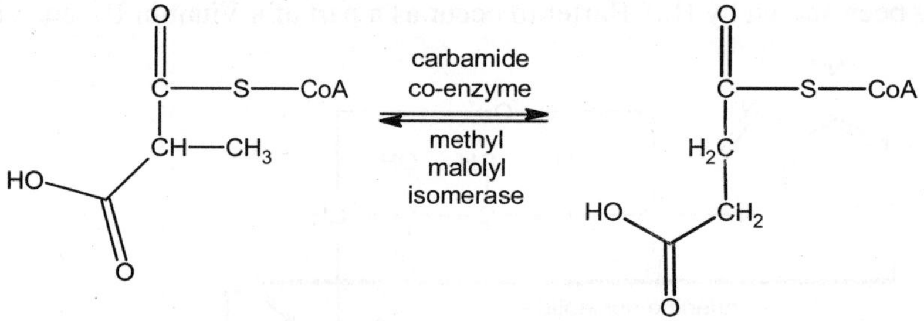

carbamide co-enzyme

methyl malolyl isomerase

methyl malolyl CoA mutase

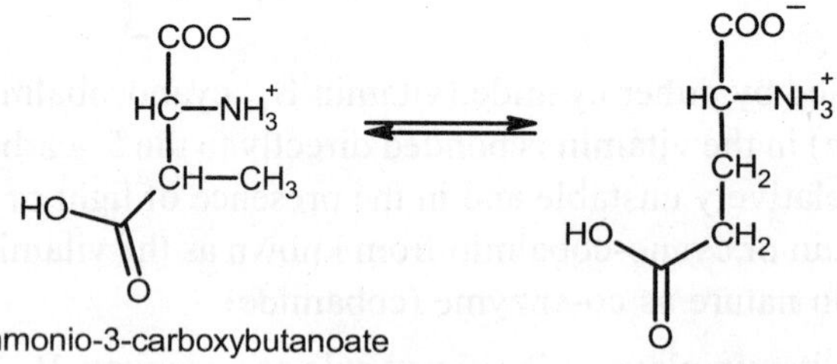

2-ammonio-3-carboxybutanoate

2-ammonio-4-carboxybutanoate

$$CH_3CHOHCH_2OH \rightarrow CH_3CH_2CHO$$

Propyleneglycol Propionaldehyde

Methylaspartase mutase

Formation of methyl group of methionine is linked to vitamin B_{12} in both E Coli extracts and mammalian systems. A reaction sequence is suggested for B_{12} folic acid and methionine interaction.

$$\text{Methyl } N^5 - FH_4 + \text{Homocysteine} \xrightarrow[\text{s-adenosyl+FADH}_2]{\text{cobala min e}} \text{Methionine} + FH_4$$

DBC plays an essential role as a hydrogen transferring agent in the cobamide dependant ribonucleotide reductase reaction. The ribose part in a ribonucleotide reaction, the ribose part in a ribonucleotide reaction is converted into deoxyribose.

In addition extracts of *Methanosarcina barkeri* convert converts cobalmin to methane in the presence of pyruvate. There is a considerable evidence on an in vitro basis that a biochemical relation exists between vitamin B_{12} and the total synthesis of acetate from CO_2 by extracts of *clostridium thermoaceticum*. The extracts of these organisms convert the methyl of B_{12} - $^{14}CH_3$ to $^{14}C_3$ - COOH with CO_2 presumably contributing the carboxyl group directly and the methyl carbon indirectly.

The B_{12} co-enzymes have been isolated both from several bacterial cultures as well as form the liver of various animals.

DEFICIENCY

Vitamin B_{12} deficiency is a reduction in vitamin B_{12} from inadequate dietary intake or impaired absorption. The condition is commonly asymptomatic, but can also present as anemia characterized by enlarged blood corpuscles with characteristic changes in neutrophils, known as megaloblastic anemia.

In serious cases deficiency can potentially cause severe and irreversible damage to the nervous system, including subacute combined degeneration of spinal cord.

The anemia is thought to be due to problems in DNA synthesis, specifically in the synthesis of thymine, which is dependent on products of the MTR reaction. Other blood cell types such as white blood cells and platelets are often also low. Bone marrow examination may show megaloblastic hemopoiesis. The anemia responds completely to vitamin B_{12}; the neurological symptoms (if any) respond partly or completely, depending on prior severity and duration. Early and even fairly pronounced deficiency does not always cause distinct or specific symptoms. Common early symptoms are tiredness, a decreased mental work capacity, weakened concentration and memory, and irritability and depression.

Sleep disturbances may occur, because B_{12} may be involved in the regulation of the sleep wake cycle by the pineal gland (through melatonin).

Neurological signs of B_{12} deficiency, which can occur with or without anemia, include sensory disturbances (due to damage to peripheral nerves caused by demyelination) and irreversible nerve cell death. Symptoms include numbness and/or tingling of the extremities, altered proprioception, impaired sense of smell, loss of appetite (anorexia), disturbed coordination and, if not treated in time, an ataxic gait especially in the dark when there is less visual reference. In extreme cases, B_{12} deficiency can lead to a syndrome known as subacute combined degeneration of spinal cord.

B_{12} deficiency can also cause symptoms of mania, psychosis, fatigue, memory impairment, irritability, depression and personality changes.

B_{12} status may be associated with the onset and course of Alzheimer's disease. Some studies have found no relationship, while several recent studies indicate a relationship between B_{12}, homocysteine, and Alzheimer's. B_{12} status is routinely measured at the time of Alzheimer's diagnosis, and there is some indication that ongoing measurements may be useful to detect the development of a severe deficiency. In addition to checking serum B_{12}, checking the levels of other compounds (particularly methylmalonic acid) may be necessary to accurately detect a deficiency state, because serum levels do not necessarily correlate with efficient utilization of B_{12}.

Studies showing a relationship between clinical depression levels and deficient B_{12} blood levels in elderly people are documented in the clinical literature.

Bipolar disorder appears to genetically co-segregate with the hereditary B_{12}-deficiency disorder pernicious anemia.

Science Daily reported that "a deficiency of B-vitamins may cause vascular cognitive impairment, according to a new study by the Jean Mayer USDA Human Nutrition Research Center on Aging (HNRCA) at Tufts University." Aron Troen, PhD, said that: "The vascular changes occurred in the absence of neurotoxic or degenerative changes. Metabolic impairments induced by a diet deficient in three B-vitamins—folate, B_{12} and B_6—caused cognitive dysfunction and reductions in brain capillary length and density in our mouse model."

The following symptoms have been recorded, beginning with those perhaps most likely to be indicative of B_{12} deficiency:

1. Itchy or tingling tongue. The tongue suddenly itches from time to time without warning. This occurs on the edge of the tongue, along one side or the other or at the tip. Some individuals experience stinging, pain, or tingling instead of itching.
2. White spots in the skin, resulting from melanin becoming absent in the area. These often occur on the outside of the forearm, but may occur in other places. The longer these spots are there, the whiter they get. As time goes by, the spots become very dry and flaky to the extent that small raw spots of skin may be exposed.
3. Sharp stabbing, tingling pain in the palm of one or both hands. This occurs suddenly and for no apparent reason in a spot directly below the ring finger, approximately where the first palm crease is. If B_{12} deficiency is not treated, a tingling pain may begin to occur along the outside edge of the hand, starting from the wrist. This pain occurs when the wrist is extended.
4. Sores at the corners of the mouth, sometimes extending along the edge of the lip. These are raw spots, not blisters, and they tend to come and go.
5. Nerve shock in the side of the body. It can be felt coming on a few seconds before it hits, and then it hits almost like a mild but deep electric shock and quickly subsides. It can occur at the side of either hip or on either side of the upper body, along the ribs. It can occur consecutively in at least two or three locations, one right after the other.
6. Shortness of breath, but without chest pain. This can occur when walking just a few yards.
7. Hemifacial (one side of the face) spasm usually manifesting in the eye. Can occur in multiple spots on the face such as the lip, nose, eyelids, and eyebrow.

8. Facial pain, usually on only one side of the face at a time. This pain varies so much that it would be difficult to describe all the possibilities. This can present as a dull pain in the cheek bone right underneath an eye, or a sharp shooting pain across the forehead, sometimes coming downward from the scalp to the edge of the nose by the eye.
9. Tingling along the back of one or both thighs, starting at the hips and shooting downward.
10. Memory loss and/or disorientation.
11. Migraine headaches. These may be preceded by a temporary blind spot in the center of the field of vision, usually lasting about ten minutes, and sometimes followed by facial pain under the eyes. After the blind spot vanishes, there may be zigzag streaks through the vision that may last up to hours. Even in the same person, there may be extreme variations in the headaches themselves. They may be quite severe with nausea or they may be virtually nonexistent.

    Migraines of most individuals have causes other than B_{12} deficiency, but migraines of certain individuals diminish or stop completely after they are treated for B_{12} deficiency.
12. Irritability and/or depression and/or personality changes.

VITAMIN C ASCORBIC ACID

DISCOVERY

The need to include fresh plant food or raw animal flesh in the diet to prevent disease was known from ancient times. Native people living in marginal areas incorporated this into their medicinal lore. For example, spruce needles were used in temperate zones in infusions, or the leaves from species of drought-resistant trees in desert areas. In 1536, the French explorer Jacques Cartier, exploring the St. Lawrence River, used the local natives' knowledge to save his men who were dying of scurvy. He boiled the needles of the arbor vitae tree to make a tea that was later shown to contain 50 mg of vitamin C per 100 grams. Throughout history, the benefit of plant food to survive long sea voyages has been occasionally recommended by authorities.

John Woodall, the first appointed surgeon to the British East India Company, recommended the preventive and curative use of lemon juice in his book, *The Surgeon's Mate*, in 1617. The Dutch writer, Johann Bachstrom, in 1734, gave the firm opinion that *"scurvy is solely owing to a total abstinence from fresh vegetable food, and greens, which is alone the primary cause of the disease."* While the earliest documented case of scurvy was described by Hippocrates around the year 400 BC, the first attempt to give scientific basis for the cause of this disease was by a ship's surgeon in the British Royal Navy, James Lind.

Scurvy was common among those with poor access to fresh fruit and vegetables, such as remote, isolated sailors and soldiers. While at sea in May 1747, Lind provided some crew members with two oranges and one lemon per day, in addition to normal rations, while others continued on cider, vinegar, sulfuric acid or seawater, along with their normal rations. In the history of science this is considered to be the first occurrence of a controlled experiment comparing results on two populations of a factor applied to one group only with all other factors the same. The results conclusively showed that citrus fruits prevented the disease.

Lind published his work in 1753 in his *Treatise on the Scurvy.*

Lind's work was slow to be noticed, partly because his *Treatise* was not published until six years after his study, and also because he recommended a lemon juice extract known as "rob". Fresh fruit was very expensive to keep on board, whereas boiling it down to juice allowed easy storage but destroyed the vitamin (especially if boiled in copper kettles). Ship captains concluded wrongly that Lind's other suggestions were ineffective because those juices failed to prevent or cure scurvy. It was 1795 before the British navy adopted lemons or lime as standard issue at sea.

Limes were more popular, as they could be found in British West Indian Colonies, unlike lemons, which were not found in British Dominions, and were therefore more expensive. This practice led to the American use of the nickname "limey" to refer to the British. Captain James Cook had previously demonstrated and proven the principle of the advantages of carrying "Sour krout" on board, by taking his crews to the Hawaiian Islands and beyond without losing any of his men to scurvy. For this otherwise unheard of feat, the British Admiralty awarded him a medal.

The name "antiscorbutic" was used in the eighteenth and nineteenth centuries as general term for those foods known to prevent scurvy, even though there was no understanding of the reason for this. These foods included but were not limited to: lemons, limes, and oranges; sauerkraut, cabbage, malt, and portable soup. Even before the antiscorbutic substance was identified, there were indications that it was present in amounts sufficient to prevent scurvy, in nearly all fresh (uncooked and uncured) foods, including raw animal-derived foods. In 1928 the Arctic anthropologist Vilhjalmur Stefansson attempted to prove his theory of how the Eskimos are able to avoid scurvy with almost no plant food in their diet, despite the disease striking European Arctic explorers living on similar high-meat cooked diets.

Stefansson theorised that the natives get their vitamin C from fresh meat that is minimally cooked. Starting in February 1928, for one year he and a colleague lived on an exclusively minimally-cooked meat diet while under medical supervision; they remained healthy. Later studies done after vitamin C could be quantified in mostly-raw traditional food diets of the Yukon, Inuit, and Métís of the Northern Canada, showed that their daily intake of vitamin C averaged between 52 and 62 mg/day, an amount approximately the dietary reference intake (DRI), even at times of the year when little plant-based food were eaten.

In 1907, the needed biological-assay model to isolate and identify the antiscorbutic factor was discovered. Axel Holst and Theodor Frølich, two Norwegian physicians studying shipboard beriberi contracted aboard ship's crews in the Norwegian Fishing Fleet, wanted a small test mammal to substitute for the pigeons then used in beriberi research. They fed guinea pigs their test diet of grains and flour, which had earlier produced beriberi in their pigeons, and were surprised when classic scurvy resulted instead.

This was a serendipitous choice of model. Until that time, scurvy had not been observed in any organism apart from humans, and had been considered an exclusively human disease. (Pigeons, as seed-eating birds, were also later found to make their own vitamin C.) Holst and Frølich found they could cure the disease in guinea pigs with the addition of various fresh foods and extracts.

This discovery of a clean animal experimental model for scurvy, made even before the essential idea of "vitamins" in foods had even been put forward, has been called the single most important piece of vitamin C research. In 1912, the Polish-American biochemist Casimir Funk, while researching beriberi in pigeons, developed the concept of vitamins to refer to the non-mineral micro-nutrients that are essential to health. The name is a blend of "vital", due to the vital role they play biochemically, and "amines" because Funk thought that all these materials were chemical amines. Although the "e" was dropped after skepticism that all these compounds were amines, the word vitamin remained as a generic name for them. One of the "vitamins" was thought to be the anti-scorbutic factor in foods discovered by Holst and Frølich.

In 1928 this vitamin was referred to as "water-soluble C," although its chemical structure had still not been determined. From 1928 to 1933, the Hungarian research team of Joseph L. Svirbely and Albert Szent-Györgyi and the American worker Charles Glen King, first identified the anti-scorbutic factor, calling it "ascorbic acid" for its vitamin activity. Szent-Györgyi had isolated the chemical hexuronic acid from animal adrenal glands at the Mayo clinic, and suspected it to be the antiscorbutic factor, but could not prove it without a biological assay.

At the same time, for five years King's laboratory at the University of Pittsburgh had been trying to isolate the antiscorbutic factor in lemon juice, using the model of scorbutic guinea pigs, which developed scurvy when not fed fresh foods, but were cured by lemon juice. They had also considered hexuronic acid, but had been put off the trail when a coworker made the explicit (and mistaken) experimental claim that this substance was not the antiscorbutic substance.

Finally, in late 1931, Szent-Györgyi gave Svirbely, a former worker in King's lab who had recently joined Szent-Györgyi's lab, the last of this hexuronic acid, with the suggestion that it might be the anti-scorbutic factor. By the spring of 1932, King's laboratory had proven this, but published the result without giving Szent-Györgyi credit for it, leading to a bitter dispute over priority claims (in reality it had taken a teamwork effort by both groups, since Szent-Györgyi was unwilling to do the difficult and messy animal studies).

By 1932, Szent-Györgyi's group had discovered that paprika peppers, a common spice in the Hungarian diet, was a rich source of hexuronic acid, the antiscorbutic factor, by then named ascorbic acid, in honour of its activity against scurvy. Ascorbic acid turned out *not* to be an amine, nor even to contain any nitrogen.For his accomplishment, Szent-Györgyi was alone awarded the 1937 Nobel Prize in Medicine "for his discoveries in connection with the biological combustion processes, with special reference to vitamin C and the catalysis of fumaric acid".

Between 1933 and 1934, the British chemists Sir Walter Norman Haworth and Sir Edmund Hirst and, independently, the Polish chemist Tadeus Reichstein, succeeded in synthesizing the vitamin, making it the first to be artificially produced. This made possible the cheap mass-production of what was by then known as vitamin C. Only Haworth was awarded the 1937 Nobel Prize in Chemistry for this work, but the Reichstein process, a combined chemical and bacterial fermentation sequence still used today to produce vitamin C, retained Reichstein's name.

In 1934 Hoffmann–La Roche, which bought the Reichstein process patent, became the first pharmaceutical company to mass-produce and market synthetic vitamin C, under the brand name of Redoxon. In 1957 the American J.J. Burns showed that the reason some mammals were susceptible to scurvy was the inability of their liver to produce the active enzyme L-gulonolactone oxidase, which is the last of the chain of four enzymes that synthesize vitamin C. American biochemist Irwin Stone was the first to exploit vitamin C for its food preservative properties. He later developed the theory that humans possess a mutated form of the L-gulonolactone oxidase coding gene.

In 2008 researchers at the University of Montpellier discovered that in humans and other primates the red blood cells have evolved a mechanism to more efficiently utilize the vitamin C present in the body by recycling oxidized L-dehydroascorbic acid (DHA) back into ascorbic acid, which can be reused by the body. The mechanism was not found to be present in mammals that synthesize their own vitamin C.

THE VITAMIN C

Vitamin C or L-ascorbic acid or L-ascorbate is an essential nutrient for humans and certain other animal species, in which it functions as a vitamin. In living organisms, ascorbate is an anti-oxidant, since it protects the body against oxidative stress. It is also a cofactor in at least eight enzymatic reactions, including several collagen synthesis reactions that cause the most severe symptoms of scurvy when they are dysfunctional. In animals, these reactions are especially important in wound-healing and in preventing bleeding from capillaries.

Ascorbate (an ion of ascorbic acid) is required for a range of essential metabolic reactions in all animals and plants. It is made internally by almost all organisms; notable mammalian group exceptions are most or all of the order chiroptera (bats), and one of the two major primate suborders, the Anthropoidea (Haplorrhini) (tarsiers, monkeys and apes, including human beings). Ascorbic acid is also not synthesized by guinea pigs and some species of birds and fish. All species that do not synthesize ascorbate require it in the diet. Deficiency in this vitamin causes the disease scurvy in humans. It is also widely used as a food additive.

THE CHEMISTRY OF VITAMIN C

L-ascorbic acid, or **vitamin C** is a natural compound, whose peculiar antioxidant properties are used in biological systems and for the conservation of several different manufacts.

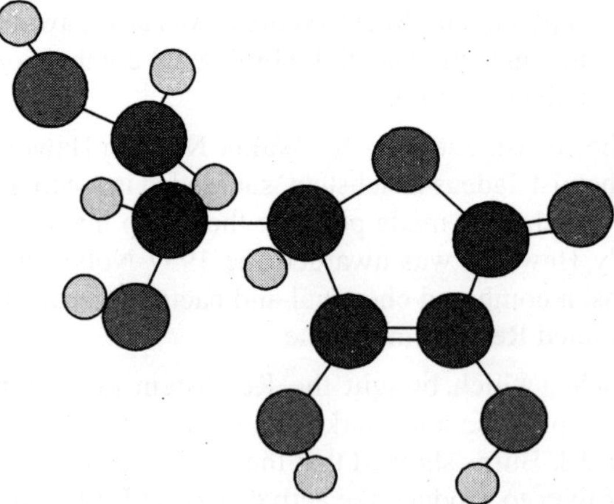

The ascorbic acid molecule (see Fig above) contains four hydroxyl groups in positions 2, 3, 5 and 6; the -OH group in position 3 is acidic ($pK_{a,3}$=4.2), the hydroxyl in position 2 has $pK_{a,2}$=11.6, while those in position 5 and 6 behave as a secondary and primary alcoholic residue respectively

Figures below illustrates the tautomeric equilibrium (see Fig. below) where the C_1=O and C_3-OH groups interchange with the shift of the double bond.

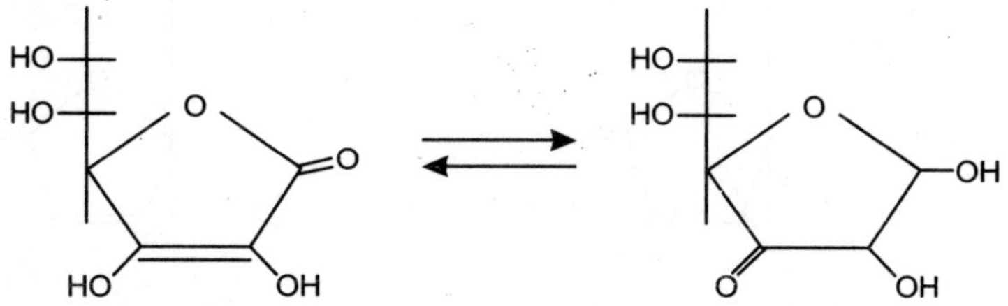

tautomerism

Vitamin C is very sensitive to even slight heating, to the light, and to the action of oxidizing agents and metal ions. Vitamin C is readily oxidized, especially in aqueous solutions, by reacting with atmospheric oxygen, and behaves as a two-electron donor:

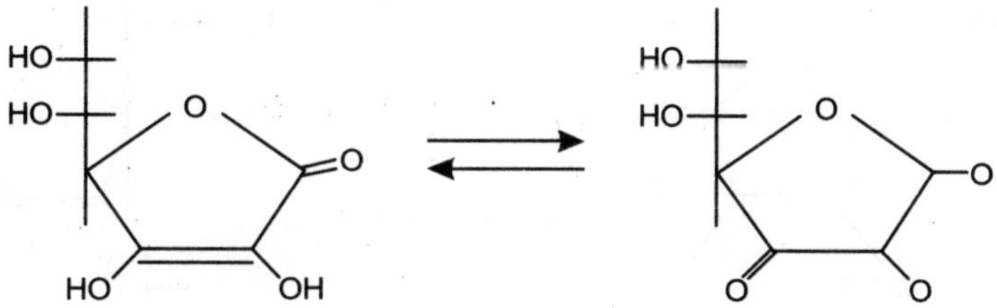

Oxidation process

Ascorbic acid is active in many biological processes; it maintains prolyl hydroxylase in an active form by preserving its iron atom in the reduced Fe^{2+} state, and it keeps collagen in a sufficiently hydroxylated form preventing skin lesions and blood-vessel fragility that are so prominent in scurvy. The biological activity of several important compounds is dramatically affected by the state in which they are found. For instance many enzymes-such as porcine pancreatic lipase, soybean lipoxygenase and phospholipase A_2-perform their activity more rapidly when they react with substrates dispersed in supramolecular aggregates such as micelles and microemulsions. This is why the study of the aggregation properties of bio-active molecules is important in order to better understand their activity.

Among the possible amphiphilic vitamin C derivatives that can be synthesized and that retain its antioxidant properties there are:

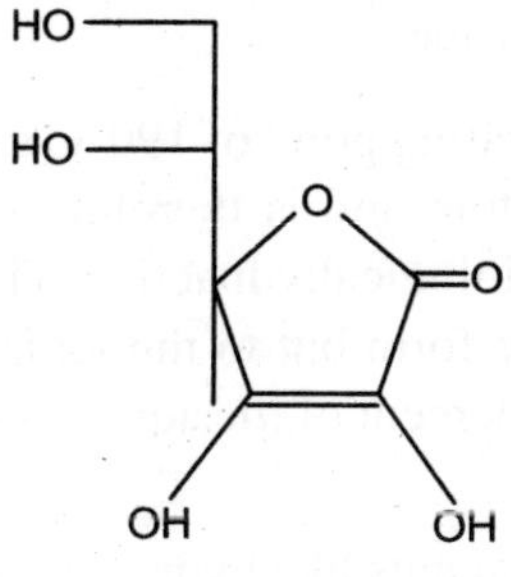

Ascorbic Acid

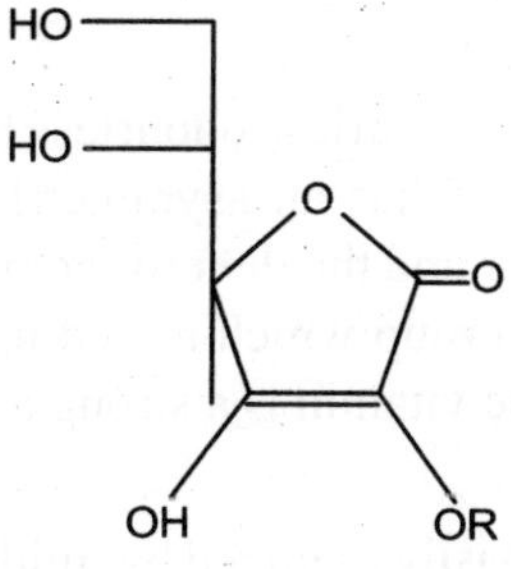

2-O-alkyl-ascorbic acid ether

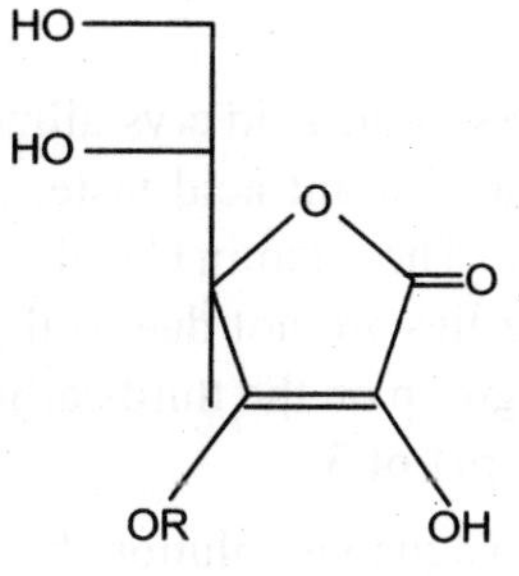

3-O-alkyl-ascorbic acid ether

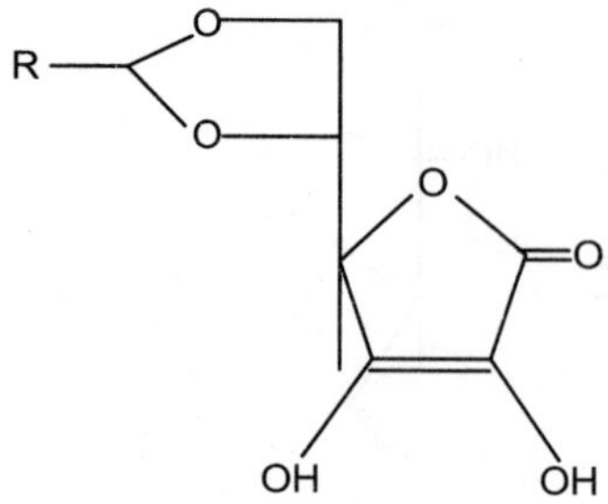

5,6-O-alkylidene-ascorbic acid

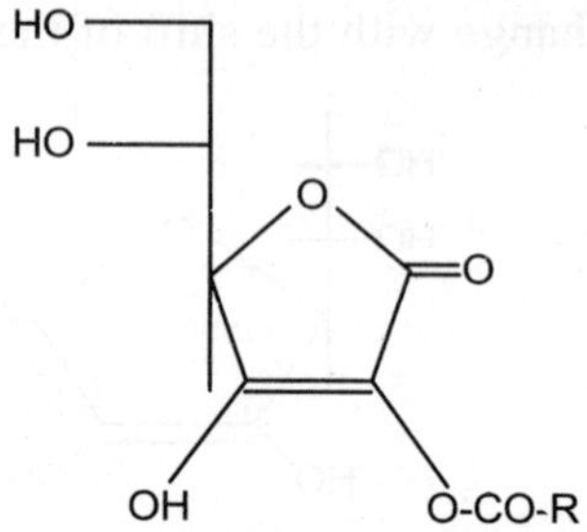

2-O-alkanoyl-ascorbic acid

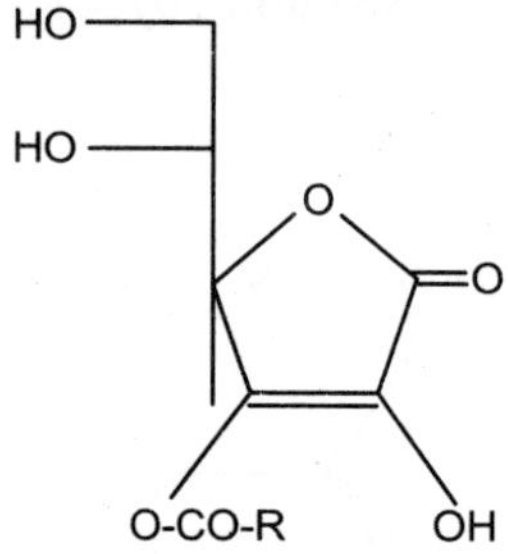

3-O-alkanyol-ascorbic acid

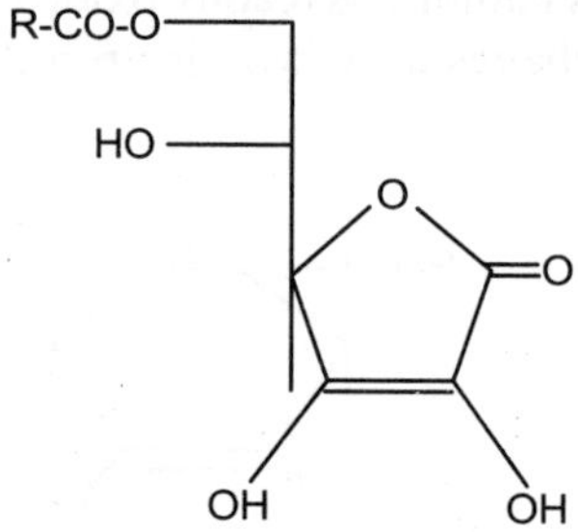

6-O-alkanoyl-ascorbic acid

The structural formal of ascorbic acid is shown below the empirical formula is $C_6H_8O_6$. It is soluble in water and alcohol but insoluble in fats, it is easily oxidized to dehydroascorbic acid.

$-2(H_2O_2)$ / $+2H(H_2S)$

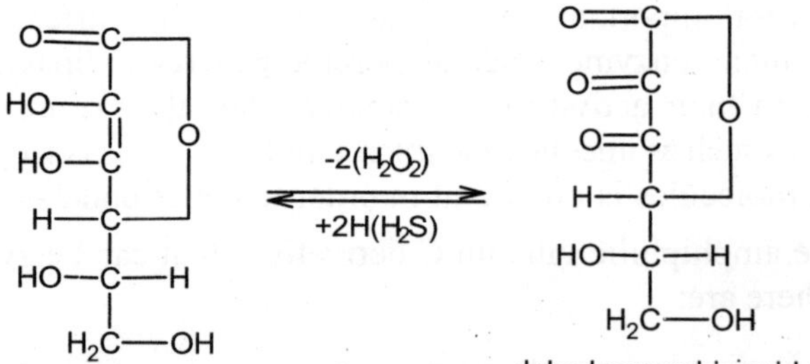

l - ascorbic acid

dehydroascorbic acid

Ascorbic acid crystallizes in white, colourless odourless having a melting point of 190°C to 192°C. It is a pleasant acid taste; Ascorbic acid has an asymmetric carbon atom and is therefore optically active. The vitamin C is the lævo isomer and the dexo(d) compound is biologically inactive. The acidic properties are not due to the carboxyl group which is tied up in lactone form but to the ionization of enol group on the third carbon atom, the vitamin is a strong acid, a half percent of the aqueous solution has a pH of 3.

In aqueous solution the vitamin is easily oxidized by mild oxidizing agents like hydrogen peroxide, ferric chloride, and quinone, iodine in acid or neutral solution to dehydroascorbic acid. Dehydroascorbic acid retains the biological activity of the vitamin and is reduced in animal tissues; a reaction is which sulphydryl compounds like glutathione play an important role. The reduction of dehydroascorbic acid can be effected by hydrogen sulphide. The dehydroascorbic acid undergoes a molecular rearrangement at pH above 5 to form diketogluconic acid in which the lactone ring is broken with the loss of its biological activity and becomes non-reducible.

Treatment with hydrogen iodide converts diketogulonic acid to dehydroascorbic acid. Once the lactone ring has opened the molecule undergoes, further oxidation and is degraded to oxalic acid. The oxidation of ascorbic acid by molecular oxygen even more strongly to form dehydroascorbic acid. Ascorbic acid is a strong reducing agent, it reduces silver nitrate, iodine, ferricyanide, methylene blue and 2, 6 dichlorophenol-indophenol, this served as a means of its histological detection and chemical estimation. The vitamin is destroyed by light specially in presence of flavins. The vitamin reacts with niacin (nicotinamide) forming coloured compounds.

SYNTHESIS OF VITAMIN C

The commercial synthesis of ascorbic acid starts with glucose. The glucose is reduced to sorbitol which is oxidized (by selective fermentation to keto sugar L-sorbose. Oxidation of the sorbose converts it into 2 ketogulonic acid lactonizes into *l*-ascorbic acid.

xylose —phenylhydrazine→ 4,5-bis(2-phenylhydrazinylidene)pentane-1,2,3-triol (xylosazone) —hydrolysis→ 3,4,5,6-tetrahydroxy-2-oxohexanal (xylosone) —HCN→ nitrile of xylosone —8% HCl at 40°C→ vitamin C

DETERMINATION OF ASCORBIC ACID

Both chemical and biological methods are available for the determination of ascorbic acid. One chemical method is based upon the reduction of 2, 6 dichlorophenol indophenol dye in acid solution.

The vitamin is first extracted from the tissues by a mixture of acetic acid and metaphosphoric acid. Several micro methods for the determination of ascorbic acid in blood are available. One of these methods, substitutes methylene blue for 2, 6 dichlorophenol indophenol as the former has higher colour value.

L - ascorbic acid → diclorophenol indophenol → reduced indicator → dehydroascorbic acid

Another method applied to blood and urine measures the colour developed upon the addition of the 85% sulphuric acid on the 2, 4 dinitrophenyl hydrazine derivative of ascorbic acid.

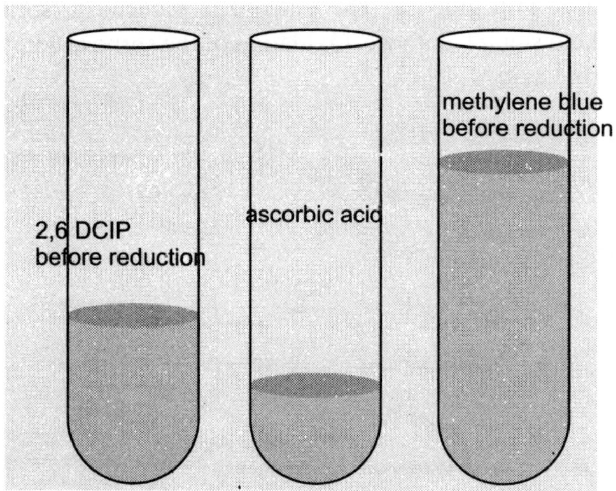

Fig. 12.5 : The DCIP and Methylene Blue and Methylene Blue before reduction

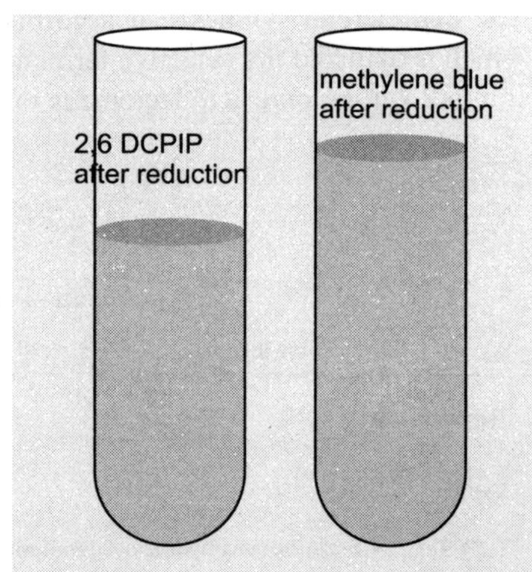

Fig. 12.6 : The DCIP after reduction

THE BIOSYNTHESIS OF VITAMIN C

The biosynthetic pathway proved to be a comparatively simple one. D-glucuronate (formed from glucose) is converted to L-gluconate and then to L-gulono-gamma-lactone, which in turn is further reduced (via L-xylo-hexulonolactone) to L-ascorbic acid (2-oxo-L-gulono-gamm-lactone). The final enzymatic step is catalyzed by L-gulonolactone oxidase (EC 1.1.3.8.)—in the liver in evolutionarily "advanced" species such as the cow, goat, rat, rabbit, and sheep and in the kidney in other species such as the frog, snake, toad, and tortoise—and it is this enzyme that is lacking in those species unable to synthesize vitamin C.

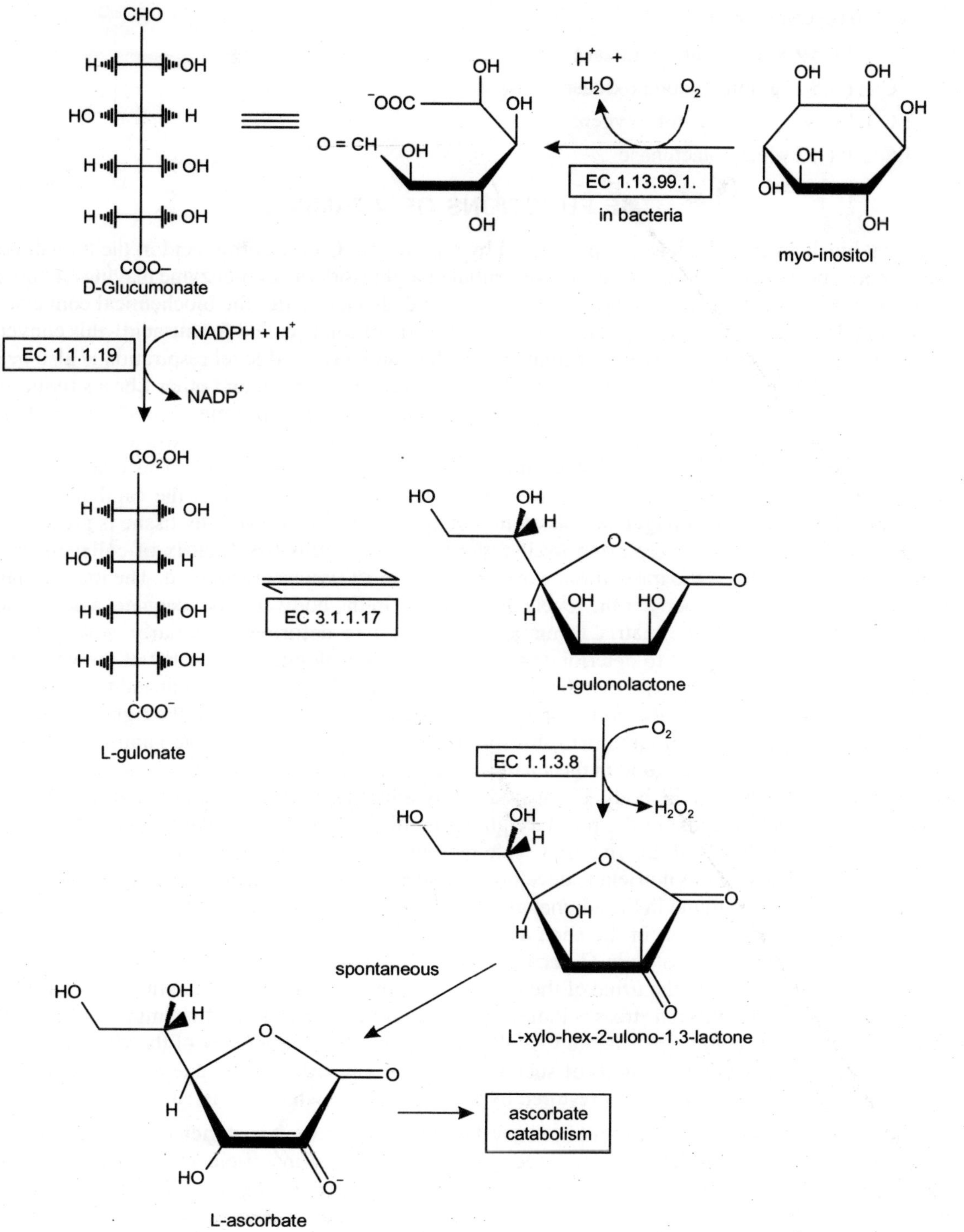

CHO
H OH
HO H
H OH
H OH
COO–
D-Glucumonate

^{-}OOC OH OH O = CH OH OH

H^+ + H_2O O_2
EC 1.13.99.1.
in bacteria

OH OH OH OH OH OH
myo-inositol

EC 1.1.1.19
NADPH + H^+
$NADP^+$

CO_2OH
H OH
HO H
H OH
H OH
COO^-
L-gulonate

EC 3.1.1.17

HO OH H O O H OH HO
L-gulonolactone

EC 1.1.3.8
O_2
H_2O_2

HO OH H O O H OH O
L-xylo-hex-2-ulono-1,3-lactone

spontaneous

HO OH H O O H HO O^-
L-ascorbate

ascorbate catabolism

ascorbate catabolism

EC 1.1.1.19 glucuronate reductase

EC 1.1.3.8 L-gulonolactone oxidase

EC 1.13.99.1 *myo*-inositol oxygenase

EC 3.1.1.17 gluconolactonase

THE FUNCTIONS OF VITAMIN C

The vital biochemical functions are performed by the vitamin C or ascorbic acid in the human body - these actions occur at the cellular level. This essential vitamin works as a co-enzyme playing an important role in the metabolism of amino acids. The ascorbic acid also facilitates the biochemical conversion of the B complex vitamin folic acid (folate) to the active form of folate called folinic acid - this conversion is required for the proper utilization of folate by the human body. Cell level respiration is also heavily influenced by the presence of the vitamin C. The formation of collagen and other fibrous tissues is by far primary role performed by vitamin C in the human body. The proteins found in the skin, the tendons, the bones, the teeth, the cartilaginous tissues and general connective tissues is composed of collagen - this protein plays a supportive functional role in the human body. The vitamin C is also required for the maintenance of proper structural and functional integrity in the capillary walls and larger blood vessels. The collagen acts as an intercellular matrix wherever any tissue is growing, is in the developing process or is repairing itself following injury - this base matrix of collagen serves a structural function and is the main framework on which most tissues are formed. The matrix may not be set down or formed properly if the levels of vitamin C in the body are not adequate, a deficiency of collaged may also leave the matrix formation incomplete or deformed. Similarly, tissue structures already constructed will start to deteriorate in the event of a prolonged deficiency of the vitamin C in the body. Therefore, it can be said that the proper growth, the correct development and the maintenance of just about all the tissues in the body depend on the presence of vitamin C in the right amounts at the cellular level. The pathway of vitamin C absorption goes like this, almost all the vitamin C in the body is absorbed from the digested food in the small intestine, the nutrient is then circulated in the blood and stored in various tissues in the body. Compared to any other tissue, high amounts of the vitamin C are stored in the adrenal glands, in the pituitary gland, the thymus and in the corpus luteum of females. Higher than normal levels of the vitamin C is to be found in all metabolically active tissues due to the increased requirement of this nutrient in developing tissues and organ systems.The highest concentrations of the vitamin C can be found in the adrenal glands among all storage sites for this vitamin in the human body. Vitamin C mobilization in the adrenal glands peaks whenever the organism is stressed by any factor, this mobilization of vitamin C occurs from all other tissues as well and large amounts of the vitamin C start to appear in the urine of the animal. The importance of the vitamin C in promoting the ability of the body to withstand stress is indicated by the rapid mobilization of vitamin C during times of stress. The ability to synthesize vitamin C is also rapidly loss by a deficiency of the vitamin A, at least in laboratory rats, the adrenal glands of such rats also starts to dysfunction due to this factor. Normal functioning in the adrenal glands is restored by supplements of vitamin C in such cases.

Vitamin C deficiency symptoms in many humans - symptomatic deficiency disorders such as scurvy - can be similar to the symptoms seen in cases of adrenal insufficiency including generalized and persistent fatigue, generalized muscle weakness, the presence of digestive disorders and a greatly reduced capability to tolerate stress.The anti-oxidant function of vitamin C is another beneficial role played by the vitamin in the body, as an anti-oxidant the vitamin helps protect the cells and the various tissues from the damaging effects of oxidation from free radicals released in the cells as part of normal metabolic reactions.

As more research is being undertaken on oxidation by free radicals as the basis for many human disease states, the anti-oxidant role of vitamin C's is of extreme importance in the treatment of many well known diseases. The requirement of the vitamin C in the tissues is also raised by the stimulation of muscle tissues during physical activity, as exercised muscle increases the rate at which vitamin is utilized in the metabolism of the cells. The reason for generalized muscle weakness in cases of vitamin C deficiency could be the linked to this increased rate of vitamin C use in the muscles.Simply put the amount of vitamin C needed in the body to prevent scurvy from affecting otherwise healthy individuals is the RDA for vitamin C.

The acute form of the disease is what scientists who set the standard RDA mean when they say scurvy - one of the oldest recognized vitamin deficiency disorders. However, the onset of scurvy may not be simple and several doctors and clinical researchers believed that cases of scurvy can come about with a much more insidious development of symptoms, or that a whole range of possible disorders can begin from localized scurvy affecting certain areas or tissues of the human body. That chronic or latent type of scurvy is prevalent in the modern world is a suggestion made by many scientists who base this belief on clinical evidence gathered with respect to vitamin C and disease states. What this implies is that the ability to resist diseases as insignificant as bleeding gums and easy bruising or even as severe and life changing as cancer and heart disease of millions of people around the world is seriously impaired and weakened.

An increased requirement for the vitamin C or a vitamin C deficiency in the body can be brought about by many different factors. Individuals who do not like eating acidic foods for example, tend to be affected by vitamin C deficiency at some stage in their life-symptoms seen in these individuals tends to be minor. Compared to people who do not smoke, all smokers in general, tend to have lower tissue levels of the vitamin C. The vitamin C levels in the body are known to be directly depleted by smoking. One factor contributing to the higher then average death rate from cancer and heart disease of smokers could be this depletion of vitamin C stores in the body.

People affected by liver disease have lower vitamin C levels then healthy people do, this deficiency or low level of the vitamin C can be directly attributed to the sudden increase in the toxic effects of the different medications used in the treatment of the liver disorder. Low tissue levels of the vitamin C also tends to be evident in patients suffering from hyperthyroid problems.

A deficiency of the vitamin C can also be induced in the human body by the long term use of several types of medications and certain classes of drugs. These medications include compounds like the adrenal corticosteroids - these can in fact produce scurvy like symptoms in the person. A vitamin C deficiency is also induced by long term use of estrogen bearing medications like different oral contraceptives and all menopausal drugs used by women. Drugs like the barbiturates and antibiotics like the compound tetracycline can also induce deficiency states. The urinary rate of excretion of the vitamin C is increased by a factor of three through the use of aspirin.The presence of any kind of disorder that results in an increase in the total blood levels of the trace mineral copper also increases the requirement for vitamin C in the body. Significant amounts of the trace mineral copper can and does enter the human body through the water that is piped through plumbing pipes made from copper - this is one way in which copper comes into the diet. The detoxification of excess levels of copper in the body is apparently influenced by vitamin C in some way.

A wide range of doses types are available for supplemental vitamin C, these doses range from a few milligrams per dose to doses that can be in excess of 1000 mg (or a gram). One should make sure to carefully examine all the vitamin C tablets before purchase for dosage and other requirements, particularly products advertised as natural or rose hips or acerola must be checked for product information. The usual mode by which these supplemental forms of the vitamin C are made involves using some synthetic vitamin C and adding small quantities of acerola berry or rose hips to mask the synthetic nature of the product. Vitamin C which is "synthetic" is similar in all respects with the naturally occurring vitamin C in plant and animal sources. The synthetic form is produced by fermenting glucose, a process that is essentially similar to the way in which vitamin C is produced in natural sources. The cost is the deciding factors as the so called natural vitamin C tablets may be sold by clever marketers for a much higher price than the same dose type of tablet which has been plainly labeled "ascorbic acid"- this is just a marketing gimmick and natural sources of vitamin C are in no way better than synthesized forms of the same vitamin. The sodium ascorbate compound form of the supplemental vitamin C is also available in many drug stores. The powdered form of vitamin C is also available in many drug stores; this supplemental form of the vitamin is the cheapest way to supplement the vitamin.

Vitamin C is found in very high amounts in plant foods such as all citrus fruits and their juices, in fruits like strawberries, in cantaloupes, in all raw vegetables such as peppers, herbs like parsley, broccoli and cauliflower, in Kale, in Brussels sprouts and turnip greens, in vegetables like cabbage, tomatoes and potatoes and bean sprouts among others. The method employed in growing, storing and preparing these vegetables decides about the level of vitamin C found in these plant foods. The vitamin C content of any plant food is determined by the amount of sunlight the plant receives, the greater the amount of incident sunlight, the more vitamins produced in the plant. Additionally, the vitamin C is chemically susceptible to the damaging actions of oxidation - a significant amount of the vitamin can be lost by exposing the vitamin C to ambient air.

Severe losses of the vitamin from food can ensue if the food is steamed for long periods of time, if it is washed again and again, or if it is soaked or canned in liquids - the reason is that vitamin C is soluble in water. A complete loss of the vitamin C content of citrus juice may result when the product is stored at warm temperatures for a long period of time.While societies in the West are do not suffer from scurvy - a serious deficiency symptom of vitamin - and the disease is rare these days, the amounts of vitamin C consumed by most people is not at the optimal rate of intake according to most nutritional scientists. The physical symptoms include generalized and persistent fatigue, easily bruised and bleeding gums - these symptoms are aggravated as scurvy becomes full blown. Low levels of vitamin C are also evident in smokers and such individuals normally need a much higher daily intake of the vitamin just to maintain vitamin C levels at levels that are normal for non-smokers.

DEFICIENCY OF THE VITAMIN C

Scurvy is an avitaminosis resulting from lack of vitamin C, since without this vitamin, the synthesized collagen is too unstable to perform its function. Scurvy leads to the formation of liver spots on the skin, spongy gums, and bleeding from all mucous membranes. The spots are most abundant on the thighs and legs, and a person with the ailment looks pale, feels depressed, and is partially immobilized. In advanced scurvy there are open, suppurating wounds and loss of teeth and, eventually, death. The human body can store only a certain amount of vitamin C and so the body soon depletes itself if fresh supplies are not consumed. It has been shown that smokers who have diets poor in vitamin C are at a higher risk of lung-borne diseases than those smokers who have higher concentrations of vitamin C in the blood.

Nobel Prize winner Linus Pauling and Dr. G. C. Willis have asserted that chronic long term low blood levels of vitamin C (chronic scurvy) is a cause of atherosclerosis. Western societies generally consume sufficient Vitamin C to prevent scurvy. In 2004, a Canadian Community health survey reported that Canadians of 19 years and above have intakes of vitamin C from food of 133 mg/d for males and 120 mg/d for females, that is higher than the RDA recommendation. Notable human dietary studies of experimentally-induced scurvy have been conducted on conscientious objectors during WW II in Britain, and on Iowa state prisoner "volunteers" in the late 1960s.

These studies both found that all obvious symptoms of scurvy previously induced by an experimental scorbutic diet with extremely low vitamin C content could be completely reversed by additional vitamin C supplementation of only 10 mg a day. In these experiments, there was no clinical difference between men given 70 mg vitamin C per day (which produced blood level of vitamin C of about 0.55 mg/dl (about 1/3 of tissue saturation levels), and those given 10 mg per day. Men in the prison study developed the first signs of scurvy about 4 weeks after starting the vitamin C free diet, whereas in the British study, six to eight months were required, possibly due to the pre-loading of this group with a 70 mg/day supplement for six weeks before the scorbutic diet was fed. Men in both studies on a diet devoid or nearly devoid of vitamin C had blood levels of vitamin C too low to be accurately measured when they developed signs of scurvy, and in the Iowa study, at this time were estimated (by labeled vitamin C dilution) to have a body pool of less than 300 mg, with daily turnover of only 2.5 mg/day.

Moderately higher blood levels of vitamin C measured in healthy persons have been found to be prospectively correlated with decreased risk of cardiovascular disease and ischemic heart disease, and an increase life expectancy. The same study found an inverse relationship between blood vitamin C levels and cancer risk in men, but not women. An increase in blood level of 20 micromol/L of vitamin C (about 0.35 mg/dL, and representing a theoretical additional 50 grams of fruit and vegetables per day) was found epidemiologically to reduce the all-cause risk of mortality, four years after measuring it, by about 20%. However, because this was not an intervention study, causation could not be proven, and vitamin C blood levels acting as a proxy market for other differences between the groups could not be ruled out.

However, the prospective nature of the study did rule out vitamin-C lowering effects of terminal illness or end-of-life poor health. Studies with much higher doses of vitamin C, usually between 200 and 6000 mg, for the treatment of infections and wounds have shown inconsistent results. While combinations of antioxidants seem to improve wound healing, this effect cannot be achieved with vitamin C alone.

VITAMIN D CALCIFEROL

Vitamin D is a group of fat-soluble secosteroids, the two major physiologically relevant forms of which are vitamin D_2 (ergocalciferol) and vitamin D_3 (cholecalciferol). Vitamin D without a subscript refers to either D_2 or D_3 or both. Vitamin D_3 is produced in the skin of vertebrates after exposure to ultraviolet B light from the sun or artificial sources, and occurs naturally in a small range of foods. In some countries staples such as milk, flour and margarine are artificially fortified with vitamin D, and it is also available as a supplement in pill form. Vitamin D is carried in the bloodstream to the liver, where it is converted into the prohormone calcidiol. Circulating calcidiol may then be converted into calcitriol, the biologically active form of vitamin D, either in the kidneys or by monocyte-macrophages in the immune system.

When synthesized by monocyte-macrophages, calcitriol acts locally as a cytokine, defending the body against microbial invaders. When synthesized in the kidneys, calcitriol circulates as a hormone, regulating, among other things, the concentration of calcium and phosphate in the bloodstream, promoting the healthy mineralization, growth and remodeling of bone, and the prevention of hypocalcemic tetany. Vitamin D insufficiency can result in thin, brittle, or misshapen bones, while sufficiency prevents rickets in children and osteomalacia in adults, and, together with calcium, helps to protect older adults from osteoporosis. Vitamin D also modulates neuromuscular function, reduces inflammation, and influences the action of many genes that regulate the proliferation, differentiation and apoptosis of cells.

FORMS OF VITAMIN D

Name	Chemical composition	Structure
Vitamin D_1	molecular compound of ergocalciferol with lumisterol, 1:1	
Vitamin D_2	ergocalciferol (made from ergosterol)	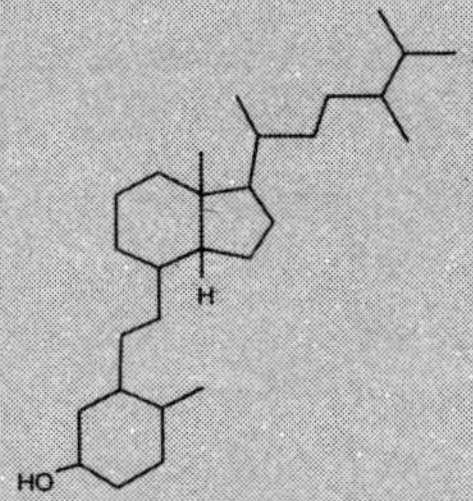 H, HO
Vitamin D_3	cholecalciferol (made from 7-dehydrocholesterol in the skin).	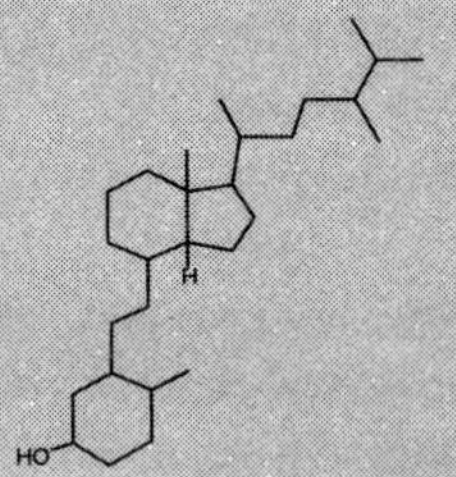 H, HO
Vitamin D_4	22-dihydroergocalciferol	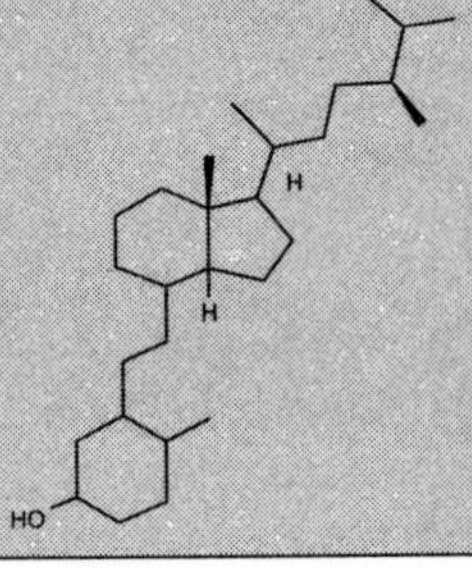 H, H, HO

Name	Chemical composition	Structure
Vitamin D_5	sitocalciferol (made from 7-dehydrositosterol)	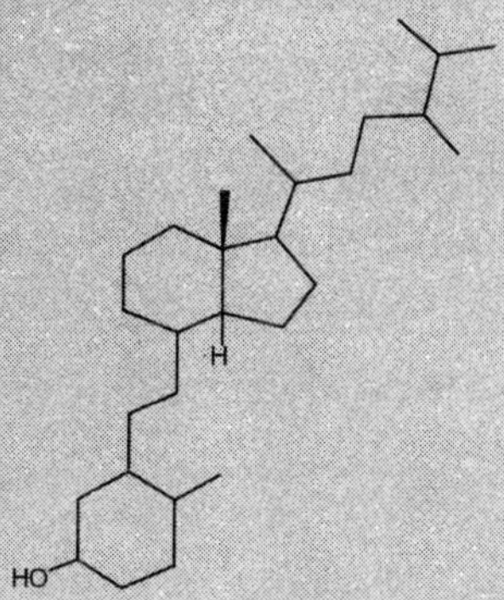 HO H

Several forms (vitamers) of vitamin D have been discovered (see table above). The two major forms are vitamin D_2 or ergocalciferol, and vitamin D_3 or cholecalciferol. These are known collectively as calciferol. Vitamin D_2 was chemically characterized in 1932. In 1936 the chemical structure of vitamin D_3 was established and resulted from the ultraviolet irradiation of 7-dehydrocholesterol. Chemically, the various forms of vitamin D are secosteroids; i.e., steroids in which one of the bonds in the steroid rings is broken. The structural difference between vitamin D_2 and vitamin D_3 is in their side chains.

The side chain of D_2 contains a double bond between carbons 22 and 23, and a methyl group on carbon 24.Vitamin D_2 (made from ergosterol) is produced by invertebrates, fungus and plants in response to UV irradiation; it is not produced by vertebrates. Little is known about the biologic function of vitamin D2 in nonvertebrate species. Because ergosterol can more efficiently absorb the ultraviolet radiation that can damage DNA, RNA and protein it has been suggested that ergosterol serves as a sunscreening system that protects organisms from damaging high energy ultraviolet radiation. Vitamin D_3 is made in the skin when 7-dehydrocholesterol reacts with UVB ultraviolet light at wavelengths between 270-300 nm, with peak synthesis occurring between 295-297 nm. These wavelengths are present in sunlight when the UV index is greater than 3. At this solar elevation, which occurs daily within the tropics, daily during the spring and summer seasons in temperate regions, and almost never within the arctic circles, vitamin D3 can be made in the skin. Depending on the concentration of UVB rays and the minutes of exposure, equilibrium can develop in the skin, and vitamin D degrades as fast as it is generated.

THE CHEMISTRY AND SYNTHESIS OF VITAMIN D

The structures of vitamin D_2 (ergocalciferol) and vitamin D_3 (cholecalciferol) and their provitamins are presented in Figure 1 below. Vitamin D is a generic term and indicates a molecule of the general structure shown for rings A, B, C, and D with differing side chain structures. The A, B, C, and D ring structure is derived from the cyclopentanoperhydrophenanthrene ring structure for steroids. Technically vitamin D is classified as a seco-steroid. Seco-steroids are those in which one of the rings has been broken; in vitamin D, the 9,10 carbon-carbon bond of ring B is broken, and it is indicated by the inclusion of "9,10-seco" in the official nomenclature.

Vitamin D (calciferol) is named according to the revised rules of the International Union of Pure and Applied Chemists (IUPAC). Because vitamin D is derived from a steroid, the structure retains its numbering from the parent compound cholesterol. Asymmetric centers are designated by using the R,S notation; the configuration of the double bonds are notated E for "entgegen" or *trans*, and Z for "zuzammen" or *cis*. Thus the official name of vitamin D_3 is 9,10-seco(5Z,7E)-5,7,10(19)cholestatriene-3b-ol, and the official name of vitamin D_2 is 9,10-seco(5Z,7E)-5,7,10(19), 22-ergostatetraene-3b-ol.

Synthesis Mechanism (form 3)

7-dehydrocholesterol, a derivative of cholesterol, is photolyzed by ultraviolet light in 6-electron conrotatory electrocyclic reaction. The product is pre-vitamin D_3.

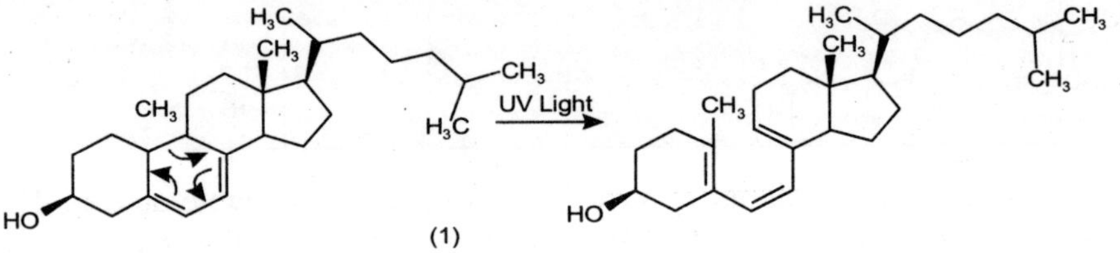

UV Light

(1)

Pre-vitamin D_3 then spontaneously isomerizes to Vitamin D_3 in a antarafacial hydride [1,7] Sigmatropic shift. At room temperature the transformation of previtamin-D_3 to vitamin D_3 takes about 12 days to complete.

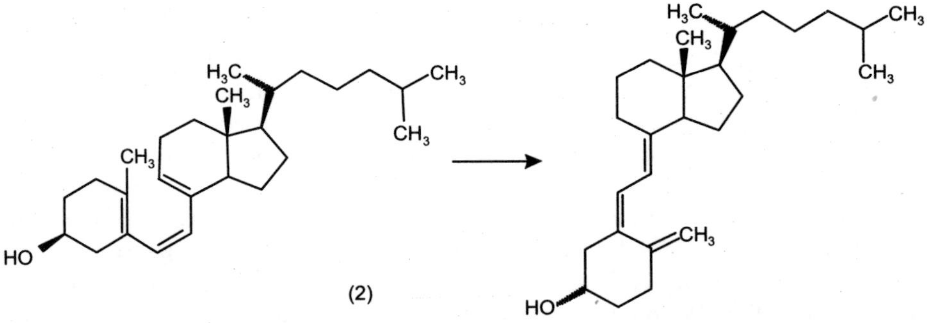

(2)

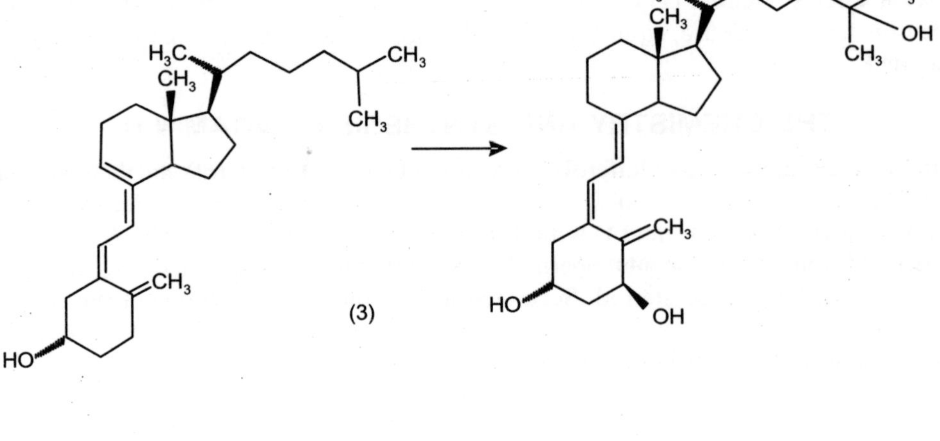

(3)

Whether it is made in the skin or ingested, vitamin D_3 (cholecalciferol) is then hydroxylated in the liver to 25-hydroxycholecalciferol (25(OH)D_3 or calcidiol) by the enzyme 25-hydroxylase produced by hepatocytes. This hydroxylation reaction occurs in the endoplasmic reticulum and requires NADPH, O_2 and Mg^{2+} yet it is not a cytochrome P450 enzyme. Once made the product is stored in the hepatocytes until it is needed and then can be released into the plasma where it will be bound to an α-globulin.

25-hydroxycholecalciferol is then transported to the proximal tubules of the kidneys where it can be hydroxylated by one of two enzymes to different forms of vitamin D, one of which is active

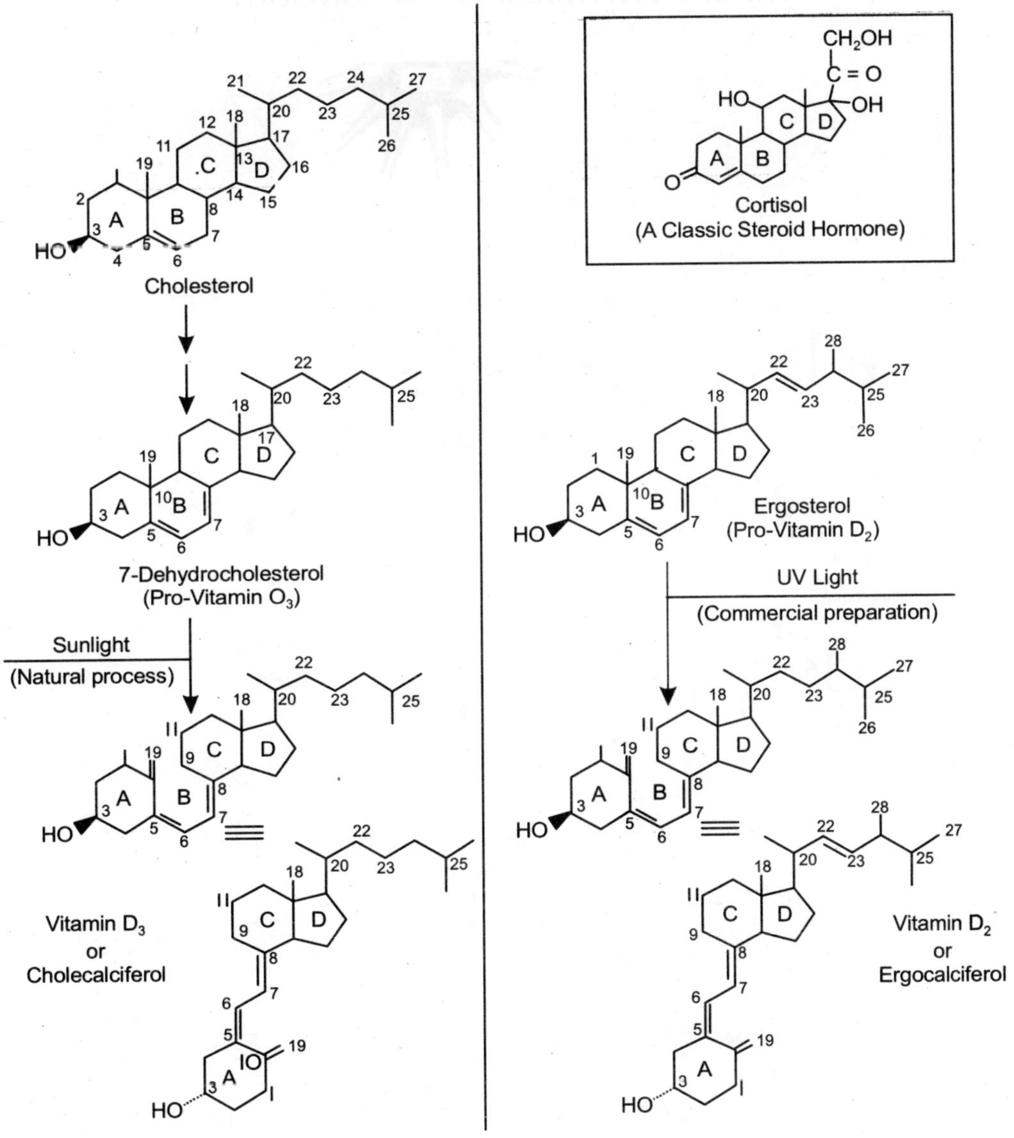

Cholesterol

7-Dehydrocholesterol
(Pro-Vitamin O_3)

Sunlight
(Natural process)

Vitamin D_3
or
Cholecalciferol

Cortisol
(A Classic Steroid Hormone)

Ergosterol
(Pro-Vitamin D_2)

UV Light
(Commercial preparation)

Vitamin D_2
or
Ergocalciferol

vitamin D (1, 25-OH D) and another which is inactive vitamin D (24, 25-OH D). The enzyme 1α-hydroxylase which is activated by parathyroid hormone (and additionally by low calcium or phosphate) forms the main biologically active vitamin D hormone with a C1 hydroxylation forming 1, 25-dihydroxycholecalciferol (1,25$(OH)_2D_3$, also known as calcitriol). A separate enzyme hydroxylates the C24 atom forming 24R,25$(OH)_2D_3$ when 1α-hydroxylase is not active, this inactivates the molecule from any biological activity. Calcitriol is represented below right (hydroxylated Carbon 1 is on the lower ring at right, hydroxylated Carbon 25.

THE BIOSYNTHESIS FROM SUNLIGHT

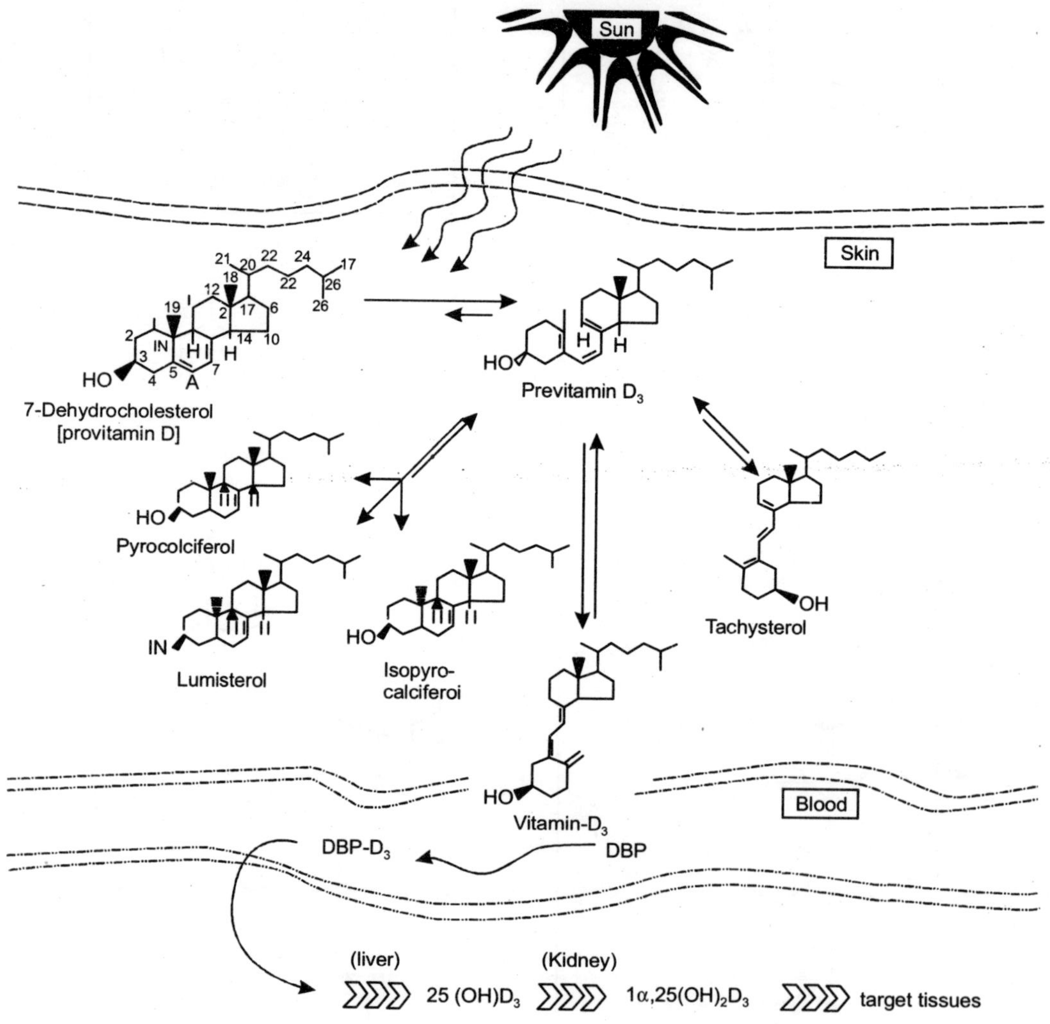

Fig. 12.7 : The Biosynthesis of Vitamin D from Sunlight

Vitamin D_3 can be produced photochemically by the action of sunlight or ultraviolet light from the precursor sterol 7-dehydrocholesterol which is present in the epidermis or skin of most higher animals. The chief structural prerequisite of a provitamin D is that it be a sterol with a D diene double bond system in ring B. The conjugated double bond system in this specific location of the molecule allows the absorption of light quanta at certain wavelengths in the UV range; this can readily be provided in most geographical locations by natural sunlight (or UV-B). This initiates a complex series of transformations that ultimately results in the appearance of vitamin D_3. Thus, it is important to appreciate that vitamin D_3 can be endogenously produced and that as long as the animal (or human) has access on a regular basis to sunlight there is no dietary requirement for this vitamin.

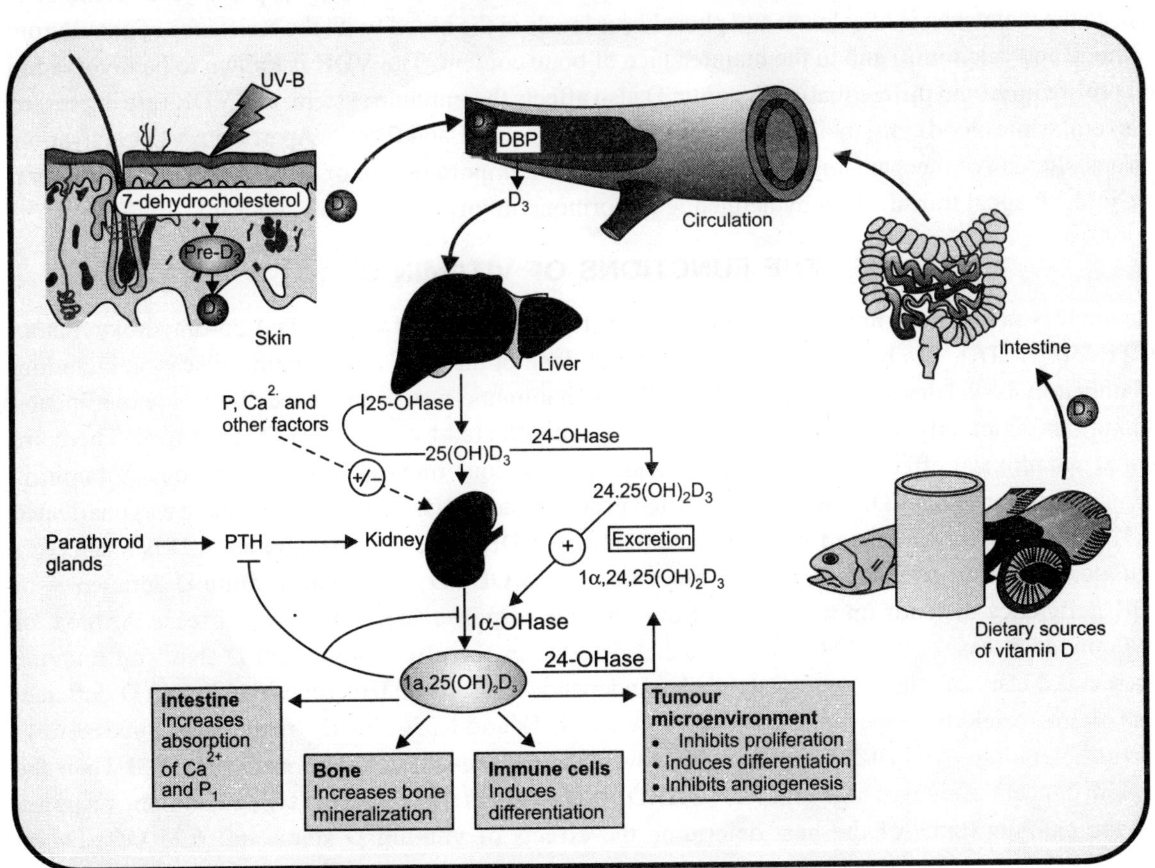

Fig. 12.8 : The Over All Biosynthesis of Vitamin D

MECHANISM OF ACTION

Following the final converting step in the kidney, calcitriol (the physiologically active form of vitamin D) is released into the circulation. By binding to vitamin D-binding protein (VDBP), a carrier protein in the plasma, calcitriol is transported to various target organs. Calcitriol mediates its biological effects by binding to the vitamin D receptor (VDR), which is principally located in the nuclei of target cells. The binding of calcitriol to the VDR allows the VDR to act as a transcription factor that modulates the gene expression of transport proteins (such as TRPV6 and calbindin), which are involved in calcium absorption in the intestine.The vitamin D receptor belongs to the nuclear receptor superfamily of steroid/thyroid hormone receptors, and VDRs are expressed by cells in most organs, including the brain, heart, skin, gonads, prostate, and breast. VDR activation in the intestine, bone, kidney, and parathyroid gland cells leads to the maintenance of calcium and phosphorus levels in the blood (with the assistance of parathyroid hormone and calcitonin) and to the maintenance of bone content. The VDR is known to be involved in cell proliferation and differentiation. Vitamin D also affects the immune system, and VDRs are expressed in several white blood cells, including monocytes and activated T and B cells. Apart from VDR activation, various alternative mechanisms of action are known. An important one of these is its role as a natural inhibitor of signal transduction by hedgehog (a hormone involved in morphogenesis).

THE FUNCTIONS OF VITAMIN D

Vitamin D is an important immune system regulator. The active form of vitamin D, 1,25-dihydroxyvitamin D_3 [$1,25(OH)_2D_3$], has been shown to inhibit the development of autoimmune diseases, including inflammatory bowel disease (IBD). Paradoxically, other immune system-mediated diseases (experimental asthma) and immunity to infectious organisms were unaffected by $1,25(OH)_2D_3$ treatment. There are similar paradoxical effects of vitamin D deficiency on various immune system functions. Vitamin D and vitamin D receptor (VDR) deficiency resulted in accelerated IBD. Experimental asthma was unaffected by $1,25(OH)_2D_3$ treatment and was less severe among VDR-deficient mice. Vitamin D is a selective regulator of the immune system, and the outcome of $1,25(OH)_2D_3$ treatment, vitamin D deficiency, or VDR deficiency depends on the nature of the immune response (e.g., infectious disease, asthma, or autoimmune disease). An additional factor that determines the effect of vitamin D status on immune function is dietary calcium. Dietary calcium has independent effects on IBD severity. Vitamin D-deficient mice on low-calcium diets developed the most severe IBD, and $1,25(OH)_2D_3$ treatment of mice on low-calcium diets improved IBD symptoms. However, the best results for IBD were observed when the calcium concentration was high and $1,25(OH)_2D_3$ was administered. Both the type of immune response and the calcium status of the host determine the effects of vitamin D status and $1,25(OH)_2D_3$ on immunity.

Vitamin D, which helps the body absorb calcium and is obtained from sunlight, may prevent parathyroid hormone production, which leads to osteoporosis.

Thus, the addition of the vitamin to a diet also has the potential to help the brain by assisting the brain-pituitary-thyroid axis. With news of vitamin D's benefits, recommended daily dose has risen significantly and thyroid problems have decreased. But of course there is always the option of taking alternate supplements you can pick up at the local pharmacy; these have also been shown to help alleviate some deficiency symtoms due to lack of Vitamin D intake.

When a study was conducted with patients who had their thyroid's removed, those given the combination of vitamin D and calcium showed significantly fewer signs of patients who had undergone the surgery. Similarly, this healing/therapy process contributed to a lower number of people who experienced low amounts of calcium in the blood, formally known as hypercalcaemia. Meanwhile, the patients with no treatment or only oral calcium, showed no improvement. Despite the exact conclusive studies, it is obvious that vitamin D helps more than it hurts.

To reap the benefits of vitamin D improved thyroid function, there is no need to bask in the sunlight for days. Instead, eat foods fortified or rich in the vitamin. Milk often comes fortified with the healthy vitamin while fishes like salmon and tuna are naturally equip.

VITAMIN D OSSIFICATION AND BONE

Because bone is made up of minerals and is hard, many people think that it is not living material. But a bone in a living animal consists of both **living tissue** and **non-living substances**. Within the "alive bone" are blood vessels, nerves, collagen, and living cells including:

- **osteoblasts** (cells that help **form** bone), and
- **osteoclasts** (cells that help **eat away** old bone).

In addition, bone contains cells called **osteocytes**, which are mature osteoblasts that have ended their bone-forming careers. These cells engage in metabolic exchange with the blood that flows through the bones. The **non-living**, but very important, substances in bone are the minerals and salts.

Besides the metabolically active cellular portion of bone tissue, bone is also made up of a **matrix** (a bonding of multiple fibers and chemicals) of different materials, including primarily **collagen fibers** and **crystalline salts**. The crystalline salts deposited in the matrix of bone are composed principally of **calcium** and **phosphate**, which are combined to form **hydroxyapatite crystals**. As you can see, the **chemical formula** for hydroxyapatite crystals includes molecules of calcium (Ca), phosphate (PO_4), and hydroxide (OH):

$Ca_{10}[PO_4]_6[OH]_2$

In particular, it is the collagen fibers and the calcium salts that help to strengthen bone. In fact, the collagen fibers of bone have great **tensile strength** (the strength to endure stretching forces), while the calcium salts, which are similar in physical properties to marble, have great **compressional strength** (the strength to endure squeezing forces). These combined properties, plus the degree of bondage between the collagen fibers and the crystals, provide a bony structure that has both extreme tensile and compressional strength (Figure 12.9).

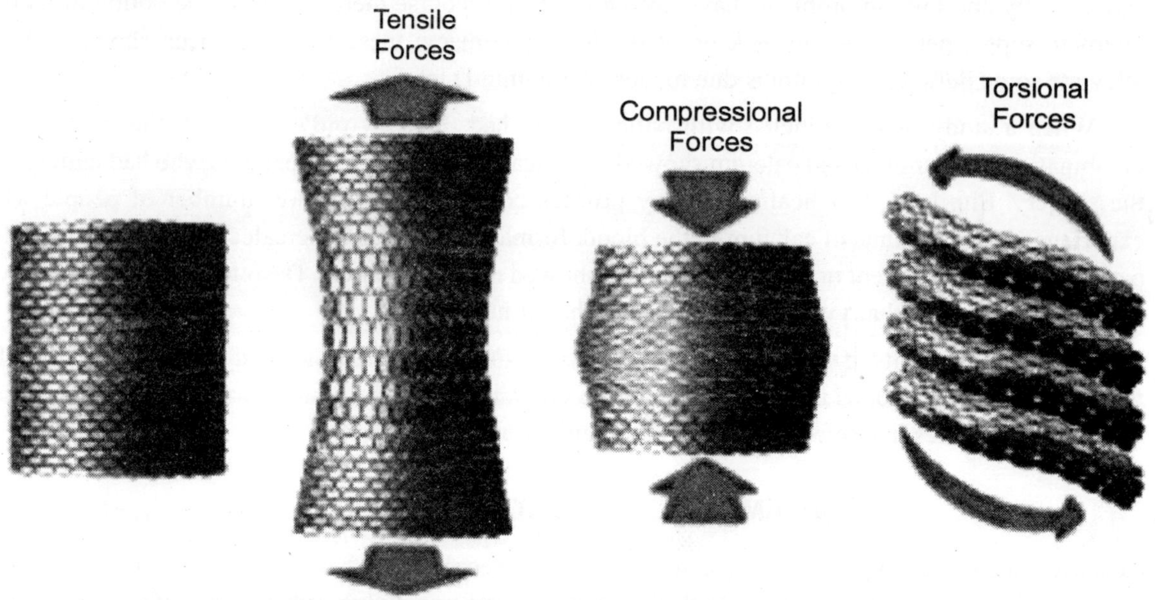

Fig. 12.9: Examples of compressional forces (squeezing), tensile forces (stretching), and torsional forces(twisting)

Thus, bones are constructed in exactly the same way that **reinforced concrete** is constructed. The steel of reinforced concrete provides the tensile strength, while the cement, sand, and rock provide the compressional strength. However, the **compressional strength of bone is greater than that of even the best reinforced concrete, and the tensile strength approaches that of reinforced concrete**. But, even with their great compressional and tensile strengths, neither bone nor concrete has a very high level of **torsional strength** (the strength to endure twisting). In fact, bone fractures often occur as a result of torsional forces that are exerted on an arm or a leg. First, let's examine the major structural components of our bones and then let's briefly discuss how bone develops.

Bones can be classified according to their shapes as: **long** (including the arm and leg bones), **short** (including the bones of the wrists and ankles), **flat** (including the ribs and the bones of the skull), and **irregular** (including the vertebrae along our spine). In describing the general structure of bone, a long bone will be used as an example (Figure 4). At each end of such a bone there is an expanded portion called an **epiphysis**, which forms a joint with another bone. The shaft of the bone, which is located between the epiphyses, is called the **diaphysis**. Except for the **articular** cartilage that covers the very ends of each epiphysis, the bone is completely enclosed by a tough covering called the **periosteum**. Within the periosteum lies a bony layer called **compact bone**, which is solid, strong, and resistant to bending. The epiphyses are composed largely of **spongy (cancellous) bone**, which **provides the greatest amount of elastic strength** since the epiphyses are subjected to the greatest forces of compression.

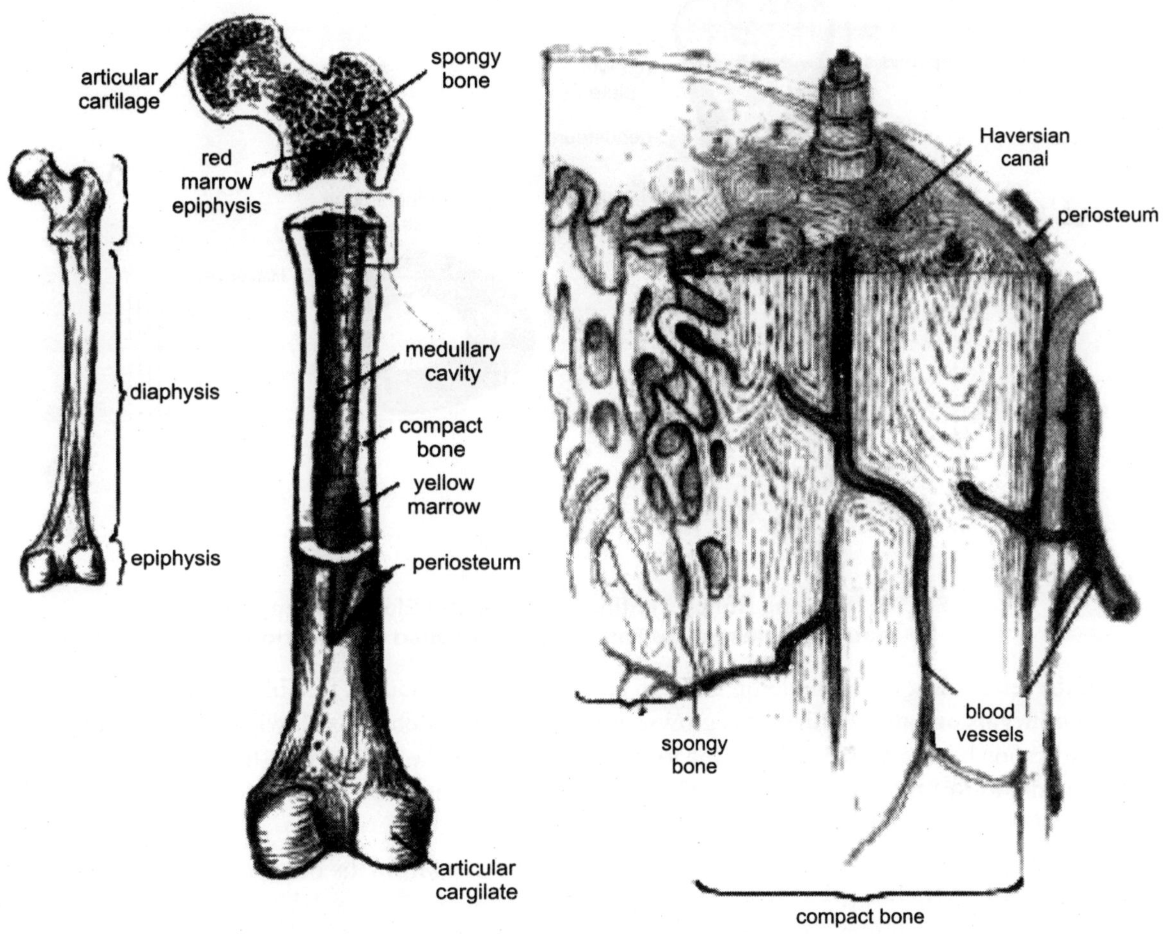

Fig. 12.10 : The Major Parts of a Long Bone

The compact bone in the diaphysis of a long bone forms a rigid tube with a hollow chamber called the **medullary cavity**. This cavity is continuous with the spaces of the spongy bone and is filled with a specialized type of connective tissue called **marrow**. The marrow in the medullary cavity of an adult bone is usually of a type called **yellow marrow**, which functions as fat storage tissue. The marrow in the spaces of spongy bone is likely to be **red marrow**, which functions to produce various types of blood cells. In an earlier chapter, we discussed how the red bone marrow participates in the formation of red blood cells (RBCs) through the process of erythropoiesis. But since we are focusing on the bones in our bodies, let's examine the process of bone formation, or **ossification**.

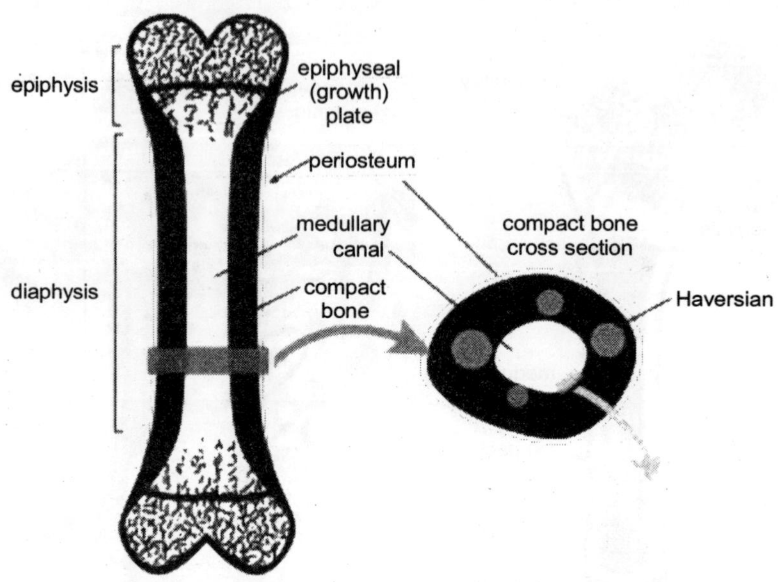

Fig. 12.11 : Long Bones Continue to Grow and Elongate (lengthen) though Adolescence. This Process is called Ossification

In the fetus, most of the skeleton is made up of **cartilage**, a tough, flexible **connective tissue** that has **no minerals or salts**. As the fetus grows, osteoblasts and osteoclasts slowly replace cartilage cells and ossification begins. **Ossification** is the formation of bone by the activity of osteoblasts and osteoclasts and the addition of minerals and salts. Calcium compounds must be present for ossification to take place.

Osteoblasts do not make these minerals, but must take them from the blood and deposit them in the bone.

By the time we are born, many of the bones have been at least partly ossified. In long bones, the growth and elongation (lengthening) continue from birth through adolescence. **Elongation** is achieved by the activity of two cartilage plates, called **epiphyseal plates**, located between the **shaft** (the diaphysis) and the **heads** (epiphyses) of the bones (Figure 12.11). These plates expand, forming new cells, and increasing the **length** of the shaft. In this manner, the length of the shaft increases **at both ends**, and each head of the bone moves progressively apart. As growth proceeds, the thickness of the epiphyseal plates gradually decreases and this bone lengthening process ends. In humans, different bones stop lengthening at different ages, but ossification is fully complete by about age 25. During this lengthening period, **the stresses of physical activity result in the strengthening of bone tissue.** In contrast to the lengthening of bone, the **thickness and strength** of bone must continually be maintained by the body. That is, old bone must be replaced by new bone all the time. This is accomplished as bone is continually deposited by **osteoblasts**, while at the same time, it is *continually being reabsorbed (broken down and digested by the body)* by **osteoclasts** (Figure 12.11). Osteoblasts are found on the outer surfaces of the bones and in the bone cavities. A small amount of osteoblastic activity occurs continually in all living bones (on about 4% of all surfaces at any given time) so that at least some new bone is being formed constantly. Normally, in fact, except in growing bones, **the rates of bone deposition and absorption are equal to each other so that the total mass of bone remains constant.** Usually, osteoclasts exist in small but concentrated masses, and once a mass of osteoclasts begins to develop, it usually eats away at the bone for about three weeks, eating out a tunnel that may be as large as millimetre in diameter and several millimeters in length.

At the end of this time the osteoclasts disappear and the tunnel is invaded by osteoblasts instead; then new bone begins to develop, Bone deposition then continues for several months, the new bone being laid down in successive layers of concentric circles on the inner surfaces of the cavity until the tunnel is filled. Deposition of new bone ceases when the bone reaches the surface of the blood vessels supplying the area. The canal through which these blood vessels run, called the **haversion canal** (Figure 12.11), therefore, is all that remains of the original cavity.

This process continues until about age 40, when the activity of osteoblasts slows and bones become more brittle. Let's look at some important factors that are necessary to produce healthy bone. Bone development is influenced by a number of factors, including **nutrition, exposure to sunlight, hormonal secretions, and physical exercise**. For example, exposure of skin to the ultraviolet portion of sunlight is favorable to bone development, because the skin can produce **vitamin D** when it is exposed to such radiation. Vitamin D is necessary for the proper absorption of calcium in the small intestine. In the absence of this vitamin, calcium is poorly absorbed, the bone matrix is deficient in calcium, and the bones are likely to be deformed or very weak. **Vitamins A** and **C** also are needed for normal bone growth and development.

Hormones that affect bone growth and development include those secreted by the pituitary gland, thyroid gland, parathyroid glands, and the ovaries and testes (Figure 12.12).

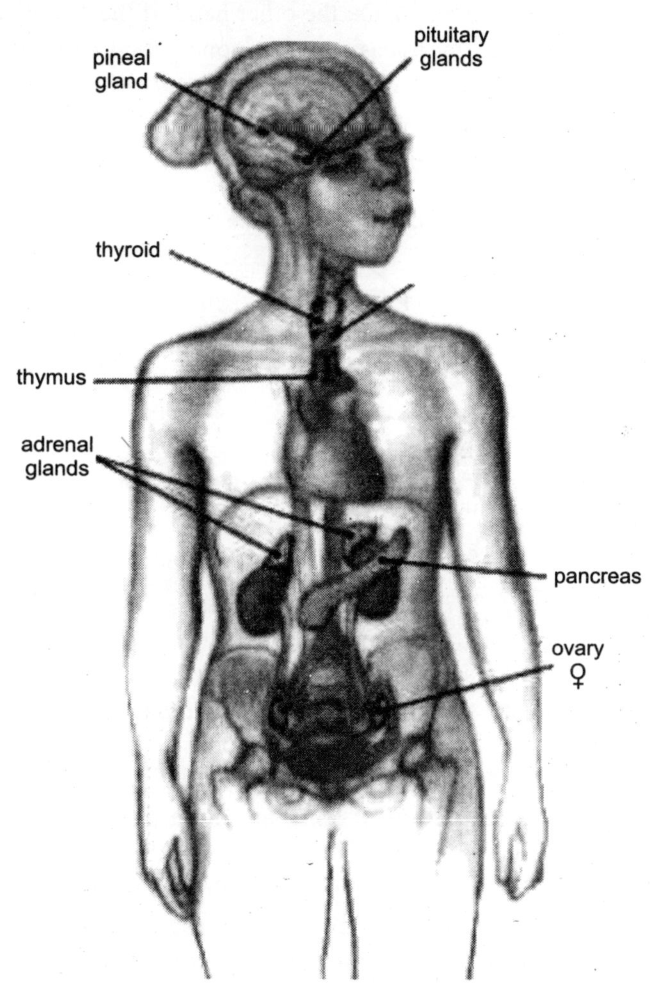

Fig. 12.12 : Certain Endocrine Glands Secrete Hormone

The pituitary gland, for instance, secretes **growth hormone (GH)**, also called **somatotropin**, which stimulates activity in the epiphyseal plates. This hormone is the main regulator of height. Somatotropin plays many roles in the body: it stimulates bone and muscle growth, maintains the normal rate of protein synthesis in all body cells, and speeds the release of fats as an energy source for growth. Other hormones play a part in maintaining the strength and health of the bone matrix by functioning to **control the level of blood calcium**. In fact, calcium is needed for a number of metabolic processes other than for bone formation, including blood clot formation, nerve impulse conduction, and muscle cell contraction. When a low blood calcium condition exists, the **parathyroid glands** respond by releasing **parathyroid hormone (PTH)**.

This hormone stimulates **osteoclasts to break down bone *tissue***, and as a result, calcium salts are released into the blood. On the other hand, if the blood calcium level is excessively high, the **thyroid gland** responds by releasing a hormone called **calcitonin**. Its effect is opposite that of parathyroid hormone; **it inhibits osteoclast activity allowing osteoblasts to form bone tissue**. As a result, the excessive calcium is stored in bone matrix. The actions of these hormones are both excellent examples of some important **negative feedback loops** present in our bodies (Figure 12.13).

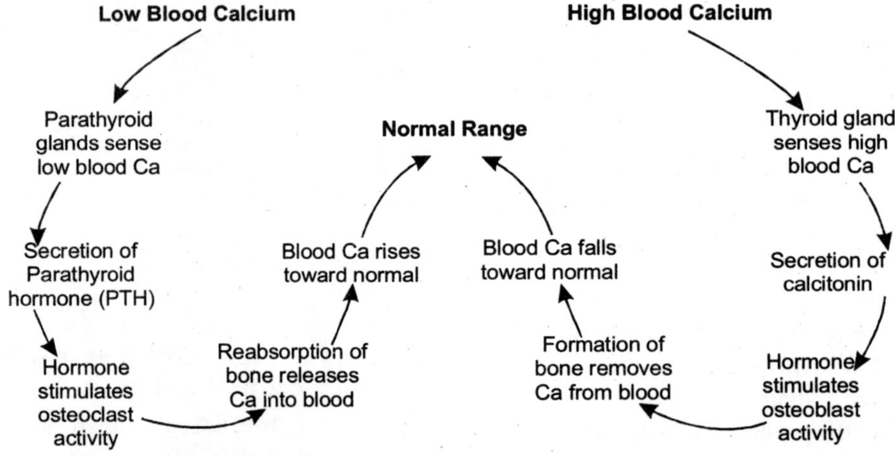

Fig. 12.13 : The Parathyroid and Thyroid Glands Function to Control the Level of Blood Calcium

Without adequate supplies of these important chemicals, the bones will not develop or grow normally.

Women, especially if slim, are more prone to bone loss after **menopause** (when women stop menstruation, age 45-60). Race is another factor that affects ones tendency to lose bone. White and Asian women are especially vulnerable to a bone-thinning disease called **osteoporosis** (which means "porous bone"), which strikes one in four American women over 60. Usually, in osteoporosis, the **osteoblastic activity in the bone is less than normal**, and consequently the rate of bone deposition is reduced. This results in brittle bones in the elderly, and, therefore, their bones are more susceptible to fracture. Older people may also develop a hump in the upper back, often referred to as a **dowager's**

hump. Some women can lose as much as **one-third of their skeletal structure by age 75**. This is twice the rate of bone loss in older men, who have about 30% more bone mass to start with. **Osteoporosis afflicts more people than any other bone disease.** There are many causes of osteoporosis:

- the lack of physical stress on the bones because of less activity;
- malnutrition, including not enough calcium in the diet, to the extent that sufficient bone matrix cannot be formed;
- lack of vitamin C, which is necessary for osteoblasts to create healthy bone matrix;
- the lack of **estrogen** secretion, which occurs in women after menopause (estrogen is also important to stimulate osteoblast activity); end,
- old age, when the body simply cannot form sufficient bone matrix.

In order to prevent the development of osteoporosis, **extra calcium** is needed in the diet, **exercise** can help by stimulating the formation of new bone, and **estrogen therapy** (estrogen is an important ovarian sex hormone in females) can curb bone loss in women after menopause. Let's briefly discuss exercise and how "loading" the bones helps them to stay strong.

Bone is deposited **in proportion to the compressional load** that the bone must carry. For instance, the bones of athletes become considerably heavier than those of nonathletes. Also, if a person has one leg in a cast but continues to walk on the opposite leg, the bone of the leg in the cast becomes thin and as much as 30% decalcified within a few weeks, while the opposite bone remains thick and normally calcified. Therefore, continual physical stress stimulates calcification and osteoblastic deposition of bone, producing stronger bones. Now that we are familiar with what bone is and how it is formed, let's examine how bones serve to support, protect, and move our bodies.

OSSIFICATION

Ossification (or **osteogenesis**) is the process of laying down new bone material by cells called osteoblasts. It is synonymous with bone tissue formation. There are two processes resulting in the formation of normal, healthy bone tissue: Intramembranous ossification is the direct laying down of bone into the primitive connective tissue (mesenchyme), while endochondral ossification involves cartilage as a precursor.

In fracture healing, endochondral osteogenesis is the most commonly occurring process, for example in fractures of long bones treated by plaster of Paris, whereas fractures treated by open reduction and stabilization by metal plate and screws may heal by intramembranous osteogenesis.

Heterotopic ossification is a process resulting in the formation of bone tissue that is often atypical, at an extraskeletal location. Calcification is often confused with ossification. Calcification is synonymous with the formation of calcium-based salts and crystals within cells and tissue. It is a process that occurs during ossification, but not *vice versa*.The exact mechanisms by which bone development is triggered remains unclear, but it involves growth factors and cytokines in some way.

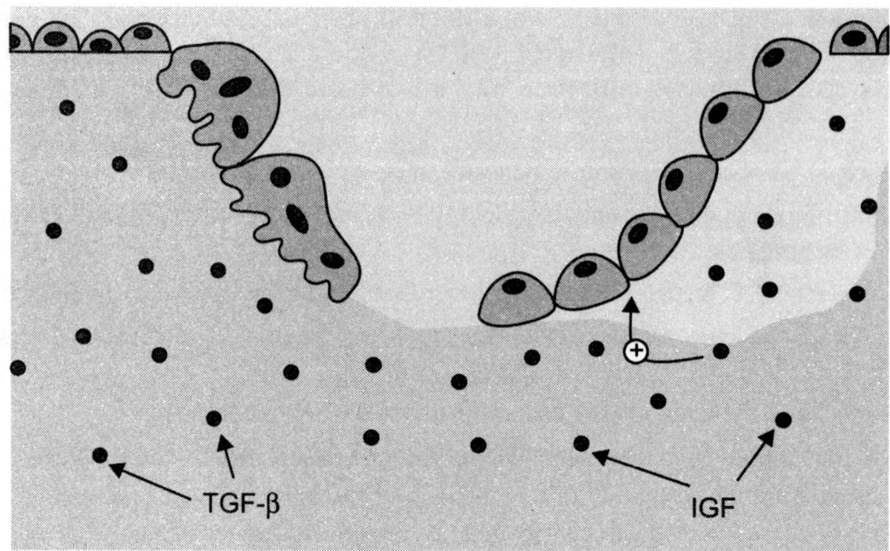

Fig. 12.14: The Osteoblasts and Osteoclasts a Diagrammatic Representation

THE OTHER FUNCTIONS OF VITAMIN D

Vitamin D is best known for its role in calcium metabolism, especially for its role in the treatment of rickets. Yet modern science is discovering that it has many more important roles, from mental health and immunity to blood sugar regulation and cancer prevention. In fact, there is some evidence that the vitamin D found in those notoriously "artery-clogging" cholesterol-rich foods may help prevent heart disease.

Cod liver oil was first used as a therapeutic agent in the 1770s, and by the mid-1800s it was well-recognized as a cure for the childhood bone disease, rickets, which is marked by an expansion of the bone's metaphyseal plate and a buildup of unmineralized bone matrix, and its adult equivalent, osteomalacia. Upon the discovery of vitamin A in 1913 as a component of cod liver oil and butterfat, researchers assumed vitamin A to be responsible for cod liver oil's ability to cure rickets. In 1921, however, a team of researchers established that there was a separate component of cod liver oil responsible for its ability to cure rickets, which came to be known as vitamin D.

In the early 1970s researchers discovered that vitamin D is actually activated into an endocrine hormone within the body, and over the course of the 1980s researchers discovered the vitamin D receptor, which is a nuclear receptor for vitamin D that enters the nucleus and alters the expression of genes once vitamin D binds to it.[12]

Since then, the scope of vitamin D research has broadened immensely. The Vitamin D Council, headed by Dr. John Jacob Cannel, MD, has compiled information on vitamin D's role in the following areas:

- Autoimmune illness
- Cancer
- Chronic pain
- Diabetes

- Heart disease
- Hyperparathyroidism
- Hypertension
- Mental illness
- Multiple sclerosis
- Muscle weakness and coordination
- Obesity
- Osteoarthritis

Dr. Cannel has proposed that a variety of the above conditions be used as indicators for "Vitamin D Deficiency Syndrome," and urges that physicians test the vitamin D levels of patients who exhibit those indicators and treat patients with deficient levels with sunlight or vitamin D supplementation.

DEFICIENCY

Low blood calcidiol (25-hydroxy-vitamin D) can result from avoiding the sun. Deficiency results in impaired bone mineralization, and leads to bone softening diseases including:

- Rickets, a childhood disease characterized by impeded growth, and deformity, of the long bones which can be caused by calcium or phosphorus deficiency as well as a lack of vitamin D; today it is largely found in low income countries in Africa, Asia or the Middle East and in those with genetic disorders such as pseudovitamin D deficiency rickets. Rickets was first described in 1650 by Francis Glisson who said it had first appeared about 30 years previously in the counties of Dorset and Somerset. In 1857 John Snow suggested the rickets then widespread in Britain was being caused by the adulteration of bakers bread with alum. The role of diet in the development of rickets was determined by Edward Mellanby between 1918–1920. Nutritional rickets exists in countries with intense year round sunlight such as Nigeria and can occur without vitamin D deficiency. Although rickets and osteomalacia are now rare in Britain there have been outbreaks in some immigrant communities in which osteomalacia sufferers included women with seemingly adequate daylight outdoor exposure wearing Western clothing Having darker skin and reduced exposure to sunshine did not produce rickets unless the diet deviated from a Western omnivore pattern characterized by high intakes of meat, fish and eggs, and low intakes of high-extraction cereals. The dietary risk factors for rickets include abstaining from animal foods. Vitamin D deficiency remains the main cause of rickets among young infants in most countries, because breast milk is low in vitamin D and social customs and climatic conditions can prevent adequate UVB exposure. In sunny countries such as Nigeria, South Africa, and Bangladesh where the disease occurs among older toddlers and children it has been attributed to low dietary calcium intakes, which are characteristic of cereal-based diets with limited access to dairy products. Rickets was formerly a major public health problem among the US population; in Denver where ultraviolet rays are approximately 20% stronger than at sea level on the same latitude almost two thirds of 500 children had mild rickets in the late 1920s. An increase in the proportion of animal protein in the 20th century American diet coupled with increased consumption of milk fortified with relatively small quantities of vitamin D coincided with a dramatic decline in the number of rickets cases.
- Osteomalacia, a bone-thinning disorder that occurs exclusively in adults and is characterized by proximal muscle weakness and bone fragility. The effects of osteomalacia are thought to contribute to chronic musculoskeletal pain, there is no persuasive evidence of lower vitamin D status in chronic pain sufferers.

Adequate vitamin D may also be associated with healthy hair follicle growth cycles. There are also associations between low 25(OH)D levels and peripheral vascular disease, certain cancers, multiple sclerosis, rheumatoid arthritis, juvenile diabetes, Parkinson's and Alzheimer's disease However these associations were found in observational studies and vitamin D vitamin supplements have not been demonstrated to reduce the risks of these diseases. Research conducted in the United Kingdom and the United States shows that dark-skinned people living in Western societies may not be getting enough vitamin D from the sun. This is particularly true among people of Asian descent, with worse levels of deficiency among Asian women. This is partly due to the fact that dark-skinned people are less efficient at making vitamin D, but among some groups it may be attributed to cultural practices of covering up exposed skin. Other groups that are at risk for vitamin D deficiency include children under five, pregnant and breast-feeding women, anyone who is housebound, and seniors who are less able to make vitamin D from their skin—particularly those who don't get out much, or people living in care homes where they don't sit outdoors.

VITAMIN E TOCOPHEROL

During feeding experiments with rats Herbert McLean Evans concluded in 1922 that besides vitamins B and C, an unknown vitamin existed. Although every other nutrition was present, the rats were not fertile. This condition could be changed by additional feeding with wheat germ. It took several years until 1936 when the substance was isolated from wheat germ and the formula $C_{29}H_{50}O_2$ was determined. Evans also found that the compound reacted like an alcohol and concluded that one of the oxygen atoms was part of an OH (hydroxyl) group. As noted in the introduction, the vitamin was given its name by Evans from Greek words meaning "to bear young" with the addition of the -ol as an alcohol. The structure was determined shortly thereafter in 1938.

Vitamin E is a generic term for tocopherols and tocotrienols. Vitamin E is a family of α-, β-, γ-, and δ- (respectively: alpha, beta, gamma, and delta) tocopherols and corresponding four tocotrienols. Vitamin E is a fat-soluble antioxidant that stops the production of reactive oxygen species formed when fat undergoes oxidation. Of these, α-tocopherol (also written as alpha-tocopherol) has been most studied as it has the highest bioavailability.

TOCOPHEROLS AND TOCOTRIENOLS

Compared with tocopherols, tocotrienols are poorly studied Less than 1% of PubMed papers on vitamin E relate to tocotrienols. Current research direction is starting to give more prominence to the tocotrienols, the lesser known but more potent antioxidants in the vitamin E family. Some studies have suggested that tocotrienols have specialized roles in protecting neurons from damage and cholesterol reduction by inhibiting the activity of HMG-CoA reductase [16-1]; δ-tocotrienol blocks processing of sterol regulatory element binding proteins (SREBPs)[16-1].Oral consumption of tocotrienols is also thought to protect against stroke-associated brain damage in vivo. Until further research has been carried out on the other forms of vitamin E, conclusions relating to the other forms of Vitamin E, based on trials studying only the efficacy of α-tocopherol, may be premature.

TYPES OF TOCOPHEROLS

Vitamin E exists in eight different forms, four tocopherols and four tocotrienols. All feature a chromanol ring, with a hydroxyl group that can donate a hydrogen atom to reduce free radicals and a hydrophobic side chain which allows for penetration into biological membranes.

Both the tocopherols and tocotrienols occur in alpha, beta, gamma and delta forms, determined by the number of methyl groups on the chromanol ring. The alpha is most highly methylated (3 methyls on the chromanol ring) with beta, gamma, and delta forms having two, one, and no methyl groups on the chromanol ring, respectively. The tocotrienols have the same methyl structure at the ring and the same Greek letter-methyl-notation, but differ from the analogous tocopherols by the presence of three double bonds in the hydrophobic side chain. The unsaturation of the tails gives tocotrienols only a single stereoisomeric carbon (and thus two possible isomers per structural formula, one of which occurs naturally), whereas tocopherols have 3 centres (and eight possible stereoisomers per structural formula, again, only one of which occurs naturally).

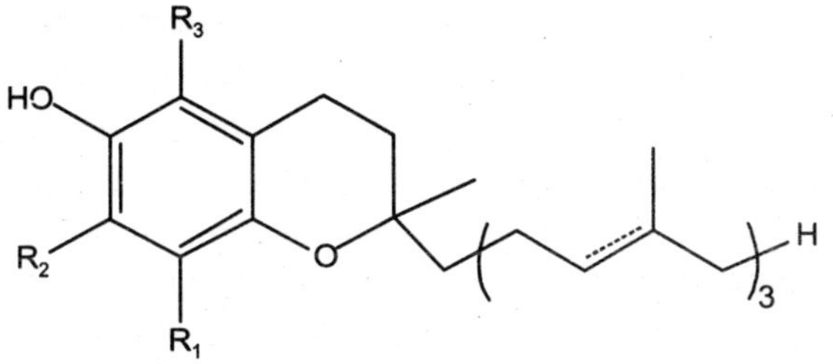

α-tocopherol, $R_1 = R_2 = R_3 = CH_3$
α-tecotrienol, $R_1 = R_2 = R_3 = CH_3$

β-tocopherol, = $R_1 = R_3 = CH_3$; $R_2 = H$
β-tecotrienol, = $R_1 = R_3 = CH_3$; $R_2 = H$

γ-tocopherol, = $R_1 = R_3 = CH_3$ $R_3 = H$
γ-tecotrienol, = $R_1 = R_3 = CH_3$ $R_3 = H$

δ-tocopherol, $R_1 = R_2 = R_3 = H$
δ-tecotrienol, $R_1 = R_2 = R_3 = H$

Each form has slightly different biological activity. In general, the unnatural l-isomers of tocotrienols lack almost all vitamin activity, and half of the possible 8 isomers of the tocopherols (those with 2S chirality at the ring-tail junction) also lack vitamin activity. Of the stereoisomers which retain activity, increasing methylation, especially full methylation to the alpha-form, increases vitamin activity. In tocopherols, this is due to the preference of the tocophrol binding protein for the alpha-tocopherol form of the vitamin.As a food additive, tocopherol is labeled with these E numbers: **E307** (α-tocopherol), **E308** (γ-tocopherol), and **E309** (δ-tocopherol).

α-Tocopherol

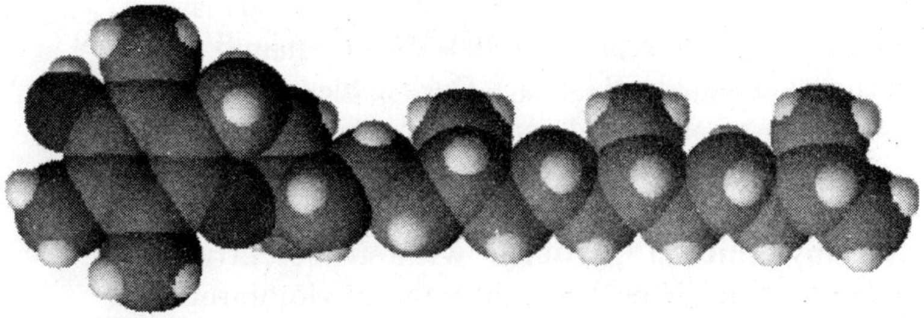

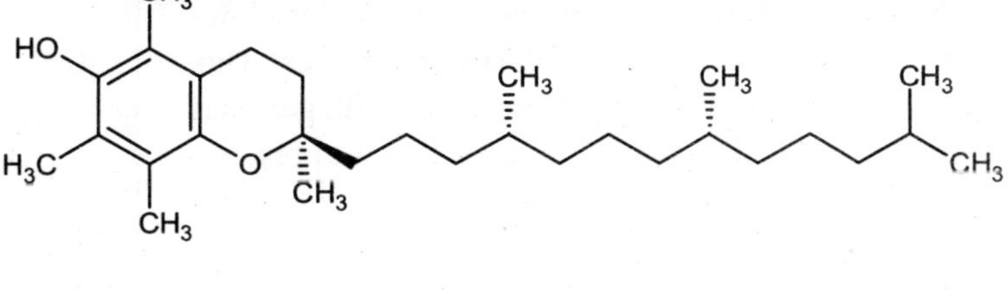

It has been claimed that α-tocopherol is the most important lipid-soluble antioxidant, and that it protects cell membranes from oxidation by reacting with lipid radicals produced in the lipid peroxidation chain reaction. This would remove the free radical intermediates and prevent the oxidation reaction from continuing. The oxidised α-tocopheroxyl radicals produced in this process may be recyclcd back to the active reduced form through reduction by other antioxidants, such as ascorbate, retinol or ubiquinol. However, the importance of the antioxidant properties of this molecule at the concentrations present in the body are not clear and it is possible that the reason why vitamin E is required in the diet is unrelated to its ability to act as an antioxidant. Other forms of vitamin E have their own unique properties. For example, γ-tocopherol (also written as gamma-tocopherol) is a nucleophile that can react with electrophilic mutagens.However, the roles and importance of all of the various forms of vitamin E are presently unclear, and it has even been suggested that the most important function of vitamin E is as a signaling molecule, and that it has no significant role in antioxidant metabolism. So far, most studies about vitamin E have supplemented using only the synthetic alpha-tocopherol, but doing so leads to reduced serum gamma- and delta-tocopherol concentrations. Moreover, a 2007 clinical study involving synthetic alpha-tocopherol concluded that supplementation did not reduce the risk of major cardiovascular events in middle aged and older men. For more information, read article tocopherol.

γ - Tocopherol

When the term "vitamin E" is used, alpha-tocopherol is generally what comes to mind. However, vitamin E is actually a collective term that encompasses 3 other tocopherols (beta, gamma, and delta) in addition to alpha-tocopherol. These molecules are differentiated by the number and placement of methyl groups on their structure. Recent research has been revealing that gamma-tocopherol possesses some unique properties that allow it to function independently from as well as synergistically with alpha-tocopherol. Alpha-tocopherol is the major form of vitamin E that can be found in blood as well as many tissues in humans. Due to its abundance in the body research has generally focused on alpha-tocopherol, with less emphasis on the other isomers. However, gamma-tocopherol is the major form of vitamin E consumed in the diet and is found in many plant seeds and the oils made from them. It has been estimated that gamma-tocopherol represents 70% of the vitamin E consumed in the typical US diet. Once in the body, the metabolism of alpha and gamma-tocopherol differs significantly. Alpha and gamma-tocopherol are absorbed similarly from the gastrointestinal tract and secreted into chylomicron particles without selective discrimination. However, when the chylomicron remnant particles are taken up by the liver, alpha-tocopherol is preferentially incorporated into very low-density lipoprotein (VLDL) cholesterol particles. This preferential placement of alpha compared with gamma-tocopherol into VLDL particles is a result of the presence of a protein in the liver called alpha-tocopherol transfer protein.

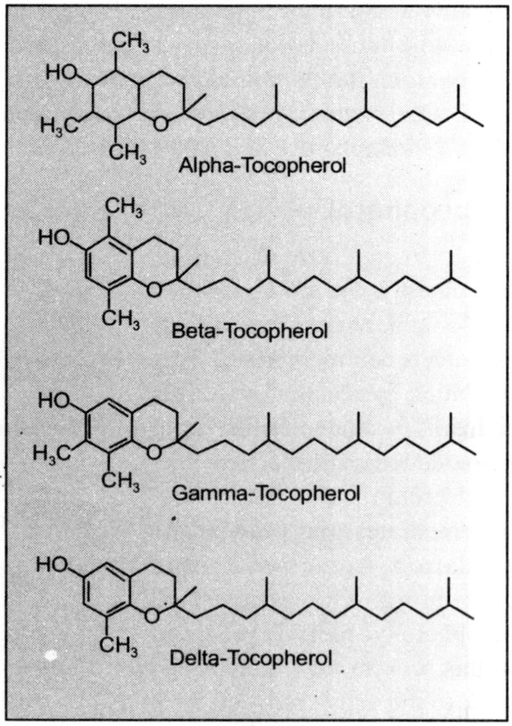

Alpha-Tocopherol

Beta-Tocopherol

Gamma-Tocopherol

Delta-Tocopherol

This protein is able to recognize the stereochemistry of the various tocopherols and has the greatest affinity for the d-alpha form. Consequently, alpha-tocopherol becomes selectively located in VLDL particles, which after circulation and metabolism are transformed in to low-density lipoprotein (LDL) particles. An interesting note is that supplementation with alpha-tocopherol in the absence of gamma-tocopherol leads to a reduction of both tissue and plasma levels of gamma-tocopherol. Nonetheless, given its presence in chylomicrons, and presence in other lipoprotein particles (albeit it at levels less than alpha-tocopherol) gamma-tocopherol does reach the circulation and plays an important role.

Alpha-tocopherol is generally thought of as the most powerful antioxidant of the various tocopherols, and due to its structure it can more readily donate electrons compared with gamma-tocopherol.

Gamma-tocopherol, however, can better quench certain dangerous reactive nitrogen species such as peroxynitrate and nitrogen dioxide, both of which have been associated with a number of degenerative diseases. In fact, in relation to alpha-tocopherol it has been reported that gamma-tocopherol is superior in detoxifying nitrogen dioxide to less harmful compounds. In addition to its antioxidant properties, gamma-tocopherol has also been shown to possess anti-inflammatory properties and can inhibit the activity of cyclooxygenase-2 (COX-2) and production of prostaglandin E2.

Given the role that inflammation plays in the pathology of cardiovascular disease this finding is particularly important. Some researchers have also reported that gamma but not alpha-tocopherol levels are lower in those with cardiovascular disease compared with control subjects. Recent work has also brought to light a relationship between gamma-tocopherol and prostate cancer. In a case-control study, the correlation between alpha-tocopherol, gamma-tocopherol, selenium intake and prostate cancer was examined. The researchers found a significant inverse correlation between the intake of gamma-tocopherol and the incidence of prostate cancer (i.e., the greater the intake of gamma-tocopherol, the lower the risk of disease). Most intriguing was that alpha-tocopherol and selenium intake was only protective when gamma-tocopherol intake was also high.

These interesting scientific findings, coupled with the fact that gamma-tocopherol, but not alpha-tocopherol levels have been shown to decline with age in humans give further credibility to the importance of supplementing with a well-rounded mixed tocopherol supplement that contains significant amounts of gamma-tocopherol.

THE CHEMISTRY OF TOCOPHEROLS

Chemical Structure and Homologues

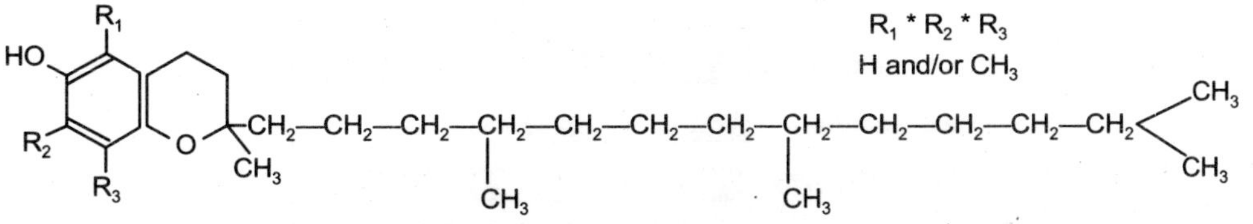

R_1 * R_2 * R_3
H and/or CH_3

Natural TOCOPHEROL exists as a mixture of 4 homologues, Alpha, Beta, Gamma, Delta - available as a mixed product known as MIXED TOCOPHEROL - which is also known as vitamin E. Riken Vitamin

Co., Ltd. supplies TOCOPHEROL products and specialties according to requirements by separating or combining the homologues using unique technology—Homologues -R1 -R2 -R3

Alpha—TOCOPHEROL $- CH_3 - CH_3 - CH_3$

Beta—TOCOPHEROL $- CH_3 - H - CH_3$

Gamma—TOCOPHEROL $- H - CH_3 - CH_3$

Delta—TOCOPHEROL $- H - H - CH_3$

Comparison of Antioxidant Activity of Homologues

Alpha-Toc has strong vitamin E activity, while Beta, Gamma, and Delta-Toc have strong antioxidant activity outside the body (in foodstuffs etc.). Therefore, Toc preparations which have a high Alpha-Toc content are suitable for health foods and vitamin E enrichment.

Biological and Antioxidant Activities of TOCOPHEROLS

Kind of TOC. -Biological activity

(IU/mg) -Antioxidant activity

(with d-Alpha-TOC as 100)

dl-Alpha-Toc acet.-1 -0

d-Alpha-Toc acet.-1.36 -0

dl-Alpha-Toc-1.1 -100

d-Alpha-Toc-1.49 -100

d-Beta-Toe-0.75 -130

d-Gamma-Toc-0.15 -200

d-Delta-Toc-0.05 -300-500

dl-form = synthetic, d-form = natural

A schale oven test was carried out to examine the antioxidant activity of each homologue on lard. As a result, when comparing the days on which the POV values reached 50, the extent of antioxidant activity was in the following order: Delta > Gamma, Beta > Alpha. It us considered that MIXED TOCOPHEROL with a high content of Delta toc or with a high content of Delta-Toc and Gamma-Toc are suitable for use as antioxidant in foods.

Comparison of Antioxidant Activity of Related Products

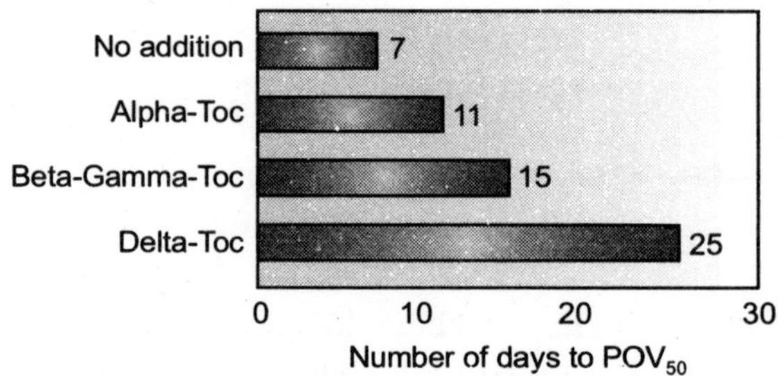

(Amount added: 210 ppm in oil 20g of oil in a 9 ø cm Petri-schale was kept in a dark place at 60°C). --Alpha-Toc has weak antioxidant activity, which is true of synthetic TOCOPHEROL (dl-Alpha-Toc). Interestingly however, it is also known that the addition of a large amount of Alpha-Toc accelerates oxidation. Care must be taken in using dl-Alpha-Toc, bind a lower ratio of Alpha-Toc is favorable also in the case of natural MIXID Toc.

Comparison of Antioxidant Activity of Related Products

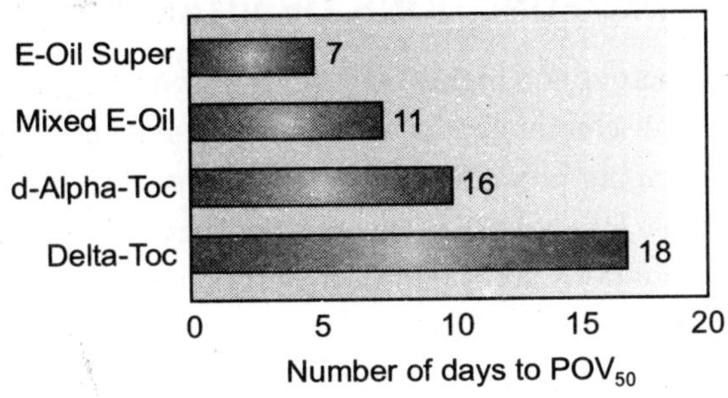

(Amount added: 210 ppm in oil 20g of oil in a 9 ø cm Petri-schale was kept in a dark place at 60°C)

THE SYNTHESIS OF TOCOPHEROL

a-Tocopherol (commonly known as vitamin E) is the mostimportant lipophilic radical-chain-breaking antioxidant in livingtissues. It also participates in the stabilization of biological membranes. The absence of vitamin E in membranes could make them highly permeable and therefore vulnerable to degradation.Vitamin E seems also to influence other important biophysicalmembrane characteristics, such as fluidity, in a manner similar tothat of cholesterol. Earlier investigations showed that the accumulation of carotenoidsduring chloroplast-to-chromoplast differentiation in thefruit pericarp of *Capsicum annuum* L. (yellow pepper) is accompaniedby an increasing a-tocopherol content.

Prenylase | Methyltransferase I | Cyclase | Methyltransferase II

Hemogentisate; Phytyl-PP; Prenylation (CO_2); 2-Me-6-PhOH$_2$; 2-Methylation (SAM → SAH); 2,3-Me$_2$-5-PhOH$_2$; Cyclization; γ-Tocopherol; 2-Methylation (SAM → SAH); α-Tocopherol

About 80% of the label was found in 2,3-dimethylphytylquinol and 20% in a-tocopherol, while c-tocopherol formation was at the detection limit, which agrees withthe results of previous investigations.

Both methylation capacities have been recently separated by fractionation of carefully shocked chromoplasts on continuous sucrose gradients. We suggest, therefore, that α-tocopherol biosynthetic enzymes in these chromoplasts are localized on different domains (peripheral or integral) of chromoplast membranes. The distinct membrane location of these enzymes and the hydrophobic character of the tocopherol cyclase, therefore, requires distinct assay conditions in chromoplast membrane preparations.

THE FUNCTIONS OF VITAMIN E

The E vitamin functions are numerous and highly beneficial to your health and healthy living. Vitamin E is a powerful antioxidant and anti-carcinogen as well as anti-aging supplement. Some studies suggest Vitamin E may slow down the aging process & prevent premature aging by protecting our cells and extending their useful life.Vitamin E is utilized by the most of the tissues in our body. Vitamin E is stored muscles, fat tissue, high concentrations are found in the pituitary gland, adrenal gland and testes. The E vitamin functions by using the Vitamin E stored in these tissue to fight off the free radicals we encounter every day. Due to its antioxidant capabilities Vitamin E protects our body from carbon, lead, mercury, tetrachloride, the ozone, nitrous oxide and other carcinogens.

The E vitamin functions by preventing the formation of nitrosamines from nitrates and nitrates found in cured meats, cigarette smoke and polluted air. Vitamin E is amazing; it's able to prevent the rancidity caused by the oxidation of polyunsaturated fatty acids (PUFAs for short) and thus preventing the free radical damage that can result from ingesting PUFAs. There is definitely a synergy with selenium *(another antioxidant)* and E vitamin functions to protect Vitamin A and C from oxidizing, thus keeping them potent in our body. Vitamin E also helps to increase the levels of some free radical fighting enzymes in our body.

Studies indicate Vitamin E appears to protect form lung, esophageal, and colorectal *(relating to or involving both the colon and the rectum)* cancer and possibly cancer of the cervix and breast.

In the blood Vitamin E rejuvenate the blood and strengthens capillary walls, adequate Vitamin E in the body means the body is getting all the oxygen needed for healthy functioning. Vitamin E helps to thin the blood without interfering with the normal clotting activities. There are many studies indicating the benefits of vitamin E for heart disease prevention and how it allows the blood to flow easier making it less prone to the plaque buildups associated with a heart attack or stroke. The E vitamin functions to help skin and scar tissue heal, many stories are out there about how burn victims are given large does of Vitamin E and the results are amazing. It is also used for internal healing after surgery and been proven to reduce scar tissue after surgery. It has been shown to reduce old scars to almost nothing. Vitamin E is amazing. Vitamin E is capable of keeping your LDL *('BAD')* cholesterol from being oxidized, less oxidized cholesterol means less plaque, resulting in lower heart disease risk. Vitamin E has also been shown to promote proper growth of the placenta in pregnant women, thus lessening the chance of miscarriages.

THE DEFICIENCY OF VITAMIN E

The deficiency of this vitamin causes neurological problems due to poor nerve conduction. These include neuromuscular problems such as spinocerebellar ataxia and myopathies. Deficiency can also cause anemia, due to oxidative damage to red blood cells.Vitamin E deficiency is rare in humans and is almost never caused by a poor diet. Instead, there are three specific situations when a vitamin E deficiency is likely to occur. It is seen in persons who cannot absorb dietary fat, has been found in premature, very low birth weight infants (birth weights less than 1500 grams, or 3.5 pounds), and is seen in individuals with rare disorders of fat metabolism. Individuals who cannot absorb fat may require a vitamin E supplement because some dietary fat is needed for the absorption of vitamin E from the gastrointestinal tract. Anyone diagnosed with cystic fibrosis, individuals who have had part or all of their stomach removed, and individuals with malabsorptive problems such as Crohn's disease, liver disease or pancreatic insufficiency may not absorb fat and should discuss the need for supplemental vitamin E with their physician. People who cannot absorb fat often pass greasy stools or have chronic diarrhea and bloating. Very low birth weight infants may be deficient in vitamin E. A neonatologist, a pediatrician specializing in the care of newborns, typically evaluates the nutritional needs of premature infants.Abetalipoproteinemia is a rare inherited disorder of fat metabolism that results in poor absorption of dietary fat and vitamin E. The vitamin E deficiency associated with this disease causes problems such as poor transmission of nerve impulses, muscle weakness, and degeneration of the retina that can cause blindness. Individuals with abetalipoproteinemia may be prescribed special vitamin E supplements by a physician to treat this disorder. In addition, there is a rare genetic condition termed isolated vitamin E deficiency or ataxia with isolated with vitamin E deficiency, caused by mutations in the gene for the tocopherol transfer protein.[4] These individuals have an extremely poor capacity to absorb vitamin E and develop neurological complications that are reversed by high doses of vitamin

VITAMIN K

In 1929, Danish scientist Henrik Dam investigated the role of cholesterol by feeding chickens a cholesterol-depleted diet. After several weeks, the animals developed hemorrhages and started bleeding. These defects could not be restored by adding purified cholesterol to the diet. It appeared that—together with the cholesterol—a second compound had been extracted from the food, and this compound was called the coagulation vitamin. The new vitamin received the letter K because the initial discoveries were reported in a German journal, in which it was designated as *Koagulationsvitamin*. Edward Adelbert Doisy of Saint Louis University did much of the research that led to the discovery of the structure and chemical nature of vitamin K. Dam and Doisy shared the 1943 Nobel Prize for medicine for their work on vitamin K (K_1 and K_2) published in 1939. Several laboratories synthesized the compound(s) in 1939.

For several decades the vitamin K-deficient chick model was the only method of quantifying vitamin K in various foods: the chicks were made vitamin K-deficient and subsequently fed with known amounts of vitamin K-containing food. The extent to which blood coagulation was restored by the diet was taken as a measure for its vitamin K content. Three groups of physicians independently found this: Biochemical Institute, University of Copenhagen, University of Iowa Department of Pathology:

The first published report of successful treatment with vitamin K of life-threatening hemorrhage in a jaundiced patient with prothrombin deficiency was made in 1938.

Vitamin K (K from "Koagulations-Vitamin" in German and Scandinavian languages) denotes a group of lipophilic, hydrophobic vitamins that are needed for the posttranslational modification of certain proteins, mostly required for blood coagulation, but also a number of other proteins that chelate calcium ion, and are involved in bone and other tissue metabolism. They are 2-methyl-1,4-naphthoquinone derivatives. Vitamin K_1 is also known as phylloquinone or phytomenadione (also called phytonadione). Vitamin K_2 (menaquinone, menatetrenone) is normally produced by bacteria in the large intestine, and dietary deficiency is extremely rare unless the intestines are heavily damaged, are unable to absorb the molecule, or are subject to decreased production by normal flora, as seen in broad spectrum antibiotic use. There are three synthetic forms of vitamin K, vitamins K_3, K_4, and K_5, which are used in many areas including the pet food industry (vitamin K_3) and to inhibit fungal growth (vitamin K_5).

THE CHEMICAL PROPERTIES AND STRUCTURE

All members of the vitamin K group of vitamins share a methylated naphthoquinone ring structure, and vary in the aliphatic side chain attached at the 3-position (see figure below). Phylloquinone (also known as vitamin K_1) invariably contains in its side chain four isoprenoid residues, one of which is unsaturated.

Menaquinones have side chains composed of a variable number of unsaturated isoprenoid residues; generally they are designated as MK-n, where n specifies the number of isoprenoids.

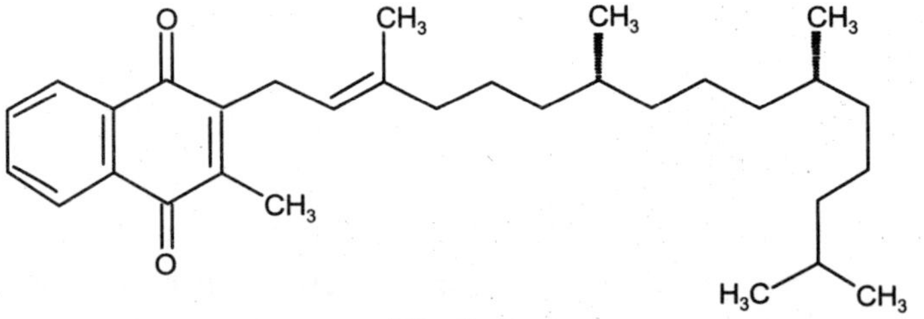

Vitamin K_1

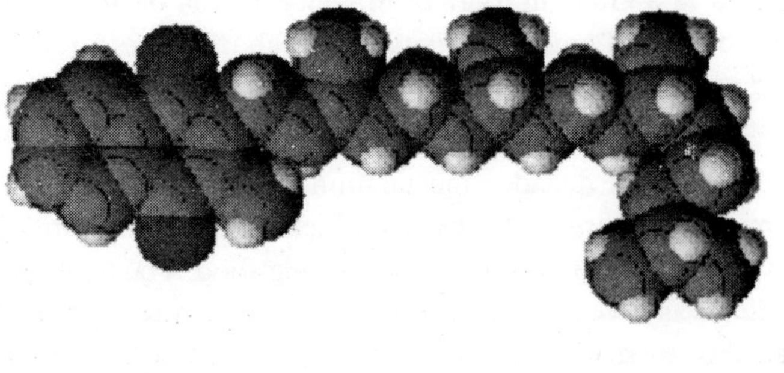

It is generally accepted that the naphthoquinone is the functional group, so that the mechanism of action is similar for all K-vitamins. Substantial differences may be expected, however, with respect to intestinal absorption, transport, tissue distribution, and bio-availability. These differences are caused by the different lipophilicity of the various side chains, and by the different food matrices in which they occur.

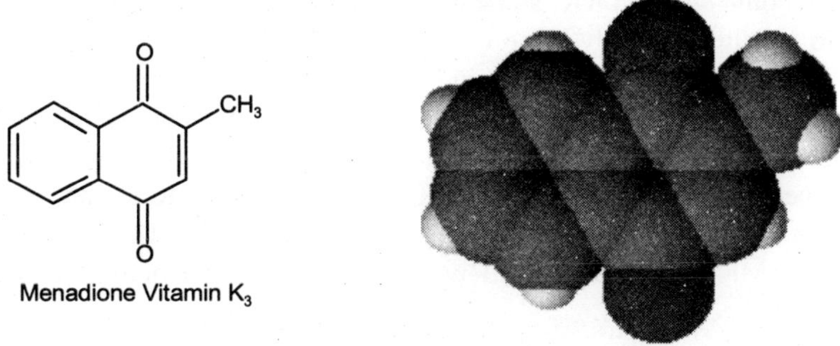

Fig. 12.15 : The Different Structures of Vitamin K

PHYSIOLOGY

Vitamin K is involved in the carboxylation of certain glutamate residues in proteins to form gamma-carboxyglutamate residues (abbreviated Gla-residues). The modified residues are often (but not always) situated within specific protein domains called Gla domains. Gla-residues are usually involved in binding calcium. The Gla-residues are essential for the biological activity of all known Gla-proteins.

At this time 14 human proteins with Gla domains have been discovered, and they play key roles in the regulation of three physiological processes:

- Blood coagulation: (prothrombin (factor II), factors VII, IX, X, protein C, protein S, and protein Z).
- Bone metabolism: osteocalcin, also called bone Gla-protein (BGP), and matrix gla protein (MGP).
- Vascular biology.

Like other liposoluble vitamins (A, D, E), vitamin K is stored in the fat tissue of the human body.

RECOMMENDED AMOUNTS

The U.S. Dietary Reference Intake (DRI) for an Adequate Intake (AI) of vitamin K for a 25-year old male is 120 micrograms/day. The Adequate Intake (AI) of this phytonutrient for adult women is 90 micrograms/day, for infants is 10–20 micrograms/day, for children and adolescents 15–100 micrograms/day. In 2002 it was found that to get maximum carboxylation of osteocalcin, one may have to take up to 1000 μg of vitamin K_1.

SOURCES

Vitamin K_1 is found chiefly in leafy green vegetables such as spinach, swiss chard, and *Brassica* (*e.g.* cabbage, kale, cauliflower, broccoli, and brussels sprouts); some fruits such as avocado and kiwifruit are also high in vitamin K. By way of reference, two tablespoons of parsley contain 153% of the recommended daily amount of vitamin K. Some vegetable oils, notably soybean, contain vitamin K, but at levels that would require relatively large caloric consumption to meet the USDA recommended levels. It is believed that phylloquinone's tight binding to the thylakoid membranes in the chloroplasts is the reason behind the poor bioavailability of vitamin K in green plants. For example, cooked spinach has a 5 percent bioavailability of phylloquinone. However when one adds fat to the spinach, the bioavailability increases to 13 percent due to the increased solubility of vitamin K in fat. Menaquinone-4 and menaquinone-7 (vitamin K_2) are found in meat, eggs, dairy, and natto]. MK-4 is synthesized by animal tissues, the rest (mainly MK-7) are synthesized by bacteria during fermentation. In natto 0% of vitamin K is from MK-4 and in cheese 2–7%.

DEFICIENCY

Average diets are usually not lacking in vitamin K and primary vitamin K deficiency is rare in healthy adults. As previously mentioned, newborn infants are at an increased risk of deficiency. Other populations with an increased prevalence of vitamin K deficiency include individuals who suffer from liver damage or disease (e.g. alcoholics), people with cystic fibrosis, inflammatory bowel diseases or those who have recently had abdominal surgeries. Groups that may suffer from secondary vitamin K deficiency include bulimics, those on stringent diets, and those taking anticoagulants. Other drugs that have been associated with vitamin K deficiency include salicylates, barbiturates, and cefamandole, although the mechanism is still unknown. There is no difference between the sexes as both males and females are affected equally. Symptoms of deficiency include heavy menstrual bleeding in women, anemia, bruising, and bleeding of the gums or nose. Osteoporosis and coronary heart disease are strongly associated with lower levels of K_2 (menaquinone). Menaquinone is not inhibited by salicylates as happens with K_1, so menaquinone supplementation can alleviate the chronic vitamin K deficiency caused by long term aspirin use.

VITAMIN P

FLAVONOIDS

Flavonoids (or **bioflavonoids**), also collectively known as **Vitamin P** and **citrin**, are a class of plant secondary metabolites. According to the IUPAC nomenclature,[2] they can be classified into:

- *flavonoids*, derived from 2-phenylchromen-4-one (2-phenyl-1,4-benzopyrone) structure (examples: quercetin, rutin).
- *isoflavonoids*, derived from 3-phenylchromen-4-one (3-phenyl-1,4-benzopyrone) structure
- *neoflavonoids*, derived from 4-phenylcoumarine (4-phenyl-1,2-benzopyrone) structure.

The three flavonoid classes above are all ketone-containing compounds, and as such, are flavonoids and flavonols. This class was the first to be termed "bioflavonoids." The terms flavonoid and **bioflavonoid** have also been more loosely used to describe non-ketone polyhydroxy polyphenol compounds which are more specifically termed flavanoids, flavan-3-ols, or catechins (although catechins are actually a subgroup of flavanoids).

BIOLOGICAL ROLES

Flavonoids are widely distributed in plants fulfilling many functions.

Flavonoids are the most important plant pigments for flower coloration producing yellow or red/blue pigmentation in petals designed to attract pollinator animals.

Flavonoids secreted by the root of their host plant help *Rhizobia* in the infection stage of their symbiotic relationship with legumes like peas, beans, clover, and soy. Rhizobia living in soil are able to sense the flavonoids and this triggers the secretion of Nod factors, which in turn are recognized by the host plant and can lead to root hair deformation and several cellular responses such as ion fluxes and the formation of a root nodule.They also protect plants from attacks by microbes, fungi and insects.

POTENTIAL FOR BIOLOGICAL ACTIVITY

Flavonoids (specifically flav**a**noids such as the catechins) are "the most common group of polyphenolic compounds in the human diet and are found ubiquitously in plants". Flavonols, the original bioflavonoids such as quercetin, are also found ubiquitously, but in lesser quantities. Both sets of compounds have evidence of health-modulating effects in animals which eat them.

The widespread distribution of flavonoids, their variety and their relatively low toxicity compared to other active plant compounds (for instance alkaloids) mean that many animals, including humans, ingest significant quantities in their diet. Resulting from experimental evidence that they may modify allergens, viruses, and carcinogens, flavonoids have potential to be biological "response modifiers", such as anti-allergic, anti-inflammatory, anti-microbial and anti-cancer activities shown from in vitro studies.[7]

ANTIOXIDANT ACTIVITY IN VITRO

Flavonoids (both flav**o**nols and flav**a**nols) are most commonly known for their antioxidant activity in vitro.Consumers and food manufacturers have become interested in flavonoids for their possible medicinal properties, especially their putative role in prevention of cancers and cardiovascular diseases. Although physiological evidence is not yet established, the beneficial effects of fruits, vegetables, and tea or even red wine have sometimes been attributed to flavonoid compounds rather than to known micronutrients, such as vitamins and dietary minerals. Alternatively, research conducted at the Linus Pauling Institute and evaluated by the European Food Safety Authority indicates that, following dietary intake, flavonoids themselves are of little or no direct antioxidant value. As body conditions are unlike controlled test tube conditions, flavonoids and other polyphenols are poorly absorbed (less than 5%), with most of what is absorbed being quickly metabolized and excreted. The increase in antioxidant capacity of blood seen after the consumption of flavonoid-rich foods is not caused directly by flavonoids themselves, but most likely is due to increased uric acid levels that result from metabolism of flavonoids. According to Frei, "we can now follow the activity of flavonoids in the body, and one thing that is clear is that the body sees them as foreign compounds and is trying to get rid of them."

OTHER POTENTIAL HEALTH BENEFITS

Cancer

Physiological processing of unwanted flavonoid compounds induces so-called Phase II enzymes that also help to eliminate mutagens and carcinogens, and therefore may be of value in cancer prevention. Flavonoids could also induce mechanisms that may kill cancer cells and inhibit tumor invasion. UCLA cancer researchers have found that study participants who ate foods containing certain flavonoids, such as catechins found in strawberries and green and black teas; kaempferol from brussel sprouts and apples; and quercetin from beans, onions and apples, may have reduced risk of obtaining lung cancer.

Research also indicated that only small amounts of flavonoids may be needed for possible benefits. Taking large dietary supplements likely provides no extra benefit and may pose risks. However, certainty of neither a benefit nor a risk has been proven yet in large-scale human intervention trials.

Difference Between Vitamins and Hormones

Hormones	Vitamins
1. A hormone is defined as a chemical compound produced in certeian specialized cells usually in ductless glands which is delivered directly into the blood streem and exerts only a single physiological change upon cells	1. Vitamins are organic nutrients that required in small amounts for a variety of biochemical functions and which generally cannot be synthesized
2. Hormones are secreated inside the body	2. Vitamins are mainly taken trough diet
3. Hormones are directly secreted in blood	3. Vitamins are taken orally and then transferred to blood
4. Hormones do not initiale any biocelmical process but they accelerate or deccelarate any biochemical process	4. Vitamins actually partipate in the metabolic functions in their proper amounts
5. Endocrine hormone molecules are secreted (released) directly into the bloodstream, while exocrine hormones (or ectohormones) are secreted directly into a duct, and from the duct they either flow into the bloodstream or they flow from cell to cell by diffusion in a process known as paracrine signalling.	5. Vitamins have diverse biochemical functions, including function as hormones (e.g. vitamin D), antioxidants (e.g. vitamin E), and mediators of cell signaling and regulators of cell and tissue growth and differentiation (e.g. vitamin A). The largest number of vitamins (e.g. B complex vitamins) function as precursors for enzyme cofactor bio-molecules (coenzymes), that help act as catalysts and substrates in metabolism. When acting as part of a catalyst, vitamins are bound to enzymes and are called prosthetic groups.
6. Hormones control the fuctions of the different organs	6. Vitamin does not control the functions of organs

Diarrhea

A study done at Children's Hospital & Research Center Oakland, in collaboration with scientists at Heinrich Heine University in Germany, has shown that epicatechin, quercetin and luteolin can inhibit the development of fluids that result in diarrhea by targeting the intestinal cystic fibrosis transmembrane conductance regulator Cl– transport inhibiting cAMP-stimulated Cl-secretion in the intestine.

Capillary Stabilizing Agents

Bioflavonoids like rutin, monoxerutin, diosmin, troxerutin and hidrosmin have potential vasoprotective proprieties still under experimental evaluation.

HORMONES

A **hormone** is a chemical released by a cell in one part of the body, that sends out messages that affect cells in other parts of the organism. Only a small amount of hormone is required to alter cell metabolism. It is essentially a chemical messenger that transports a signal from one cell to another. All multicellular organisms produce hormones; plant hormones are also called phytohormones. Hormones in animals are often transported in the blood. Cells respond to a hormone when they express a specific receptor for that hormone. The hormone binds to the receptor protein, resulting in the activation of a signal transduction mechanism that ultimately leads to cell type-specific responses.Endocrine hormone molecules are secreted (released) directly into the bloodstream, while exocrine hormones (or ectohormones) are secreted directly into a duct, and from the duct they either flow into the bloodstream or they flow from cell to cell by diffusion in a process known as paracrine signalling.

HORMONES AS A SIGNAL

Hormonal signaling involves the following:

1. **Biosynthesis** of a particular hormone in a particular tissue
2. **Storage and secretion** of the hormone
3. **Transport** of the hormone to the target cell(s)
4. **Recognition** of the hormone by an associated cell membrane or intracellular receptor protein.
5. **Relay and amplification** of the received **hormonal** signal via a signal transduction process: This then leads to a cellular response. The reaction of the target cells may then be recognized by the original hormone-producing cells, leading to a down-regulation in hormone production. This is an example of a homeostatic negative feedback loop.
6. **Degradation** of the hormone.

Hormone cells are typically of a specialized cell type, residing within a particular endocrine gland, such as thyroid gland, ovaries, and testes. Hormones exit their cell of origin via exocytosis or another means of membrane transport. The hierarchical model is an oversimplification of the hormonal signaling process. Cellular recipients of a particular hormonal signal may be one of several cell types that reside within a number of different tissues, as is the case for insulin, which triggers a diverse range of systemic physiological affects. Different tissue types may also respond differently to the same hormonal signal. Because of this, hormonal signaling is elaborate and hard to dissect.

INTERACTIONS WITH RECEPTORS

Most hormones initiate a cellular response by initially combining with either a specific intracellular or cell membrane associated receptor protein. A cell may have several different receptors that recognize the same hormone and activate different signal transduction pathways, or alternatively different hormones and their receptors may invoke the same biochemical pathway.

For many hormones, including most protein hormones, the receptor is membrane associated and embedded in the plasma membrane at the surface of the cell. The interaction of hormone and receptor typically triggers a cascade of secondary effects within the cytoplasm of the cell, often involving phosphorylation or dephosphorylation of various other cytoplasmic proteins, changes in ion channel permeability, or increased concentrations of intracellular molecules that may act as secondary messengers (e.g. cyclic AMP). Some protein hormones also interact with intracellular receptors located in the cytoplasm or nucleus by an intracrine mechanism.

For hormones such as steroid or thyroid hormones, their receptors are located intracellularly within the cytoplasm of their target cell. To bind their receptors these hormones must cross the cell membrane. They can do so because they are lipid-soluble. The combined hormone-receptor complex then moves across the nuclear membrane into the nucleus of the cell, where it binds to specific DNA sequences, effectively amplifying or suppressing the action of certain genes, and affecting protein synthesis. However, it has been shown that not all steroid receptors are located intracellularly some are associated with the plasma membrane.

An important consideration, dictating the level at which cellular signal transduction pathways are activated in response to a hormonal signal is the effective concentration of hormone-receptor complexes that are formed. Hormone-receptor complex concentrations are effectively determined by three factors:

1. The number of hormone molecules available for complex formation
2. The number of receptor molecules available for complex formation and
3. The binding affinity between hormone and receptor.

The number of hormone molecules available for complex formation is usually the key factor in determining the level at which signal transduction pathways are activated. The number of hormone molecules available being determined by the concentration of circulating hormone, which is in turn influenced by the level and rate at which they are secreted by biosynthetic cells. The number of receptors at the cell surface of the receiving cell can also be varied as can the affinity between the hormone and its receptor.

PHYSIOLOGY OF HORMONES

Most cells are capable of producing one or more molecules, which act as signaling molecules to other cells, altering their growth, function, or metabolism. The classical hormones produced by cells in the endocrine glands mentioned so far in this article are cellular products, specialized to serve as regulators at the overall organism level. However they may also exert their effects solely within the tissue in which they are produced and originally released.

The rate of hormone biosynthesis and secretion is often regulated by a homeostatic negative feedback control mechanism. Such a mechanism depends on factors that influence the metabolism and excretion of hormones. Thus, higher hormone concentration alone cannot trigger the negative feedback mechanism. Negative feedback must be triggered by overproduction of an "effect" of the hormone.

Hormone secretion can be stimulated and inhibited by:

- Other hormones (*stimulating*- or *releasing*-hormones)
- Plasma concentrations of ions or nutrients, as well as binding globulins
- Neurons and mental activity
- Environmental changes, e.g., of light or temperature

One special group of hormones is the tropic hormones that stimulate the hormone production of other endocrine glands. For example, thyroid-stimulating hormone (TSH) causes growth and increased activity of another endocrine gland, the thyroid, which increases output of thyroid hormones.

A recently identified class of hormones is that of the "hunger hormones" - ghrelin, orexin and PYY 3-36 - and "satiety hormones" - e.g., leptin, obestatin, nesfatin-1.

To release active hormones quickly into the circulation, hormone biosynthetic cells may produce and store biologically inactive hormones in the form of pre- or prohormones. These can then be quickly converted into their active hormone form in response to a particular stimulus.

GENERAL EFFECTS OF HORMONE

Hormones have the following effects on the body:

- stimulation or inhibition of growth
- mood swings
- induction or suppression of apoptosis (programmed cell death)
- activation or inhibition of the immune system
- regulation of metabolism
- preparation of the body for mating, fighting, fleeing, and other activity
- preparation of the body for a new phase of life, such as puberty, parenting, and menopause
- control of the reproductive cycle
- hunger cravings

A hormone may also regulate the production and release of other hormones. Hormone signals control the internal environment of the body through homeostasis.

CLASSIFICATION OF HORMONES

Hormones can be classified in many ways here we are going to classify according two ways the first from the organs they are secreted and second chemically

The Classification of Hormones by their Origin of Secretion

1. Anterior Pituitary Hormones
2. Posterior Pituitary Hormones
3. Adrenal Cortical Hormones
4. Sex Hormones

The Classification of Hormones according to their Chemical Properties

1. Amines or Amino Acids
   (a) Epinephrine from adrenal medualla
   (b) Tyroxine from thyroid Glands
2. Proteinaceous or Peptide Like
   (a) Anterior Pituitary Hormones
   (b) Posterior Pituitary Hormones
   (c) Insulin from Pancreas
   (d) Parathyroid Hormones
3. Steroid Hormones
   (a) Adrenal Cortical Hormones
   (b) Ovarian Hormones

(c) Corpus Luteum Hormones

(d) Testicular hormones

ANTERIOR PITUITARY HORMONES

A major organ of the endocrine system, the **anterior pituitary**, also called the **adenohypophysis**, is the glandular, anterior lobe of the pituitary gland. The anterior pituitary regulates several physiological processes including stress, growth, and reproduction.Its regulatory functions are achieved through the secretion of various peptide hormones that act on target organs including the adrenal gland, liver, bone, thyroid gland, and gonads. The anterior pituitary itself is regulated by the hypothalamus and by negative feedback from these target organs.

Disorders of the anterior pituitary are generally classified by the presence of over- or underproduction of pituitary hormones. For example, a prolactinoma is a pituitary adenoma that overproduces prolactin. In Sheehan's syndrome of postpartum hypopituitarism, the anterior pituitary uniformly malfunctions and underproduces all hormones. Proper function of the anterior pituitary and of the organs it regulates can often be ascertained via blood tests that measure hormone levels.

MAJOR HORMONES SECRETED

Hormone	Other names	Symbol	Structure	Secretory cells	Staining	Target	Effect
Adrenocorticotropic hormone	Corticotropin	ACTH	Polypeptide	Corticotrophs	Basophil	Adrenal gland	Secretion of glucocorticoids
Beta-endorphin			Polypeptide	Corticotrophs	Basophil	Opioid receptor	Inhibit perception of pain
Thyroid-stimulating hormone	Thyrotropin	TSH	Glycoprotein	Thyrotrophs	Basophil	Thyroid gland	Secretion of thyroid hormones
Follicle-stimulating hormone	-	FSH	Glycoprotein	Gonadotrophs	Basophil	Gonads	Growth of reproductive system
Luteinizing hormone	Lutropin	LH,	Glycoprotein ICSH	Gonadotrophs	Basophil	Gonads	Sex hormone production
Growth hormone	Somatotropin	GH, STH	Polypeptide	Somatotrophs	Acidophil	Liver, adipose tissue	Promotes growth; lipid and carbohydrate metabolism
Prolactin	Lactogenic hormone	PRL	Polypeptide	Lactotrophs and Mammotrophs	Acidophil	Ovaries-mammary glands	Secretion of estrogens/ progesterone; milk production

The acidophilic cells (GH and PRL) have extensive rough endoplasmic reticulum and produce single chain polypeptides without any glycosylation or phosphorylation. Basophilic staining results from lysosome action modifying the hormones (or prohormones in the case of corticotrophs) by glycosylation.

REGULATION

Hormone secretion from the anterior pituitary gland is regulated by hormones secreted by the hypothalamus. Neuroendocrine neurons in the hypothalamus project axons to the median eminence, at the base of the brain. At this site, these neurons can release substances into small blood vessels that travel directly to the anterior pituitary gland (the hypothalamo-hypophysial portal vessels).

ADRENOCORTICOTROPIC HORMONE (APCTH)

Adrenocorticotropic hormone (ACTH), also known as 'corticotropin', is a polypeptide tropic hormone produced and secreted by the anterior pituitary gland. It is an important component of the hypothalamic-pituitary-adrenal axis and is often produced in response to biological stress (along with corticotropin-releasing hormone from the hypothalamus). Its principal effects are increased production and release of corticosteroids and, as its name suggests, cortisol from the adrenal cortex.

ACTH is synthesized from pre-pro-opiomelanocortin (pre-POMC). The removal of the signal peptide during translation produces the 267 amino acid polypeptide POMC, which undergoes a series of posttranslational modifications such as phosphorylation and glycosylation before it is proteolytically cleaved by endopeptidases to yield various polypeptide fragments with varying physiological activity. These fragments include ACTH, β-lipotropin, β-lipotropin, Melanocyte-Stimulating Hormone (MSH) and β-endorphin. POMC, ACTH and β-lipotropin are secreted from corticotropes in the anterior lobe (or adenohypophysis) of the pituitary gland in response to the hormone corticotropin-releasing hormone (CRH) released by the hypothalamus. In order to regulate the secretion of ACTH, many substances secreted within this axis exhibit slow/intermediate and fast feedback-loop activity. Glucocorticoids secreted from the adrenal cortex work to inhibit CRH secretion by the hypothalamus, which in turn decreases anterior pituitary secretion of ACTH. Glucocorticoids may also inhibit the rates of POMC gene transcription and peptide synthesis. The latter is an example of a slow feedback loop, which works on the order of hours to days, while the former works on the order of minutes.ACTH is also related to the circadian rhythm in many organisms. The half-life of ACTH in human blood is about ten minutes.

STRUCTURE

ACTH consists of 39 amino acids, the first 13 of which (counting from the N-terminus) may be cleaved to form α-melanocyte-stimulating hormone (α-MSH). (This common structure is excessively tanned skin.) After a short period of time, ACTH is cleaved into α-melanocyte-stimulating hormone (α-MSH) and CLIP, a peptide with unknown activity in humans.

Human ACTH has a molecular weight of 4,540 atomic mass units (Da).

FUNCTION

- ACTH is released in a pulsatile fashion from the pituitary following a circadian rhythm, peaking in the morning and then declining. It is released under control of hypothalamic CRH.
- The release of CRH can be affected by external influences such as stress. CRH action is potentiated by petides such as vasopressin.
- ACTH controls the production of glucocorticoids by the adrenal cortex, stimulating the conversion of cholesterol to pregnenolone - a precursor of cortisol. Cortisol feeds back to inhibit both the hypothalamus and pituitary gland.

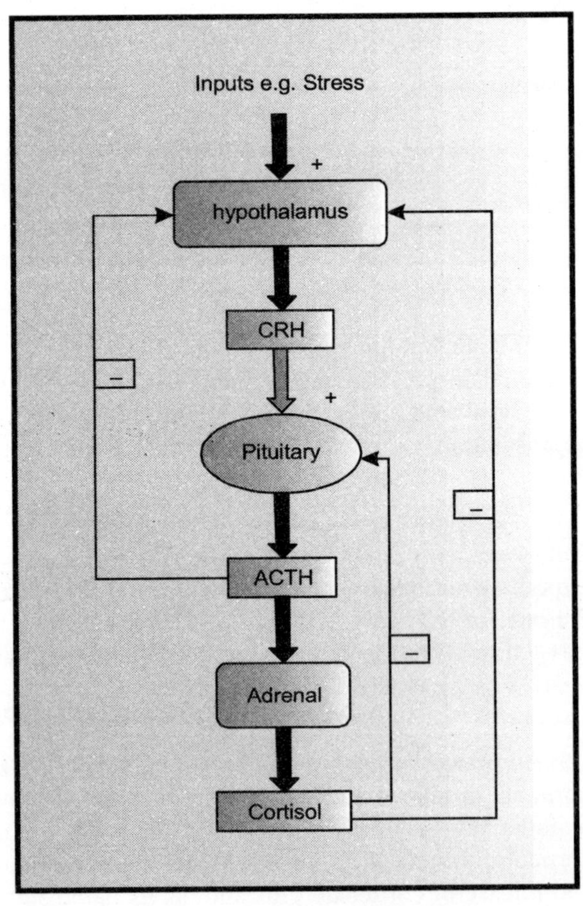

Fig. 12.16: The Flow chart of ACTH function

What can Go Wrong

Oversecretion of ACTH leads to symptoms known as Cushing syndrome. When this is a result of a feedback abnormality it is Cushing disease and is caused by ACTH and CRH secretion being suppressed at abnormally high levels of circulating cortisol. This results in an ACTH secreting pituitary tumour due to overstimulation of the corticotrophs and excess cortisol secretion secondary to this. Excess cortisol secretion may also be caused by a CRH secreting tumour Symptoms of excess cortisol include central obesity, thinning of the skin, bruising, glucose intolerance, susceptibility to infection, muscle wasting and thin bones. Hyperpigmentation of the skin may occur due to the associoated secretion of a-MSH with ACTH. When excess ACTH drives excess cortisol production adrenal androgens are also oversecreted leading to symptoms of androgen excess. Treatment is surgery to remove the tumour. Undersecretion of ACTH leads to problems of glucocorticoid deficiency, such as hypoglycaemia, and androgen deficiency, such as lack of pubic hair and libido in females. This can be treated by administering hydrocortisone.

SYNTHETIC ACTH

An active synthetic form of ACTH, consisting of the first 23 amino acids of native ACTH, was first synthesized by Klaus Hofmann at the University of Pittsburgh. ACTH is available as a synthetic derivative in the forms of cosyntropin, tradename Cortrosyn and synacthen (synthetic ACTH). Both are very rarely used in place of glucocorticoids to treat secondary adrenal insufficiency in a hospital setting, but are primarily used to conduct the ACTH stimulation test.ACTH was first synthesized as a replacement for Acthar Gel, a long-lasting animal product. Once relatively inexpensive, Acthar Gel is currently an extremely expensive pharmaceutical product.

Beta endonephrin (β-endorphin) is an endogenous opioid peptide neurotransmitter found in the neurons of both the central and peripheral nervous system.

Sequence: "Tyr Gly Gly Phe Met Thr Ser Glu Lys Ser Gln Thr Pro Leu Val Thr Leu Phe Lys Asn Ala Ile Ile Lys Asn Ala Tyr Lys Lys Gly Glu".

â-endorphin was discovered from camel pituitary extracts by C.H. Li and David Chung (Choh Hao Li and David Chung. Isolation and Structure of an Untriakontapeptide with Opiate Activity from Camel Pituitary Glands. PNAS 1976 73: 1145-1148.). The discovery was amidst a flurry of discoveries made in the opioid field of "endogenous opiates", including Met-enkephalin, Leu-enkephalin, and dynorphin. Shortly thereafter, also in C.H. Li's laboratory, the opioid effects of this peptide were characterized (Horace H. Loh, L. F. Tseng, Eddie Wei, and Choh Hao Li. β-Endorphin is a Potent Analgesic Agent. PNAS 1976 73: 2895-2898.)

Some notable early questions which were investigated included anatomical questions (where in the body? where in the brain? where in the neuron?), pharmacological questions (how similar to morphine? how does potency differ? what part of the molecule confers opioid activity?), biochemical questions (how is it biosynthesized? how is the activity regulated? how is it metabolized?), behavioral questions (what is the effect of endogenous β-endorphin on activity levels), clinical questions (do deficiencies in endogenous β endorphin lead to psychological disorders?), and therapeutic questions (is β-endorphin a viable treatment for morphine and heroin addiction, as well as chronic pain disorders?).

FORMATION

β-endorphin is a peptide, 31 amino acids long, resulting from processing of the precursor proopiomelanocortin (POMC). (Note, POMC also gives rise to other peptide hormones, including ACTH (Adrenocorticotropic hormone), as well α- and γ-MSH (Melanocyte-Stimulating Hormone), resulting from intracellular processing by internal enzymes known as prohormone convertases.)

β-endorphin is found in neurons of the hypothalamus, as well as the pituitary gland.

FUNCTION

It is an agonist of the opioid receptors, with evidence suggesting it serves as the endogenous ligand of the μ-opioid receptor, the same receptor to which the chemicals extracted from opium, such as morphine, have their analgesic and addictive effects (indeed, the ì-opioid receptor was named based on its most renowned ligand, morphine).

EFFECTS

It is used as an analgesic in the body to numb or dull pains. That is the reason why humans start to feel better immediately after an acute physical trauma even though the symptoms are still present. The reason the pain dulls is because it binds to and activates opioid receptors. β-endorphin has approximately 80 times the analgesic potency of morphine.

β-endorphin is believed to have a number of other benefits, including:

- Boosting the immune system
- Slowing the growth of cancer cells
- Promoting feeling of well-being
- Increasing relaxation

Thyroid-stimulating hormone (also known as **TSH** or **thyrotropin**) is a peptide hormone synthesized and secreted by thyrotrope cells in the anterior pituitary gland, which regulates the endocrine function of the thyroid gland.

TSH stimulates the thyroid gland to secrete the hormones thyroxine (T_4) and triiodothyronine (T_3). TSH production is controlled by thyrotropin-releasing hormone (TRH), which is manufactured in the hypothalamus and transported to the anterior pituitary gland via the superior hypophyseal artery, where it increases TSH production and release. Somatostatin is also produced by the hypothalamus, and has an opposite effect on the pituitary production of TSH, decreasing or inhibiting its release.

The level of thyroid hormones (T_3 and T_4) in the blood has an effect on the pituitary release of TSH; when the levels of T_3 and T_4 are low, the production of TSH is increased, and, on the converse, when levels of T_3 and T_4 are high, TSH production is decreased. This effect creates a regulatory negative feedback loop.

SUBUNITS OF TSH

TSH is a glycoprotein and consists of two subunits, the *alpha* and the *beta* subunit.

- The α (*alpha*) subunit (i.e., chorionic gonadotropin alpha) is identical to that of human chorionic gonadotropin (HCG), luteinizing hormone (LH), follicle-stimulating hormone (FSH).
- The β (*beta*) subunit (TSHB) is unique to TSH, and therefore determines its function.

The TSH receptor is found mainly on thyroid follicular cells. Stimulation of the receptor increases T_3 and T_4 production and secretion.

Stimulating antibodies to this receptor mimic TSH and cause Graves' disease.

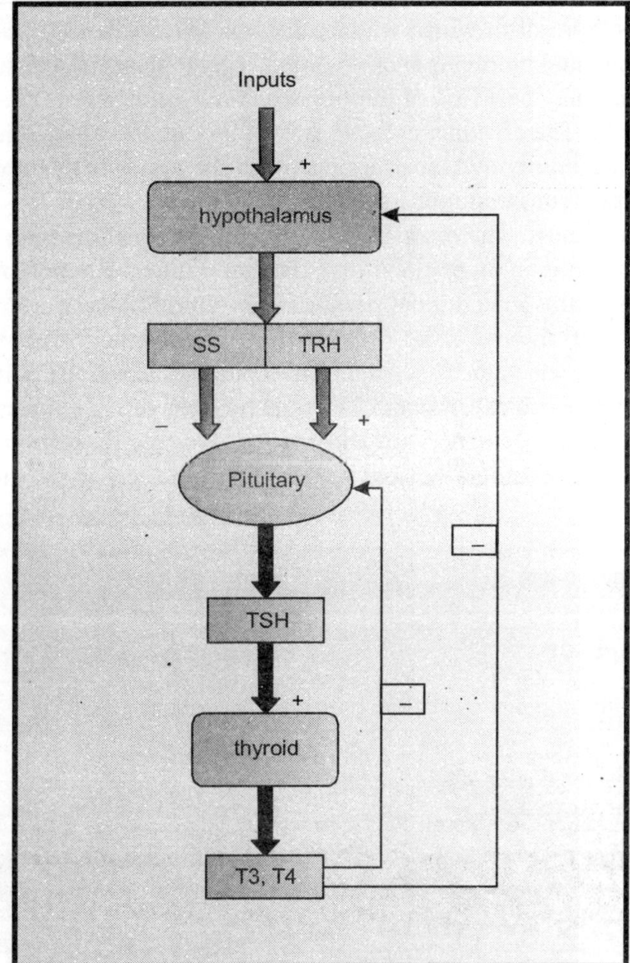

Fig. 12.17 : The Flow chart of the Function of TSH

FUNCTION

TSH is released from the anterior pituitary gland in response to thyroid releasing hormone from the hypothalamus and causes the synthesis and secretion of triiodothyronine (T_3) and thyroxine (T_4) by the thyroid gland.

- This is acheived by stimulation of the thyroid follicular cells by binding of TSH to the receptor on the basal surface of the cell and activation of adenylate cyclase.
- This leads to iodide uptake. T_3 and T_4 exert negative feedback on both the pituitary production of TSH and the hypothalamic production of TRH.
- Other factors affecting release of TRH from the hypothalamus incluse blood levels glucose and the body's metabolic rate.
- Somatostatin inhibits TSH secretion and oestrogen has been shown, in rats, to reverse the negative feedback affect of T_3 and T_4 on the TSH response to TRH.

What can Go Wrong

Problems at the level of the hypothalamo-pituitary axis can cause deficiency of TSH resulting in secondary hypothyroidism which is normally less severe than that caused by disease of the thyroid gland itself. Congenital problems such as pituitary hypoplasia of aplasia may affect any of the pituitary hormones. There may be a loss of midline structures as in septo-optic dysplasia, which also causes loss of optic nerves. The effects are due to lack of production of thyroid hormones. Treatment is by administration of oral thyroxine Hyperthroidism has the opposite symptoms to hypothyroidism including increased oxygen consumption, increased metabolism, weight loss, heat intolerance, sweating, insomnia and nervousness. The most common form is Graves' disease, an autoimmune condition in which thyroid stimulating immunoglobulins exist that mimic the action of TSH but with no normal feedback control. This results in continual production of thyroid hormones and an enlarged thyroid gland. Treatment is with anti-thyroid drugs or by thryoid gland surgery. It is rare to see overproduction of TSH due to a pituitary adenoma.The hypothalamo-pituitary axis is the unit formed by the **hypothalamus** and **pituitary gland**, which exerts control over many parts of the endocrine system. This unit functions by means of interaction of the nervous and endocrine systems whereby the nervous system regulates the endocrine system and endocrine activity modulates the activity of the CNS.

Effects of TSH deficiency

Metabolism	Body weight increasesOxygen consumption decreasesHeat production decreasesBasal metabolic rate decreases
CNS	Impaired mentalityPoor memory and concentrationDrowsiness
Motor nervous system	Activity decreases
Sympathetic nervous system	Activity decreases
Cardiovascular	BradycardiaReduced output and blood pressure
GI tract	Activity decreasedConstipation

THE HYPOTHALAMUS

The hypothalamus has many functions and is one of the major regulators of homeostasis:

- It controls the autonomic nervous system, acts with the limbic system to regulate emotional and behavioural patterns, regulates eating and drinking, controls body temperature and regulates diurnal rhythms. It also controls pituitary gland secretions.
- The hypothalamus receives input from the external and internal environment as well as having its own receptors. It receives stimuli from the somatic and visceral sense organs. These inputs travel via the medulla oblongata and reach the hypothalamus through innervation by fibres producing dopamine, adrenaline, noradrenaline, serotonin and acetylcholine as well as fibres releasing neuropeptides such as enkephalins, NPY, neurotensin, dynorphins and endorphins.

- The release of hormones from the pituitary is therefore subject to many different stimuli from higher centres' acting on the hypothalamus.In response to stimuli such as stress, pain and emotions, the hypothalamus can exert effects on the anterior and posterior pituitary gland in order to respond rapidly to environmental change as well as to feedback from internal systems. The hypothalamus is the least well understood area of the hypothalamo-pituitary axis and research into certain aspects of its function is ongoing. An example is current work involving the recently discovered peptide hormone, Leptin, which has shed light on the existance of an adipose tissue- brain endocrine axis.The hypothalmus exerts its effects on the pituitary gland in two different ways:

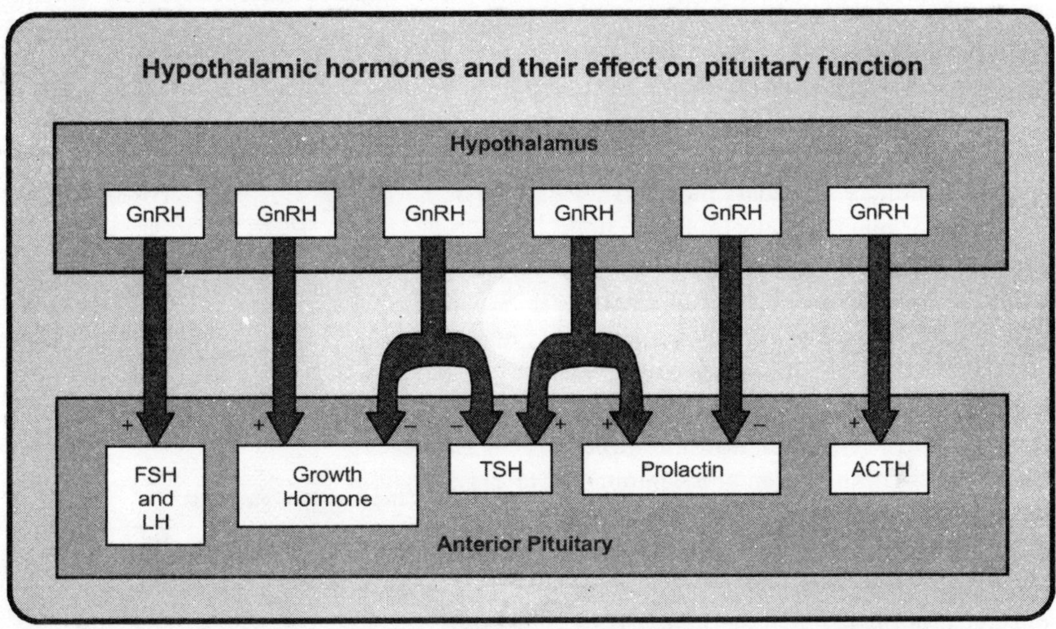

Fig. 12.18: The Hypothalamic Hormones and their Effects on Pituitary Function

The hypothalamic paraventricular and supraoptic nuclei produce secretory droplets, which are the first stage in formation of the neuropeptides vasopressin and oxytocin. The paraventricular nuclei produce mostly oxytocin and the supraoptic mostly vasopressin although both hormones may also be produced by the other type of nuclei. These hormones are packaged with the protein neurophysin into granules which then move down the axon and are stored in the posterior pituitary. Following stimulation of the hypothalamus these hormones are then released into the bloodstream. Small amounts of the hormones also enter the portal blood supply. Control of the anterior pituitary is not via a nervous link. It was discovered in the 1930s that the hypothalamus is linked to the anterior pituitary by a network of microcapillaries - the hypophyseal portal vessels. Control is maintained by release of hypothalamic hormones, some of which stimulate release and others inhibit release.

Hypothalamic hormone	Effect on anterior pituitary gland
Thyrotropin releasing hormone (TRH)	release of TSH and PRL
Gonadotropin releasing hormone (GnRH)	release of LH and FSH
Growth hormone releasing hormone (GHRH)	release of GH
Somatostatin (SS)	inhibition of GH
Corticotrophn releasing hormone (CRH)	release of ACTH
Dopamine (DA)	inhibition of PRL

These hormones are released by exocytosis from storage granules in the hypothalamic-hypophyseotropic nuclei into the capillaries of the primary plexus. From here the hormones travel in the blood through the hypophyseal portal veins into the anterior pituitary secondary plexus. They then act on anterior pituitary cells giving a rapid response. Anterior pituitary hormones are then released into the secondary plexus and anterior hypophyseal veins into the systemic circulation. Negative feedback is an important factor in controlling the hypothalamic-pituitary-target organ axis function. Once hypothalamic hormones stimulate the release or inhibition of the pituitary hormone, this may then acts at a target gland, such as the thyroid, causing release of further hormones or causing metabolic effects. The action of hypothalamic hormones may be inhibited by long feedback loops from the target gland hormone or by short feedback loops from the pituitary hormone. There may also be direct feedback from the target gland hormone to the pituitary gland.Input is also received at the hypothalamus from higher brain centres, which can be due to internal or external influences Positive feedback also plays a part in certain systems. For example, in the situation where high levels of oestradiol in the blood cause a surge in LH levels during the menstrual cycle.

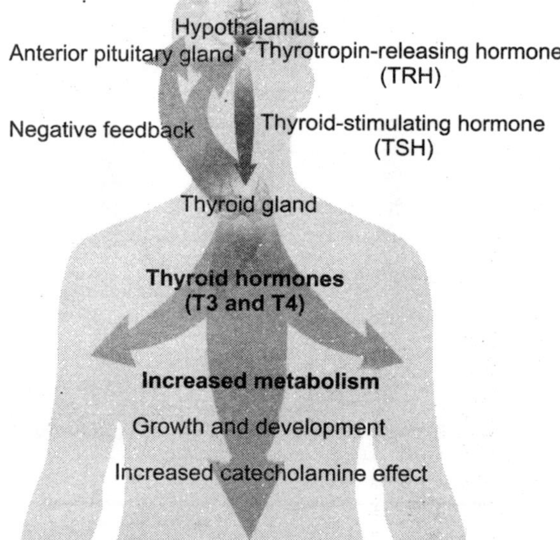

Fig. 12.19: The Thyroid System

Follicle-stimulating hormone (FSH) is a hormone found in humans and other animals. It is synthesized and secreted by gonadotrophs of the anterior pituitary gland. FSH regulates the development, growth, pubertal maturation, and reproductive processes of the body. FSH and Luteinizing hormone (LH) act synergistically in reproduction.

STRUCTURE

FSH is a glycoprotein. Each monomeric unit is a protein molecule with a sugar attached to it; two of these make the full, functional protein. Its structure is similar to those of LH, TSH, and hCG. The protein dimer contains 2 polypeptide units, labeled alpha and beta subunits. The alpha subunits of LH, FSH, TSH, and hCG are identical, and contain 92 amino acids. The beta subunits vary. FSH has a beta subunit of 118 amino acids (FSHB), which confers its specific biologic action and is responsible for interaction with the FSH-receptor. The sugar part of the hormone is composed of fucose, galactose, mannose, galactosamine, glucosamine, and sialic acid, the latter being critical for its biologic half-life. The half-life of FSH is 3-4 hours. Its molecular wt is 30000.

The gene for the alpha subunit is located on chromosome 6p21.1-23. It is expressed in different cell types. The gene for the FSH beta subunit is located on chromosome 11p13, and is expressed in gonadotropes of the pituitary cells, controlled by GnRH, inhibited by inhibin, and enhanced by activin.

FSH regulates the development, growth, pubertal maturation, and reproductive processes of the human body.

- In both *males* and *females*, FSH stimulates the maturation of germ cells.
- In *males*, FSH induces Sertoli cells to secrete inhibin and stimulates the formation of sertoli-sertoli tight junctions (zonula occludens).
- In *females*, FSH initiates follicular growth, specifically affecting granulosa cells. With the concomitant rise in inhibin B, FSH levels then decline in the late follicular phase. This seems to be critical in selecting only the most advanced follicle to proceed to ovulation. At the end of the luteal phase, there is a slight rise in FSH that seems to be of importance to start the next ovulatory cycle.

Like its partner LH, FSH release at the pituitary gland is controlled by pulses of gonadotropin-releasing hormone (GnRH). Those pulses, in turn, are subject to the oestrogen feed-back from the gonads.

EFFECTS IN FEMALES

FSH stimulates the growth and recruitment of immature Ovarian follicles in the ovary. In early (small) antral follicles, FSH is the major survival factor that rescues the follicles from apoptosis (programmed death of the somatic cells of the follicle and oocyte). In the luteal-follicle phase transition period the serum levels of progesterone and estrogen (primarily estradiol) decrease and no longer suppress the release of FSH, consequently FSH peaks at about day three (day one is the first day of menstrual flow). The cohort of small antral follicles is normally sufficiently in number to produce enough Inhibin B to lower FSH serum levels.

In addition, there is evidence that gonadotrophin surge-attenuating factor produced by small follicles during the first half of the follicle phase also exerts a negative feedback on pulsatile luteinizing hormone (LH) secretion amplitude, thus allowing a more favorable environment for follicle growth and preventing premature luteinization.

(As a woman nears perimenopause the number of small antral follicles recruited in each cycle diminishes and consequently insufficient Inhibin B is produced to fully lower FSH and the serum level of FSH begins to rise.)

When the follicle matures and reaches 8-10 mm in diameter it starts to secrete significant amounts of estradiol. Normally in humans only one follicle becomes dominant and survives to grow to 18-30 mm in size and ovulate, the remaining follicles in the cohort undergo atresia. The sharp increase in estradiol production by the dominant follicle (possibly along with a decrease in gonadotrophin surge-attenuating factor) cause a positive effect on the hypothalamus and pituitary and rapid GnRH pulses occur and an LH surge results.

The increase in serum estradiol levels cause a decrease in FSH production by inhibiting GnRH production in the hypothalamus. The decrease in serum FSH level causes the smaller follicles in the current cohort to undergo atresia as they lack sufficient sensitivity to FSH to survive. Occasionally two follicles reach the 10 mm stage at the same time by chance and as both are equally sensitive to FSH both survive and grow in the low FSH environment and thus two ovulations can occur in one cycle possibly leading to non identical (dizygotic) twins.

EFFECTS IN MALES

FSH stimulates maturation of seminiferous tubules and spermatogenesis.

FSH enhances the production of androgen-binding protein by the Sertoli cells of the testes by binding to FSH receptors on their basolateral membranes, and is critical for the initiation of spermatogenesis.

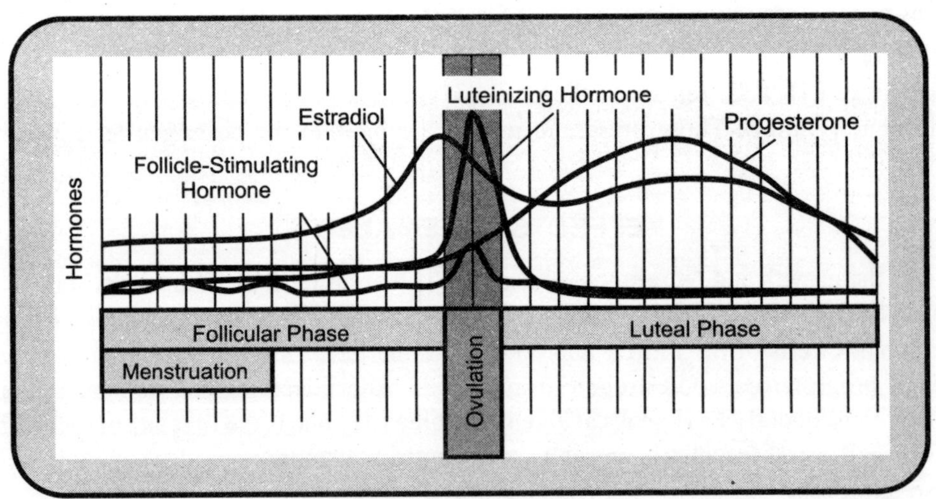

Fig. 12.20 : FSH levels (red line) during the menstrual cycle

Follicle stimulating hormone is typically measured on day three of a woman's cycle when the levels of estradiol (E2) and progesterone are at the lowest point of the menstrual cycle.

FSH levels are normally low during childhood and, in females, high after menopause.

High FSH Levels

The most common reason for high serum FSH concentration is in a female who is undergoing or has recently undergone menopause. High levels of Follicle-Stimulating Hormone indicate that the normal restricting feedback from the gonad is absent, leading to an unrestricted pituitary FSH production.

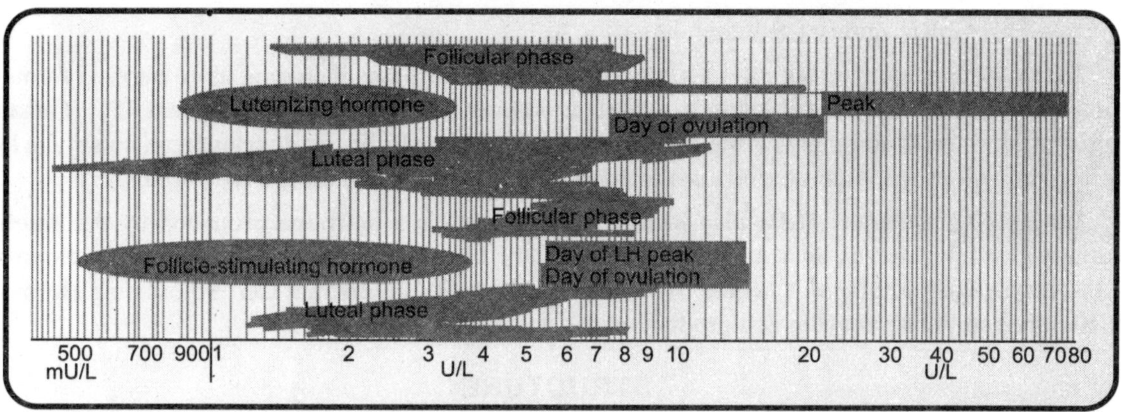

Fig. 12.21 : Reference ranges for Luteinizing Hormone and *follicle-stimulating hormone* in the Menstrual Cycle, Expressed in International Units. The Scale is Logarithmic

If high FSH levels occur during the reproductive years, it is abnormal.

1. Premature menopause also known as Premature Ovarian Failure
2. Poor ovarian reserve also known as Premature Ovarian Aging
3. Gonadal dysgenesis, Turner syndrome
4. Castration
5. Swyer syndrome
6. Certain forms of CAH
7. Testicular failure.

Most of these conditions are associated with subfertility and/or infertility. Therefore high FSH levels are an indication of subfertility and/or infertility.

Low FSH Levels

Diminished secretion of FSH can result in failure of gonadal function (hypogonadism). This condition is typically manifested in males as failure in production of normal numbers of sperm. In females, cessation of reproductive cycles is commonly observed. Conditions with very low FSH secretions are:

1. Polycystic Ovarian Syndrome
2. Polycystic Ovarian Syndrome + Obesity + Hirsutism + Infertility
3. Kallmann syndrome

4. Hypothalamic suppression
5. Hypopituitarism
6. Hyperprolactinemia
7. Gonadotropin deficiency
8. Gonadal suppression therapy
   (a) GnRH antagonist
   (b) GnRH agonist (downregulation).

FSH is available mixed with LH activity in various menotropins including more purified forms of urinary gonadotropins such as Menopur, as well as without LH activity as recombinant FSH (Gonal F, Follistim). It is used commonly in infertility therapy to stimulate follicular development, notably in IVF therapy, as well as with interuterine insemination (IUI). (See Gonadotropin Preparations.)

Luteinizing hormone (LH, also known as lutropin) is a hormone produced by the anterior pituitary gland. In *females*, an acute rise of LH called the *LH surge* triggers ovulation and development of the corpus luteum. In *males*, where LH had also been called Interstitial Cell Stimulating Hormone (ICSH), ruit stimulates Leydig cell production of testosterone.

STRUCTURE

LH is a heterodimeric glycoprotein. Each monomeric unit is a glycoprotein molecule; one alpha and one beta subunit make the full, functional protein.Its structure is similar to the other glycoprotein hormones, follicle-stimulating hormone (FSH), thyroid-stimulating hormone (TSH), and human chorionic gonadotropin (hCG). The protein dimer contains 2 glycopeptidic subunits, labeled alpha and beta subunits, that are non-covalently associated (i.e. without any disulfide bridge linking them). :

- The *alpha subunits* of LH, FSH, TSH, and hCG are identical, and contain 92 amino acids in human but 96 amino acids in almost all other vertebrate species (glycoprotein hormones do not exist in invertebrates).
- The *beta subunits* vary. LH has a beta subunit of 121 amino acids (LHB) that confers its specific biologic action and is responsible for the specificity of the interaction with the LH receptor. This beta subunit contains an amino acid sequence that exhibits large homologies with that of the beta subunit of hCG and both stimulate the same receptor, however, the hCG beta subunit contains an additional 24 amino acids, and the two hormones differ in the composition of their sugar moieties.

The different composition of these oligosaccharides affects bioactivity and speed of degradation. The biologic half-life of LH is 20 minutes, shorter than that of FSH (3–4 hours) or hCG (24 hours).

The gene for the *alpha subunit* is located on chromosome 6q12.21.

The luteinizing hormone *beta subunit* gene is localized in the LHB/CGB gene cluster on chromosome 19q13.32. In contrast to the alpha gene activity, beta LH subunit gene activity is restricted to the pituitary gonadotropic cells. It is regulated by the gonadotropin releasing hormone from the hypothalamus. Inhibin, activin, and sex hormones do not affect genetic activity for the beta subunit production of LH.

In both males and females, LH is essential for reproduction.

- In *females*, at the time of menstruation, FSH initiates follicular growth, specifically affecting granulosa cells. With the rise in oestrogens, LH receptors are also expressed on the maturing follicle that produces an increasing amount of estradiol. Eventually at the time of the maturation of the follicle, the oestrogen rise leads via the hypothalamic interface to the "positive feed-back" effect, a release of LH over a 24-48 hour period. This 'LH surge' triggers ovulation thereby not only releasing the egg, but also initiating the conversion of the residual follicle into a corpus luteum that, in turn, produces progesterone to prepare the endometrium for a possible implantation. LH is necessary to maintain luteal function for the first two weeks. In case of a pregnancy luteal function will be further maintained by the action of hCG (a hormone very similar to LH) from the newly established pregnancy. LH supports thecal cells in the ovary that provide androgens and hormonal precursors for estradiol production.
- In the male, LH acts upon the Leydig cells of the testis and is responsible for the production of testosterone, an androgen that exerts both endocrine activity and intratesticular activity on spermatogenesis.

The release of LH at the pituitary gland is controlled by pulses of gonadotropin-releasing hormone (GnRH) from the hypothalamus. Those pulses, in turn, are subject to the oestrogen feedback from the gonads.

Normal Levels

LH levels are normally low during childhood and, in women, high after menopause. As LH is secreted as pulses, it is necessary to follow its concentration over a sufficient period of time to get a proper information about its blood level.

During the reproductive years typical levels are between 1-20 IU/L. Physiologic high LH levels are seen during the LH surge (v.s.); typically they last 48 hours.

Ovulation Predictor Kit (LH kit)

The detection of the Luteinising hormone surge indicates impending ovulation. LH can be detected by urinary ovulation predictor kits (OPK, also LH-kit) that are performed, typically daily, around the time ovulation may be expected. The conversion from a negative to a positive reading would suggest that ovulation is about to occur within 24-48 hours, giving women two days to engage in sexual intercourse or artificial insemination with the intentions of conceiving.

Tests may be read manually using a colour-change paper strip, or digitally with the assistance of reading electronics. Tests for Luteinising hormone may be combined with testing for estradiol in tests such as the Clearblue fertility monitor. The sensitivity of LH tests are measured in milli international unit, with tests commonly available in the range 10-40 m.i.u. As sperm can stay viable in the woman for several days, LH tests are not recommended for contraceptive practices, as the LH surge typically occurs after the beginning of the fertile window.

Disease States

Relative Elevations

In children with precocious puberty of pituitary or central origin, LH and FSH levels may be in the reproductive range instead of the low levels typical for their age.

During the reproductive years, relatively elevated LH is frequently seen in patients with the polycystic ovary syndrome; however it would be unusual for them to have LH levels outside of the normal reproductive range.

High LH levels

Persistently high LH levels are indicative of situations where the normal restricting feedback from the gonad is absent, leading to a pituitary production of both LH and FSH. While this is typical in the menopause, it is abnormal in the reproductive years. There it may be a sign of:

1. Premature menopause
2. Gonadal dysgenesis, Turner syndrome
3. Castration
4. Swyer syndrome
5. Polycystic Ovary Syndrome
6. Certain forms of CAH
7. Testicular failure

Deficient LH Activity

Diminished secretion of LH can result in failure of gonadal function (hypogonadism). This condition is typically manifest in males as failure in production of normal numbers of sperm. In females, amenorrhea is commonly observed. Conditions with very low LH secretions are:

1. Kallmann syndrome
2. Hypothalamic suppression
3. Hypopituitarism
4. Eating disorder
5. Female athlete triad
6. Hyperprolactinemia
7. Gonadotropin deficiency
8. Gonadal suppression therapy
   (a) GnRH antagonist
   (b) GnRH agonist (downregulation)

Availability

LH is available mixed with FSH in the form of Pergonal, and other forms of urinary gonadotropins . More purified forms of urinary gonadotropins may reduce the LH portion in relation to FSH. Recombinant LH is available as lutropin alfa (Luveris). All these medications have to be given parenterally. They are commonly used in infertility therapy to stimulate follicular development, notably in IVF therapy.

Often, hCG medication is used as an LH substitute because it activates the same receptor. Medically used hCG is derived from urine of pregnant women, less costly, and has a longer half-life than LH.

Growth hormone (GH) is a protein-based poly-peptide hormone. It stimulates growth, cell reproduction and regeneration in humans and other animals. It is a 191-amino acid, single-chain polypeptide hormone that is synthesized, stored, and secreted by the somatotroph cells within the lateral wings of the anterior pituitary gland.

Somatotropin refers to the growth hormone produced natively and naturally in animals, whereas the term **somatropin** refers to growth hormone produced by recombinant DNA technology, and is abbreviated "rhGH" in humans.Growth hormone is used clinically to treat children's growth disorders and adult growth hormone deficiency. In recent years, replacement therapies with human growth hormones (hGH) have become popular in the battle against aging and weight management. Reported effects on GH deficient patients (but not on healthy people) include decreased body fat, increased muscle mass, increased bone density, increased energy levels, improved skin tone and texture, increased sexual function and improved immune system function. At this time hGH is still considered a very complex hormone and many of its functions are still unknown. In its role as an anabolic agent, HGH has been used by competitors in sports since the 1970s, and it has been banned by the IOC and NCAA.

Traditional urine analysis could not detect doping with hGH, so the ban was unenforceable until the early 2000s when blood tests that could distinguish between natural and artificial hGH were starting to be developed. Blood tests conducted by WADA at the 2004 Olympic Games in Athens, Greece primarily targeted hGH.

Genes for human growth hormone, known as growth hormone 1 (somatotropin) and growth hormone 2, are localized in the q22-24 region of chromosome 17 and are closely related to human chorionic somatomammotropin (also known as placental lactogen) genes. GH, human chorionic somatomammotropin, and prolactin (PRL) are a group of homologous hormones with growth-promoting and lactogenic activity. The identification, purification and later synthesis of growth hormone is associated with Choh Hao Li. Genentech pioneered the first use of recombinant human growth hormone for human therapy in 1981.

Prior to its production by recombinant DNA technology, growth hormone used to treat deficiencies was extracted from the pituitary glands of cadavers. Attempts to create a wholly synthetic HGH failed. Limited supplies of HGH resulted in the restriction of HGH therapy to the treatment of idiopathic short stature. Furthermore, growth hormone from other primates was found to be inactive in humans. In 1985, unusual cases of Creutzfeldt-Jacob disease were found in individuals that had received cadaver-derived HGH ten to fifteen years previously. Based on the assumption that infectious prions causing the disease were transferred along with the cadaver-derived HGH, cadaver-derived HGH was removed from the market. In 1985, biosynthetic human growth hormone replaced pituitary-derived human growth hormone for therapeutic use in the U.S. and elsewhere. As of 2005, recombinant growth hormones available in the United States (and their manufacturers) included Nutropin (Genentech),

Humatrope (Lilly), Genotropin (Pfizer), Norditropin (Novo), and Saizen (Merck Serono). In 2006, the U.S. Food and Drug Association (FDA) approved a version of rhGH called Omnitrope (Sandoz). A sustained-release form of growth hormone, Nutropin Depot (Genentech and Alkermes) was approved by the FDA in 1999, allowing for fewer injections (every 2 or 4 weeks instead of daily); however, the product was discontinued in 2004.

STRUCTURE

The major isoform of the human growth hormone is a protein of 191 amino acids and a molecular weight of 22,124 daltons. The structure includes four helices necessary for functional interaction with the GH receptor. It appears that, in structure, GH is evolutionarily homologous to prolactin and chorionic somatomammotropin. Despite marked structural similarities between growth hormone from different species, only human and primate growth hormones have significant effects in humans.

Several molecular isoforms of GH circulate in the plasma. A percentage of the growth hormone in the circulation is bound to a protein (growth hormone binding protein, GHBP) which is the truncated part of the growth hormone receptor, and an acid labile subunit (ALS).

Function

Growth hormone release from the anterior pituitary is under the control of two hypothalamic hormones.

- GHRH acts to stimulate release of GH while somatostatin acts to inhibit release of GH.
- GH exerts its effects directly and also indirectly through insulin-like growth factors (IGFs) 1 and 2. The effects include promotion of growth of bone, soft tissue and viscera as well as having affects on protein synthesis, lipolysis and glucose transport and metabolism. IGF1 provides negative feedback to the pituitary and hypothalamus.

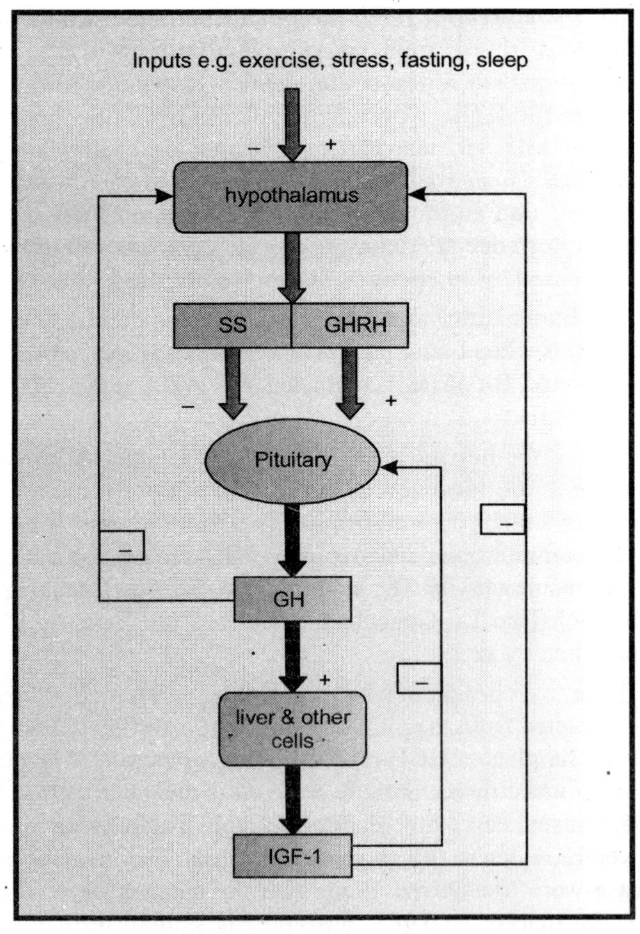

Fig. 12.22 : The Function of Groth hormone

- Other factors affecting GHRH and somatostatin secretion include sleep, exercise, stress and blood glucose levels. Oestradiol also acts to increase sensitivity of tissues to GH.

Regulation

Peptides released by neurosecretory nuclei of the hypothalamus (Growth hormone-releasing hormone/ *somatocrinin* and Growth hormone-inhibiting hormone/*somatostatin*) into the hypophyseal portal venous blood surrounding the pituitary are the major controllers of GH secretion by the somatotropes. However, although the balance of these stimulating and inhibiting peptides determines GH release, this balance is affected by many physiological stimulators (e.g., exercise, nutrition, sleep) and inhibitors of GH secretion (e.g., Free fatty acids).

Stimulators of GH secretion include:

- peptide hormones
  - Growth hormone releasing hormone (GHRH) through binding to the growth hormone releasing hormone receptor (GHRHR)
  - ghrelin through binding to growth hormone secretagogue receptors (GHSR)
- sex hormones
  - increased androgen secretion during puberty (in males from testis and in females from adrenal cortex)
  - estrogen
- clonidine and L-DOPA by stimulating GHRH release
- hypoglycaemia, arginine and propranolol by inhibiting somatostatin release
- deep sleep
- fasting
- vigorous exercise

Inhibitors of GH secretion include:

- somatostatin from the periventricular nucleus
- circulating concentrations of GH and IGF-1 (negative feedback on the pituitary and hypothalamus)
- hyperglycemia
- glucocorticoids
- dihydrotestosterone

In addition to control by endogenous and stimulus processes, a number of foreign compounds (xenobiotics such as drugs and endocrine disruptors) are known to influence GH secretion and function.

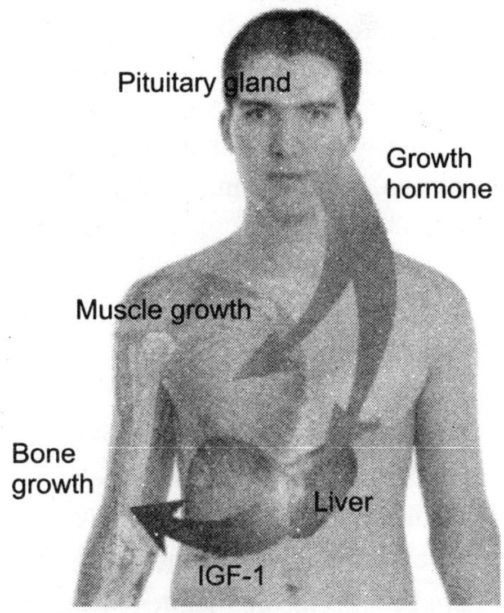

Fig. 12.23 : Action of Growth Hormone

HGH is synthesized and secreted from the anterior pituitary gland in a pulsatile manner throughout the day; surges of secretion occur at 3- to 5-hour intervals. The plasma concentration of GH during these peaks may range from 5 to even 45 ng/mL. The largest and most predictable of these GH peaks occurs about an hour after onset of sleep. Otherwise there is wide variation between days and individuals. Nearly fifty percent of HGH secretion occurs during the third and fourth REM sleep stages. Between the peaks, basal GH levels are low, usually less than 5 ng/mL for most of the day and night. Additional analysis of the pulsatile profile of GH described in all cases less than 1 ng/ml for basal levels while maximum peaks were situated around 10-20 ng/mL.

A number of factors are known to affect HGH secretion, such as age, gender, diet, exercise, stress, and other hormones. Young adolescents secrete HGH at the rate of about 700 μg/day, while healthy adults secrete HGH at the rate of about 400 μg/day.

FUNCTIONS OF GH

Effects of growth hormone on the tissues of the body can generally be described as anabolic (building up). Like most other protein hormones, GH acts by interacting with a specific receptor on the surface of cells.

Increased height during childhood is the most widely known effect of GH. Height appears to be stimulated by at least two mechanisms:

1. Because polypeptide hormones are not fat-soluble, they cannot penetrate sarcolemma. Thus, GH exerts some of its effects by binding to receptors on target cells, where it activates the MAPK/ERK pathway. Through this mechanism GH directly stimulates division and multiplication of chondrocytes of cartilage.
2. GH also stimulates, through the JAK-STAT signaling pathway, the production of insulin-like growth factor 1 (IGF-1, formerly known as somatomedin C), a hormone homologous to proinsulin. The liver is a major target organ of GH for this process and is the principal site of IGF-1 production. IGF-1 has growth-stimulating effects on a wide variety of tissues. Additional IGF-1 is generated within target tissues, making it what appears to be both an endocrine and an autocrine/paracrine hormone. IGF-1 also has stimulatory effects on osteoblast and chondrocyte activity to promote bone growth.

In addition to increasing height in children and adolescents, growth hormone has many other effects on the body:

- Increases calcium retention, and strengthens and increases the mineralization of bone
- Increases muscle mass through sarcomere hyperplasia
- Promotes lipolysis
- Increases protein synthesis
- Stimulates the growth of all internal organs excluding the brain
- Plays a role in fuel homeostasis
- Reduces liver uptake of glucose
- Promotes gluconeogenesis in the liver
- Contributes to the maintenance and function of pancreatic islets
- Stimulates the immune system

Excesses

The most common disease of GH excess is a pituitary tumor composed of somatotroph cells of the anterior pituitary. These somatotroph adenomas are benign and grow slowly, gradually producing more and more GH. For years, the principal clinical problems are those of GH excess. Eventually the adenoma may become large enough to cause headaches, impair vision by pressure on the optic nerves, or cause deficiency of other pituitary hormones by displacement.

Prolonged GH excess thickens the bones of the jaw, fingers and toes. Resulting heaviness of the jaw and increased size of digits is referred to as acromegaly. Accompanying problems can include sweating, pressure on nerves (e.g., carpal tunnel syndrome), muscle weakness, excess sex hormone binding globulin (SHBG), insulin resistance or even a rare form of type 2 diabetes, and reduced sexual function.

GH-secreting tumors are typically recognized in the fifth decade of life. It is extremely rare for such a tumor to occur in childhood, but, when it does, the excessive GH can cause excessive growth, traditionally referred to as pituitary gigantism.

Surgical removal is the usual treatment for GH-producing tumors. In some circumstances, focused radiation or a GH antagonist such as pegvisomant may be employed to shrink the tumor or block function. Other drugs like octreotide (somatostatin agonist) and bromocriptine (dopamine agonist) can be used to block GH secretion because both somatostatin and dopamine negatively inhibit GHRH-mediated GH release from the anterior pituitary.

DEFICIENCIES

The effects of growth hormone deficiency vary depending on the age at which they occur. In children, growth failure and short stature are the major manifestations of GH deficiency, with common causes including genetic conditions and congenital malformations. It can also cause delayed sexual maturity. In adults, deficiency is rare, with the most common cause a pituitary adenoma, and others including a continuation of a childhood problem, other structural lesions or trauma, and very rarely idiopathic GHD.

Adults with GHD present with non-specific problems including truncal obesity with a relative decrease in muscle mass and, in many instances, decreased energy and quality of life.

Diagnosis of GH deficiency involves a multiple-step diagnostic process, usually culminating in GH stimulation tests to see if the patient's pituitary gland will release a pulse of GH when provoked by various stimuli.

Treatment with exogenous GH is indicated only in limited circumstances, and needs regular monitoring due to the frequency and severity of side-effects. GH is used as replacement therapy in adults with GH deficiency of either childhood-onset (after completing growth phase) or adult-onset (usually as a result of an acquired pituitary tumour). In these patients, benefits have variably included reduced fat mass, increased lean mass, increased bone density, improved lipid profile, reduced cardiovascular risk factors, and improved psychosocial well-being.

TREATMENTS UNRELATED TO DEFICIENCY

GH can be used to treat conditions that produce short stature but are not related to deficiencies in GH. However, results are not as dramatic when compared to short stature that is solely due to deficiency of GH. Examples of other causes of shortness often treated with GH are Turner syndrome, chronic renal failure, Prader–Willi syndrome, intrauterine growth retardation, and severe idiopathic short stature. Higher ("pharmacologic") doses are required to produce significant acceleration of growth in these conditions, producing blood levels well above normal ("physiologic"). Despite the higher doses, side-effects during treatment are rare, and vary little according to the condition being treated.

GH treatment improves muscle strength and slightly reduces body fat in Prader-Willi syndrome, which are significant concerns beyond the need to increase height. GH is also useful in maintaining muscle mass in wasting due to AIDS. GH can also be used in patients with short bowel syndrome to lessen the requirement for intravenous total parenteral nutrition.

GH can also be used for conditions that do not cause short stature. Typically, growth hormone treatment for conditions unrelated to stature is controversial and experimental. GH has been used for remission of multiple sclerosis, to reverse the effects of aging in older adults (see below), to enhance weight loss in obesity, as well as fibromyalgia, heart failure, Crohn's disease and ulcerative colitis, burns and bodybuilding or athletic enhancement.

Anti-aging Agent

Claims for GH as an anti-aging treatment date back to 1990 when the *New England Journal of Medicine* published a study wherein GH was used to treat 12 men over 60. At the conclusion of the study, all the men showed statistically significant increases in lean body mass and bone mineral, while the control group did not. The authors of the study noted that these improvements were the opposite of the changes that would normally occur over a 10- to 20-year aging period. Despite the fact the authors at no time claimed that GH had reversed the aging process itself, their results were misinterpreted as indicating that GH is an effective anti-aging agent. This has led to organizations such as the controversial American Academy of Anti-Aging Medicine promoting the use of this hormone as an "anti-aging agent".

A Stanford University School of Medicine survey of clinical studies on the subject published in early 2007 showed that the application of GH on healthy elderly patients increased muscle by about 2 kg and decreased body fat by the same amount. However, these were the only positive effects from taking GH. No other critical factors were affected, such as bone density, cholesterol levels, lipid measurements, maximal oxygen consumption, or any other factor that would indicate increased fitness. Researchers also did not discover any gain in muscle strength, which led them to believe that GH merely let the body store more water in the muscles rather than increase muscle growth. This would explain the increase in lean body mass.

Athletic Enhancement

Athletes in many sports use human growth hormone to enhance their athletic performance. Some recent studies have not been able to support claims that human growth hormone can improve the athletic performance of professional male athletes.

Side-effects

There is theoretical concern that HGH treatment may increase the risks of diabetes, especially in those with other predispositions treated with higher doses. If used for training, growth at a young age (25 or less) can cause severe symptoms. One survey of adults that had been treated with replacement cadaver GH (which has not been used anywhere in the world since 1985) during childhood showed a mildly increased incidence of colon cancer and prostate cancer, but linkage with the GH treatment was not established. Regular application of extra GH may show several negative side-effects such as joint swelling, joint pain, carpal tunnel syndrome, and an increased risk of diabetes. Other side effects can include less sleep needed after dosing. This is common initially and decreases in effect after habitual use of GH.

Prolactin (**PRL**) or **Luteotropic hormone** (**LTH**) is a peptide hormone discovered by Dr. Henry Friesen, primarily associated with lactation. In breastfeeding, the act of an infant sucking the nipple stimulates the production of prolactin, which fills the breast with milk via a process called lactogenesis, in preparation for the next feed. Oxytocin, another hormone, is also released, which triggers milk let-down.

Production and Regulation

The system of increased production is a positive feedback cycle. (That is, the effect causes more of itself.) This is one of the few instances in the body where positive feedback is normal.Pituitary prolactin secretion is regulated by neuroendocrine neurons in the hypothalamus, the most important ones being the neurosecretory tuberoinfundibulum (TIDA) neurons of the arcuate nucleus, which secrete dopamine to act on the dopamine-2 receptors of lactotrophs, causing inhibition of prolactin secretion. Thyrotropin-releasing factor (thyrotropin-releasing hormone) has a stimulatory effect on prolactin release.Vasoactive intestinal peptide and peptide histidine isoleucine help to regulate prolactin secretion in humans, but the functions of these hormones in birds can be quite different.

Effects

Prolactin has many effects including regulating lactation, orgasms, and stimulating proliferation of oligodendrocyte precursor cells.It stimulates the mammary glands to produce milk (lactation): Increased serum concentrations of prolactin during pregnancy cause enlargement of the mammary glands of the breasts and increases the production of milk. However, the high levels of progesterone during pregnancy act directly on the breasts to stop ejection of milk. It is only when the levels of this hormone fall after childbirth that milk ejection is possible. Sometimes, newborn babies (males as well as females) secrete a milky substance from their nipples. This substance is commonly known as witch's milk. This is caused by the fetus being affected by prolactin circulating in the mother just before birth, and usually stops soon after birth. Prolactin provides the body with sexual gratification after sexual acts: The hormone counteracts the effect of dopamine, which is responsible for sexual arousal. This is thought to cause the sexual refractory period. The amount of prolactin can be an indicator for the amount of sexual satisfaction and relaxation. Unusually high amounts are suspected to be responsible for impotence and loss of libido (see hyperprolactinemia symptoms). Prolactin also stimulates proliferation of oligodendrocyte precursor cells. These cells differentiate into oligodendrocytes, the cells responsible for the formation of myelin coatings on axons in the central nervous system. Prolactin also has a number of other effects including contributing to surfactant synthesis of the fetal lungs at the end of the pregnancy and immune tolerance of the fetus by the maternal organism during pregnancy; it also decreases normal levels of sex hormones—estrogen in women and testosterone in men. It is this inhibition of sex steroids that is responsible for loss of the menstrual cycle in lactating women as well as lactation-associated osteoporosis. Prolactin also enhances luteinizing hormone-receptors in Leydig cells, resulting in testosterone secretion which leads to spermatogenesis. Prolactin delays hair regrowth in mice.

Variance in Levels

There is a diurnal as well as an ovulatory cycle in prolactin secretion. In many mammals, there is also a seasonal change in prolactin release. During pregnancy, high circulating concentrations of estrogen promote prolactin production. The resulting high levels of prolactin secretion cause further maturation of the mammary glands, preparing them for lactation. After childbirth, prolactin levels fall as the internal stimulus for them is removed. Sucking by the baby on the nipple then promotes further prolactin release, maintaining the ability to lactate. The sucking activates mechanoreceptors in and around the nipple. These signals are carried by nerve fibers through the spinal cord to the hypothalamus, where changes in the electrical activity of neurons that regulate the pituitary gland cause increased prolactin secretion. The suckling stimulus also triggers the release of oxytocin from the posterior pituitary gland, which triggers milk let-down: Prolactin controls milk production (lactogenesis) but not the milk-ejection reflex; the rise in prolactin fills the breast with milk in preparation for the next feed.In usual circumstances, in the absence of galactorrhea, lactation will cease within one or two weeks of the end of demand breastfeeding.

It has also been found that compared to un-mated males, fathers and expectant fathers have increased prolactin concentrations. High prolactin levels also tend to suppress the ovulatory cycle by inhibiting the secretion of both follicle-stimulating hormone (FSH) and gonadotropic-releasing hormone (GnRH). High prolactin levels can also contribute to mental health issues.Prolactin levels peak during REM sleep, and in the early morning. Levels can rise after exercise, meals, sexual intercourse, minor surgical procedures, or following epileptic seizures.

STRUCTURE

Prolactin is a single-chain polypeptide of 198 amino acids with a molecular weight of about 24,000 daltons. Its structure is similar to that of growth hormone and placental lactogen. The molecule is folded due to the activity of three disulfide bonds. Significant heterogeneity of the molecule has been described, thus bioassays and immunoassays can give different results due to differing glycosylation, phosphorylation, sulfation, as well as degradation. The non-glycosylated form of prolactin is the dominant form of prolactin that is secreted by the pituitary gland. **Little prolactin** is apparently the result of removal of some amino acids, whereas **big prolactin** can be the product of interaction of several prolactin molecules. Pit-1 is a transcription factor that binds to the prolactin gene at several sites to allow for the production of prolactin in the pituitary gland. A key regulator of prolactin production is estrogens that enhance growth of prolactin-producing cells and stimulate prolactin production directly, as well as suppressing dopamine.Human prolactin receptors are insensitive to mouse prolactin.

FUNCTION

- PRL acts to initiate and maintain milk secretion by the mammary glands. It works with other hormones such as oxytocin, which actually causes milk ejection, and oestradiol, progesterone, glucocorticoids, GH, thyroxine and insulin, which prepare the mammary gland for milk production.
- Other functions are unclear but experimental animals have been shown to produce PRL in response to stress. PRL may also play a part in fertility and maternal behaviour.
- PRL secretion is under inhibitory control of dopamine. This means that if the link between the hypothalamus and pituitary is severed PRL secretion increases, unlike all other pituitary hormones, where production would decrease without stimulatory control of the hypothalamus.

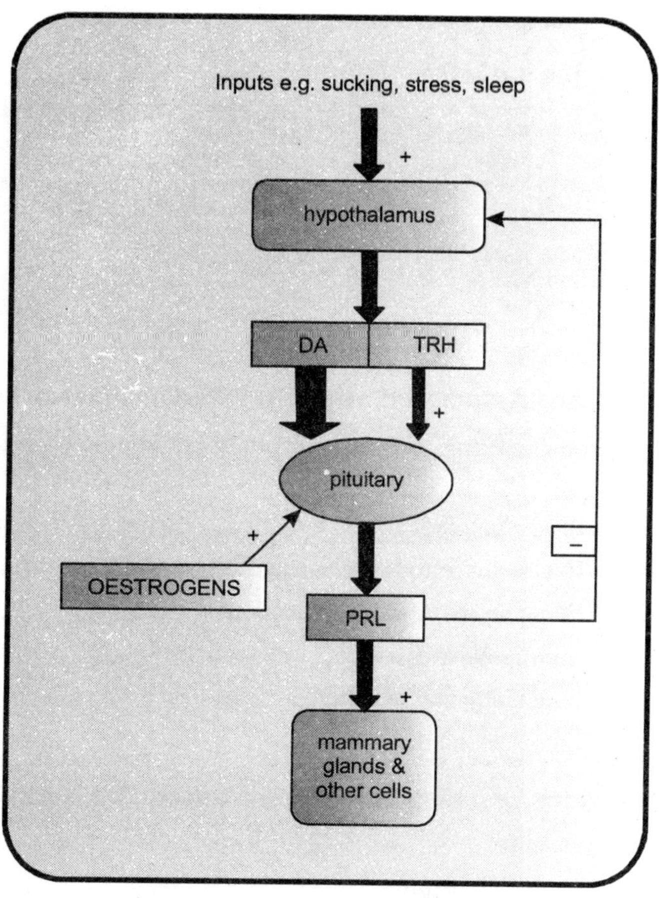

Fig. 12.24: The Function of Prolactein

- TSH has a stimulatory affect on PRL secretion. Oestradiol increases PRL production and levels of PRL rise during pregnancy and remain high during lactation.

Prolactin Receptor

Prolactin receptors are present in the mamillary glands, ovaries, pituitary glands, heart, lung, thymus, spleen, liver, pancreas, kidney, adrenal gland, uterus, skeletal muscle, skin and areas of the central nervous system. When prolactin binds to the receptor, it causes it to dimerize with another prolactin receptor. This results in the activation of Janus kinase 2, a tyrosine kinase which initiates the JAK-

STAT pathway. The activation of the prolactin receptor also results in the activation of mitogen-activated protein kinases and Src kinase. Prolactin levels may be checked as part of a sex hormone workup, as elevated prolactin secretion can suppress the secretion of FSH and GnRH, leading to hypogonadism, and sometimes causing erectile dysfunction in men.Prolactin levels may be of some use in distinguishing epileptic seizures from psychogenic non-epileptic seizures. The serum prolactin level usually rises following an epileptic seizure.

Reference Levels

	ng/mL	mIU/L	Prolactin Standard
Women	2.8-29.2	59-619	(World Health Organization [WHO]: Third International Reference Preparation [3rd IRP] of prolactin, human, lyophilized, 0.053 IU / ampoule, for immunoassay [preparation code:84/500])
Men	2.1-17.7	45-375	(WHO: 3rd IRP 84/500)

Conditions Associated with Elevated Prolactin Secretion

Hyperprolactinaemia is the term given to having too-high levels of prolactin in the blood, which can result from:

- Prolactinoma
- Excess thyrotropin-releasing hormone (TRH), usually in primary hypothyroidism
- Many antipsychotic medications
- Emotional stress
- Pregnancy and lactation.
- Some sexual disorders

Conditions associated with Decreased Prolactin

- Bulimia
- Excess dopamine

Use of breastfeeding as Contraceptive

The World Health Organization states that demand breastfeeding is more than 98% effective as a contraceptive in the first six months postpartum. This effect is said to be responsible for the natural spacing of children seen in countries where contraception is not widely available, and is thought to be an evolutionary means of ensuring adequate care is provided to each newborn. The 98% effectiveness applies only if three criteria are met:

1. The mother has had no menstrual periods at all (amenorrhea)
2. The baby is exclusively breast-fed

3. It is six months or less since birth.
   - A fourth criterion has also been mentioned: The baby does not go for longer than 5 hours between feedings (i.e. while sleeping)

If one or more of these conditions are broken, lactational amenorrhea is no longer a reliable form of birth control. This contraceptive method is highly effective as long as the three conditions stated above are fulfilled. Further, the WHO suggests that a woman who is still amenorrheic has a less-than-5% chance of getting pregnant in the first year of her baby's life, as long as she is still breastfeeding on demand

POSTERIOR PITUITARY HORMONE

The **posterior pituitary** (or **neurohypophysis**) comprises the posterior lobe of the pituitary gland and is part of the endocrine system. Despite its name, the posterior pituitary gland is not a gland, *per se*; rather, it is largely a collection of axonal projections from the hypothalamus that terminate behind the anterior pituitary gland.

HORMONES SECRETED

Hormones known classically as posterior pituitary hormones are synthesized by the hypothalamus. They are then stored and secreted by the posterior pituitary into the bloodstream.

Hormone	Other names	Symbol(s)	Target	Effect	Source
Oxytocin	-	OT	Uterus, mammary glands	Uterine contractions; lactation	supraoptic and paraventricular nuclei
Vasopressin (antidiuretic hormone)	Arginine vasopressin, argipressin, antidiuretic hormone	VP, AVP, ADH	Kidneys or Arterioles	Stimulates water retention; raises blood pressure by contracting arterioles, induces male aggression	supraoptic and paraventricular nuclei

Oxytocin (Pitocin, Syntocinon) is a mammalian hormone that acts primarily as a neurotransmitter in the brain. Also known as alpha-hypophamine (α-hypophamine), oxytocin has the distinction of being the very first polypeptide hormone to be sequenced and synthesized biochemically in 1953. Oxytocin is best known for its roles in female reproduction: 1) it is released in large amounts after distension of the cervix and vagina during labor, and 2) after stimulation of the nipples, facilitating birth and breastfeeding. Recent studies have begun to investigate oxytocin's role in various behaviors, including orgasm, social recognition, pair bonding, anxiety, and maternal behaviors. For this reason, it is sometimes referred to as the "cuddle hormone."

ACTION

Oxytocin has peripheral (hormonal) actions, and also has actions in the brain. The actions of oxytocin are mediated by specific, high affinity oxytocin receptors. The oxytocin receptor is a G-protein-coupled receptor which requires Mg^{2+} and cholesterol. It belongs to the rhodopsin-type (class I) group of G-protein-coupled receptors.

Peripheral (Hormonal) Actions

The peripheral actions of oxytocin mainly reflect secretion from the pituitary gland.

- **Letdown reflex** – in lactating (breastfeeding) mothers, oxytocin acts at the mammary glands, causing milk to be 'let down' into subareolar sinuses, from where it can be excreted via the nipple.[4] Sucking by the infant at the nipple is relayed by spinal nerves to the hypothalamus. The stimulation causes neurons that make oxytocin to fire action potentials in intermittent bursts; these bursts result in the secretion of pulses of oxytocin from the neurosecretory nerve terminals of the pituitary gland.
- **Uterine contraction** – important for cervical dilation before birth and causes contractions during the second and third stages of labor. Oxytocin release during breastfeeding causes mild but often painful uterine contractions during the first few weeks of lactation. This also serves to assist the uterus in clotting the placental attachment point postpartum. However, in knockout mice lacking the oxytocin receptor, reproductive behavior and parturition is normal.
- The relationship between oxytocin and human sexual response is unclear. At least two non-controlled studies have found increases in plasma oxytocin at orgasm – in both men and women. Plasma oxytocin levels are notably increased around the time of self stimulated orgasm and are still higher than baseline when measured 5 minutes after self arousal. The authors of one of these studies speculated that oxytocin's effects on muscle contractibility may facilitate sperm and egg transport. In a study that measured oxytocin serum levels in women before and after sexual stimulation the author suggests that oxytocin serves an important role in sexual arousal. This study found that genital tract stimulation resulted in increased oxytocin immediately after orgasm. Another study that reports increases of oxytocin during sexual arousal states that it could be in response to nipple/areola, genital, and/or genital tract stimulation as confirmed in other mammals. Researchers studying men, found that oxytocin levels were raised throughout sexual arousal and there was no acute increase at orgasm. A more recent study of men found an increase in plasma oxytocin immediately after orgasm, but only in a portion of their sample that did not reach statistical significance. The authors noted that these changes "may simply reflect contractile properties on reproductive tissue."

It is also important to note that more studies have been done to examine sexual arousal in women compared to men. Women experience longer orgasms compared to men and have a more complex reproductive endocrine system with clearly identified cycles such as, menstruation, lactation, menopause, and pregnancy. This allows more opportunities to measure and examine the hormones related to sexual arousal.

Oxytocin evokes feelings of contentment, reductions in anxiety, and feelings of calmness and security around a mate. In order to reach full orgasm, it is necessary that brain regions associated with behavioral control, fear and anxiety are deactivated; which allows individuals to let go of fear and anxiety during sexual arousal. Many studies have already shown a correlation of oxytocin with social bonding, increases in trust, and decreases in fear. One study confirmed that there was a positive correlation between oxytocin plasma levels and an anxiety scale measuring the adult romantic attachment. This suggests that oxytocin may be important for the inhibition of brain regions that are associated with behavioral control, fear, and anxiety, thus allowing orgasm to occur:

- Due to its similarity to vasopressin, it can reduce the excretion of urine slightly. In several species, oxytocin can stimulate sodium excretion from the kidneys (natriuresis), and in humans, high doses of oxytocin can result in hyponatremia.
- Oxytocin and oxytocin receptors are also found in the heart in some rodents, and the hormone may play a role in the embryonal development of the heart by promoting cardiomyocyte differentiation. However, the absence of either oxytocin or its receptor in knockout mice has not been reported to produce cardiac insufficiencies.
- Modulation of hypothalamic-pituitary-adrenal axis activity. Oxytocin, under certain circumstances, indirectly inhibits release of adrenocorticotropic hormone and cortisol and, in those situations, may be considered an antagonist of vasopressin.
- Autism. Oxytocin may play a role in autism and may be an effective treatment for autism's repetitive and affiliative behaviors. Oxytocin treatments also resulted in an increased retention of affective speech in adults with autism. Two related studies in adults, in 2003 and 2007, found that oxytocin decreased repetitive behaviors and improved interpretation of emotions. More recently, intranasal administration of oxytocin was found to increase emotion recognition in children as young as 12 who are diagnosed with autism spectrum disorders Oxytocin has also been implicated in the etiology of autism with one report suggesting that autism is correlated with genomic deletion of the gene containing the oxytocin receptor gene (OXTR). Studies involving Caucasian and Finnish samples and Chinese Han families provide support for the relationship of OXTR with autism. Autism may also be associated by an aberrant methylation of OXTR, as reported by Gregory and colleagues. After treatment with inhaled oxytocin, autistic patients exhibit more appropriate social behavior. While this research suggests some promise, further clinical trials of oxytocin are required to demonstrate potential benefit and side effects in the treatment of autism. As such, researchers do not recommend use of oxytocin as a treatment for autism outside of clinical trials.
- Increasing trust and reducing fear. In a risky investment game, experimental subjects given nasally administered oxytocin displayed "the highest level of trust" twice as often as the control group. Subjects who were told that they were interacting with a computer showed no such reaction, leading to the conclusion that oxytocin was not merely affecting risk-aversion. Nasally administered oxytocin has also been reported to reduce fear, possibly by inhibiting the amygdala (which is thought to be responsible for fear responses).. Some researchers have argued that oxytocin has a general enhancing effect on all social emotions since intranasal administration of oxytocin also increases envy and schadenfreude.

- Affecting generosity by increasing empathy during perspective taking. In a neuroeconomics experiment, intranasal oxytocin increased generosity in the Ultimatum Game by 80% but has no effect in the Dictator Game that measures altruism. Perspective-taking is not required in the Dictator Game, but the researchers in this experiment explicitly induced perspective-taking in the Ultimatum Game by not identifying to participants which role they would be in.
- Certain learning and memory functions are impaired by centrally administered oxytocin. Also, systemic oxytocin administration can impair memory retrieval in certain aversive memory tasks. Interestingly, oxytocin does seem to facilitate learning and memory specifically for social information. Healthy males administered intranasal oxytocin show improved memory for human faces, particularly happy faces. They also show improved recognition for positive social cues over threatening social cues
- Empathy in healthy males has been shown to be increased after intranasal oxytocin This is most likely due to the effect of oxytocin in enhancing eye gaze. There is some discussion about which aspect of empathy oxytocin might alter, for example cognitive vs emotional empathy

Actions within the Brain

Oxytocin secreted from the pituitary gland cannot re-enter the brain because of the blood-brain barrier. Instead, the behavioral effects of oxytocin are thought to reflect release from centrally projecting oxytocin neurons, different from those that project to the pituitary gland, or which are collaterals from them.[37] Oxytocin receptors are expressed by neurons in many parts of the brain and spinal cord, including the amygdala, ventromedial hypothalamus, septum, nucleus accumbens and brainstem.

- *Sexual arousal.* Oxytocin injected into the cerebrospinal fluid causes spontaneous erections in rats, reflecting actions in the hypothalamus and spinal cord. Centrally administrated oxytocin receptor antagonists can prevent non contact erections, which is a measure of sexual arousal. Studies using oxytocin antagonists in female rats provide data that oxytocin increases lordosis, indicating an increase in sexual receptivity.
- *Bonding.* In the Prairie Vole, oxytocin released into the brain of the female during sexual activity is important for forming a monogamous pair bond with her sexual partner. Vasopressin appears to have a similar effect in males. Oxytocin has a role in social behaviors in many species, and so it seems likely that it has similar roles in humans.
- *Maternal behaviour.* Female rats given oxytocin antagonists after giving birth do not exhibit typical maternal behavior. By contrast, virgin female sheep show maternal behavior towards foreign lambs upon cerebrospinal fluid infusion of oxytocin, which they would not do otherwise. Oxytocin is involved in the initiation of maternal behavior not its maintenance, for example, it is higher in mothers after they interact with unfamiliar children rather than their own.
- According to some studies in animals, *oxytocin inhibits* the development of tolerance to various addictive drugs (opiates, cocaine, alcohol) and reduces withdrawal symptoms.
- *Preparing fetal neurons for delivery.* Crossing the placenta, maternal oxytocin reaches the fetal brain and induces a switch in the action of neurotransmitter GABA from excitatory to

inhibitory on fetal cortical neurons. This silences the fetal brain for the period of delivery and reduces its vulnerability to hypoxic damage.

- *MDMA (ecstasy)* may increase feelings of love, empathy and connection to others by stimulating oxytocin activity via activation of serotonin 5-HT1A receptors, if initial studies in animals apply to humans. The anxiolytic Buspar (buspirone) also appears to produce some or all of its effect via 5-HT1A receptor-induced oxytocin stimulation.

Synthetic oxytocin is sold as proprietary medication under the trade names **Pitocin** and **Syntocinon** and also as generic oxytocin. Oxytocin is destroyed in the gastrointestinal tract, and therefore must be administered by injection or as nasal spray. Oxytocin has a half-life of typically about three minutes in the blood. Oxytocin given intravenously does not enter the brain in significant quantities - it is excluded from the brain by the blood-brain barrier. There is no evidence for significant central nervous system entry of oxytocin by nasal spray. Oxytocin nasal sprays have been used to stimulate breastfeeding but the efficacy of this approach is doubtful.

Injected oxytocin analogues are used for labor induction and to support labor in case of non-progression of parturition. It has largely replaced ergometrine as the principal agent to increase uterine tone in acute postpartum haemorrhage. Oxytocin is also used in veterinary medicine to facilitate birth and to increase milk production. The tocolytic agent atosiban (**Tractocile**) acts as an antagonist of oxytocin receptors; this drug is registered in many countries to suppress premature labor between 24 and 33 weeks of gestation. It has fewer side-effects than drugs previously used for this purpose (ritodrine, salbutamol and terbutaline).

Some have suggested that the trust-inducing property of oxytocin might help those who suffer from social anxieties and mood disorders, while others have noted the potential for abuse with confidence tricks and military applications.

Potential Adverse Reactions

Oxytocin is relatively safe when used at recommended doses. Potential side effects include:

- *Central nervous system:* Subarachnoid hemorrhage, seizures.
- *Cardiovascular:* Increased heart rate, decreased blood pressure, systemic venous return, cardiac output, and arrhythmias.
- *Genitourinary:* Impaired uterine blood flow, pelvic hematoma, tetanic uterine contractions, uterine rupture, postpartum hemorrhage.
- *Fetal distress:* Overstimulated uterus, too frequent contractions, resulting in reduced ability of placenta/fetus to re-oxygenate and process waste products. This increases chances of Caesarean section.

Synthesis, Storage and Release

The oxytocin peptide is synthesized as an inactive precursor protein from the *OXT* gene. This precursor protein also includes the oxytocin carrier protein neurophysin I. The inactive precursor protein is progressively hydrolyzed into smaller fragments (one of which is neurophysin I) via a series of enzymes.

The last hydrolysis which releases the active oxytocin nonapeptide is catalyzed by peptidylglycine alpha-amidating monooxygenase (PAM).

The activity of the PAM enzyme system is dependent upon ascorbate which is a necessary vitamin cofactor. By chance, it was discovered that sodium ascorbate by itself stimulated the production of oxytocin from ovarian tissue over a range of concentrations in a dose-dependent manner. Many of the same tissues (e.g. ovaries, testes, eyes, adrenals, placenta, thymus, pancreas) where PAM (and oxytocin by default) is found are also known to store higher concentrations of vitamin C.

Neural Sources

In the hypothalamus, oxytocin is made in magnocellular neurosecretory cells of the supraoptic and paraventricular nuclei and is stored in Herring bodies at the axon terminals in the posterior pituitary. It is then released into the blood from the posterior lobe (neurohypophysis) of the pituitary gland. These axons (likely, but dendrites have not been ruled out) have collaterals that innervate oxytocin receptors in the nucleus accumbens. The peripheral hormonal and behavioral brain effects of oxytocin it has been suggested are coordinated through its common release through these collaterals. Oxytocin is also made by some neurons in the paraventricular nucleus that project to other parts of the brain and to the spinal cord. Depending on the species, oxytocin-receptor expressing cells are located in other areas, including the amygdala and bed nucleus of the stria terminalis.

In the pituitary gland, oxytocin is packaged in large, dense-core vesicles, where it is bound to neurophysin I as shown in the inset of the figure; neurophysin is a large peptide fragment of the larger precursor protein molecule from which oxytocin is derived by enzymatic cleavage.

Secretion of oxytocin from the neurosecretory nerve endings is regulated by the electrical activity of the oxytocin cells in the hypothalamus. These cells generate action potentials that propagate down axons to the nerve endings in the pituitary; the endings contain large numbers of oxytocin-containing vesicles, which are released by exocytosis when the nerve terminals are depolarised.

Non-neural Sources

Outside the brain, oxytocin-containing cells have been identified in several diverse tissues including the corpus luteum, the interstitial cells of Leydig, the retina, the adrenal medulla, the placenta, the thymus and the pancreas. The finding of significant amounts of this classically "neurohypophysial" hormone outside the central nervous system raises many questions regarding its possible importance in these different tissues.

Female

Oxytocin is synthesized by corpora lutea of several species, including ruminants and primates. Along with estrogen, it is involved in inducing the endometrial synthesis of prostaglandin $F_{2\alpha}$ to cause regression of the corpus luteum.

Male

The Leydig cells in some species have also been shown to possess the biosynthetic machinery to manufacture testicular oxytocin *de novo*, specifically, in rats (who can synthesize Vitamin C endogenously), and in guinea pigs who (like humans) require an exogenous source of vitamin C (ascorbate) in their diets.

Structure and Relation to Vasopressin

Oxytocin is a peptide of nine amino acids (a nonapeptide). The sequence is *cys – tyr – ile – gln – asn – cys – pro – leu – gly* (CYIQNCPLG). The cysteine residues form a sulfur bridge. Oxytocin has a molecular mass of 1007 daltons. One international unit (IU) of oxytocin is the equivalent of about two micrograms of pure peptide.

The biologically active form of oxytocin, commonly measured by RIA and/or HPLC techniques, is also known as the octapeptide "oxytocin disulfide" (oxidized form), but oxytocin also exists as a reduced dithiol nonapeptide called oxytoceine. It has been theorized that open chain oxytoceine (the reduced form of oxytocin) may also act as a free radical scavenger (by donating an electron to a free radical); oxytoceine may then be oxidized back to oxytocin via the redox potential of dehydroascorbate <—> ascorbate.

The structure of oxytocin is very similar to that of vasopressin (*cysteine – tyrosine – phe – glu – asp – cys – pro – arg – gly*), also a nonapeptide with a sulfur bridge, whose sequence differs from oxytocin by 2 amino acids. A table showing the sequences of members of the vasopressin/oxytocin superfamily and the species expressing them is present in the vasopressin article. Oxytocin and vasopressin were isolated and synthesized by Vincent du Vigneaud in 1953, work for which he received the Nobel Prize in Chemistry in 1955.

Oxytocin and vasopressin are the only known hormones released by the human posterior pituitary gland to act at a distance. However, oxytocin neurons make other peptides, including corticotropin-releasing hormone (CRH) and dynorphin, for example that act locally. The magnocellular neurons that make oxytocin are adjacent to magnocellular neurons that make vasopressin, and are similar in many respects.

Oxytocin Receptor Polymorphism

The oxytocin receptor in humans has several alleles, which differ in their effectiveness. Individuals homozygous for the "G" allele, when compared to carriers of the "A" allele, show higher empathy, lower stress response, as well as lower prevalence of autism and of poor parenting skills.

Evolution

Virtually all vertebrates have an oxytocin-like nonapeptide hormone that supports reproductive functions and a vasopressin-like nonapeptide hormone involved in water regulation. The two genes are usually located close to each other (less than 15,000 bases apart) on the same chromosome and are transcribed in opposite directions (however, in fugu, the homologs are further apart and transcribed in the same directions).

It is thought that the two genes resulted from a gene duplication event; the ancestral gene is estimated to be about 500 million years old and is found in cyclostomes (modern members of the Agnatha).

Arginine vasopressin (AVP), also known as **vasopressin, argipressin** or **antidiuretic hormone (ADH)**, is a hormone found in most mammals, including humans. Vasopressin is a peptide hormone that controls the reabsorption of molecules in the tubules of the kidneys by affecting the tissue's permeability. It also increases peripheral vascular resistance, which in turn increases arterial blood pressure. It plays a key role in homeostasis, and the regulation of water, glucose, and salts in the blood. It is derived from a preprohormone precursor that is synthesized in the hypothalamus and stored in vesicles at the posterior pituitary. Most of it is stored in the posterior pituitary to be released into the bloodstream; however, some AVP is also released directly into the brain.

FUNCTION

One of the most important roles of AVP is to regulate the body's retention of water; it is released when the body is dehydrated and causes the kidneys to conserve water, thus concentrating the urine, and reducing urine volume. In high concentrations, it also raises blood pressure by inducing moderate vasoconstriction. In addition, it has a variety of neurological effects on the brain, having been found, for example, to influence pair-bonding in voles. The high-density distributions of vasopressin receptor AVPr1a in prairie vole ventral forebrain regions have been shown to facilitate and coordinate reward circuits during partner preference formation, critical for pair bond formation

A very similar substance, *lysine vasopressin* **(LVP)** or **lypressin**, has the same function in pigs and is often used in human therapy.

Kidney

Vasopressin has two effects by which it contributes to increased urine osmolality (increased concentration) and decreased water excretion. These are:

(1) Increase in the permeability to water of the cells of distal tubule and collecting duct in the kidney and thus allows water reabsorption and excretion of concentrated urine, i.e. antidiuresis. This occurs through insertion of water channels (Aquaporin-2) into the apical membrane of the distal tubule and collecting duct epithelial cells. The aquaporins allow water to move out of the nephron, increasing

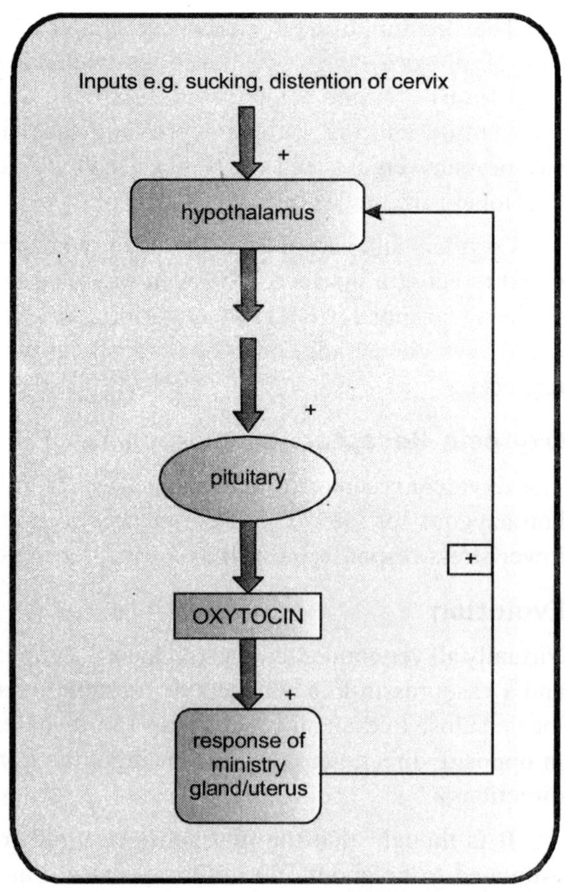

Fig. 12.25: The Function of Vasopresin

the amount of water re-absorbed from the forming urine back into the bloodstream.

V2 receptors, G protein-coupled receptors on the basolateral plasma membrane of the epithelial cells couple to the heterotrimeric G-protein G_s, which activates adenylyl cyclases III and VI to convert ATP into cAMP, plus 2 inorganic phosphates. The rise in cAMP then triggers the insertion of aquaporin-2 water channels by exocytosis of intracellular vesicles, recycling endosomes. Vasopressin also increases the concentration of calcium in the collecting duct cells, by episodic release from intracellular stores. Vasopressin, acting through cAMP also increases transcription of the aquaporin-2 gene, thus increasing the total number of aquaporin-2 molecules in collecting duct cells.

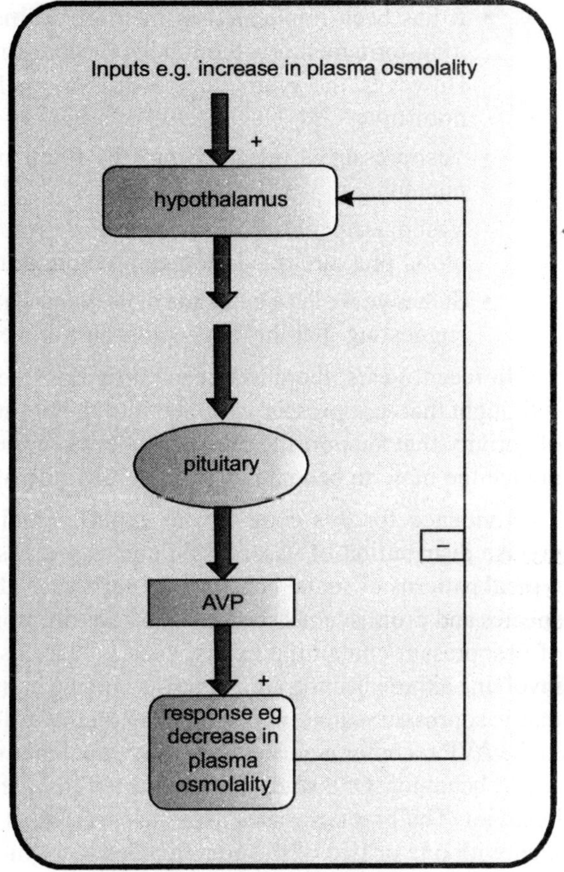

Fig. 12.26: Functions of Vasopresin

Cyclic-AMP activates protein kinase A (PKA)by binding to its regulatory subunits and allowing them to detach from the catalytic subunits. Detachment exposes the catalytic site in the enzyme, allowing it to add phosphate groups to proteins (including the aquaporin-2 protein), which alters their functions.

(2) Increase in the permeability of the inner medullary portion of the collecting duct to urea, allowing increased reabsorption of urea into the medullary interstitium, down the concentration gradient created from the removal of water in the connecting tubule, cortical collecting duct, and outer medullary collecting duct.

Cardiovascular System

Vasopressin increases peripheral vascular resistance and thus increases arterial blood pressure. This effect appears small in healthy individuals; however it becomes an important compensatory mechanism for restoring blood pressure in hypovolemic shock such as that which occurs during hemorrhage.

Central Nervous System (CNS)

Vasopressin released within the brain has many actions:

- It has been implicated in memory formation, including delayed reflexes, image, short- and long-term memory, though the mechanism remains unknown; these findings are controversial. However, the synthetic vasopressin analogue desmopressin has come to interest as a likely nootropic.
- Vasopressin is released into the brain in a circadian rhythm by neurons of the Supraoptic nucleus.
- Vasopressin released from centrally-projecting hypothalamic neurons is involved in aggression, blood pressure regulation and temperature regulation.
- Selective AVPr1a blockade in the ventral pallidum has been shown to prevent partner preference, suggesting that these receptors in this ventral forebrain region are crucial for pair bonding.

In recent years, there has been particular interest in the role of vasopressin in social behaviour. It is thought that vasopressin, released into the brain during sexual activity, initiates and sustains patterns of activity that support the pair-bond between the sexual partners; in particular, vasopressin seems to induce the male to become aggressive towards other males.

Evidence for this comes from experimental studies in several species, which indicate that the precise distribution of vasopressin and vasopressin receptors in the brain is associated with species-typical patterns of social behavior. In particular, there are consistent differences between monogamous species and promiscuous species in the distribution of AVP receptors, and sometimes in the distribution of vasopressin-containing axons, even when closely-related species are compared. Moreover, studies involving either injecting AVP agonists into the brain or blocking the actions of AVP support the hypothesis that vasopressin is involved in aggression towards other males. There is also evidence that differences in the AVP receptor gene between individual members of a species might be predictive of differences in social behavior. One study has suggested that genetic variation in male humans effects pair-bonding behavior. The brain of males uses vasopressin as a reward for forming lasting bonds with a mate, and men with one or two of the genetic alleles are more likely to experience marital discord. The partners of the men with two of the alleles affecting vasopressin reception state disappointing levels of satisfaction, affection, and cohesion. Vasopressin receptors distributed along the reward circuit pathway, to be specific in the ventral pallidum, are activated when AVP is released during social interactions such as mating, in monogamous prairie voles. The activation of the reward circuitry reinforces this behavior, leading to conditioned partner preference, and thereby initiates the formation of a pair bond.

Control

Vasopressin is secreted from the posterior pituitary gland in response to reductions in plasma volume, in response to increases in the plasma osmolality, and in response to cholecystokinin by the small intestine:

- Secretion *in response to reduced plasma volume* is activated by pressure receptors in the veins, atria, and carotids.
- Secretion *in response to increases in plasma osmotic pressure* is mediated by osmoreceptors in the hypothalamus.
- Secretion *in response to increases in plasma Cholecystokinin* is mediated by an unknown pathway.

The neurons that make AVP, in the hypothalamic supraoptic nuclei (SON) and paraventricular nuclei (PVN), are themselves osmoreceptors, but they also receive synaptic input from other osmoreceptors located in regions adjacent to the anterior wall of the third ventricle. These regions include the organum vasculosum of the lamina terminalis and the subfornical organ.

Many factors influence the secretion of vasopressin:

- Ethanol (alcohol) acts as an antagonist for AVP in the collecting ducts of the kidneys, which prevents aquaporins from binding to the collecting ducts, and prevents water reabsorption.
- Angiotensin II *may stimulate* the secretion of AVP.

Secretion

The main stimulus for secretion of vasopressin is increased osmolality of plasma. Reduced volume of extracellular fluid also has this effect, but is a less sensitive mechanism.

The AVP that is measured in peripheral blood is almost all derived from secretion from the posterior pituitary gland (except in cases of AVP-secreting tumours). However there are two other sources of AVP with important local effects:

- Vasopressin is produced in the PVN and SON and travels down the axons through the infundibulum within neurosecretory granules that are found within Herring bodies, localized swellings of the axons and nerve terminals. These carry the peptide directly to the posterior pituitary gland, where it is stored until released into the blood.
- Vasopressin is also released into the brain by several different populations of smaller neurons.

Receptors

Below is a table summarizing some of the actions of AVP at its three receptors, differently expressed in different tissues and exerting different actions:

Type	Second messenger system	Locations	Actions
1	2	3	4
AVPR1A	phosphatidylinositol/ calcium	liver, kidney, peripheral vasculature, brain	vasoconstriction, gluconeogenesis, platelet aggregation, and release of factor VIII and von Willebrand factor; social recognition, circadian tau
AVPR1B	phosphatidylinositol/ calcium	pituitary gland, brain	adrenocorticotropic hormone secretion in response to stress; social interpretation of olfactory cues
AVPR2	adenylate cyclase/ cAMP	basolateral membrane of the cells lining the collecting ducts of the	insertion of aquaporin-2 (AQP2) channels (water channels). This allows water to be reabsorbed down

(Contd...)

1	2	3	4
		kidneys (especially the cortical and outer medullary collecting ducts)	an osmotic gradient, and so the urine is more concentrated. Release of von Willebrand factor and surface expression of P-selectin through exocytosis of Weibel-Palade bodies from endothelial cells
VACM-1	phosphatidylinositol/ calcium	vascular endothelium and renal collecting tubules	Increases cytosolic calcium and acts as an inverse agonist of cAMP accumulation

STRUCTURE AND RELATION TO OXYTOCIN

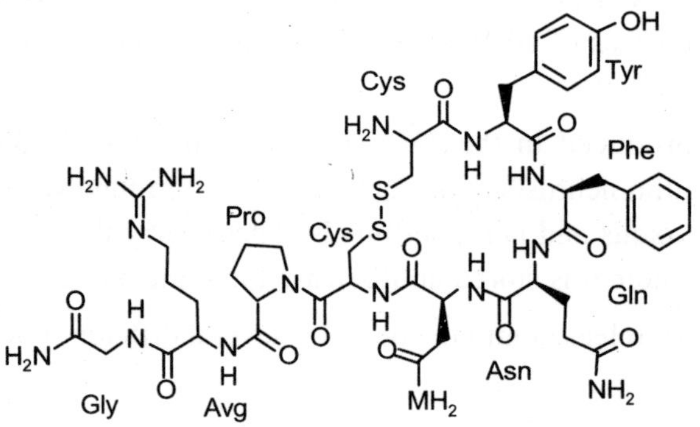

Chemical Structure of Argipressin

The vasopressins are peptides consisting of nine amino acids (nonapeptides). (NB: the value in the table above of 164 amino acids is that obtained before the hormone is activated by cleavage). The amino acid sequence of arginine vasopressin is *Cys-Tyr-Phe-Gln-Asn-Cys-Pro-Arg-Gly*, with the cysteine residues forming a sulfur bridge. Lysine vasopressin has a lysine in place of the arginine.

The structure of oxytocin is very similar to that of the vasopressins: It is also a nonapeptide with a disulfide bridge and its amino acid sequence differs at only two positions (see table below). The two genes are located on the same chromosome separated by a relatively small distance of less than 15,000 bases in most species. The magnocellular neurons that make vasopressin are adjacent to magnocellular neurons that make oxytocin, and are similar in many respects. The similarity of the two peptides can cause some cross-reactions: oxytocin has a slight antidiuretic function, and high levels of AVP can cause uterine contractions.

Here is a table showing the superfamily of vasopressin and oxytocin neuropeptides:

Superfamily of Vasopressin and Oxytocin Neuropeptides

Vertebrate Vasopressin Family		
Cys-Tyr-Phe-Gln-Asn-Cys-Pro-Arg-Gly-NH2	Argipressin (AVP, ADH)	Most mammals
Cys-Tyr-Phe-Gln-Asn-Cys-Pro-Lys-Gly-NH2	Lypressin (LVP)	Pigs, hippos, warthogs, some marsupials
Cys-Phe-Phe-Gln-Asn-Cys-Pro-Arg-Gly-NH2	Phenypressin	Some marsupials
Cys-Tyr-Ile-Gln-Asn-Cys-Pro-Arg-Gly-NH2	Vasotocin*	Non-mammals
Vertebrate Oxytocin Family		
Cys-Tyr-Ile-Gln-Asn-Cys-Pro-Leu-Gly-NH2	Oxytocin (OXT)	Most mammals, ratfish
Cys-Tyr-Ile-Gln-Asn-Cys-Pro-Ile-Gly-NH2	Mesotocin	Most marsupials, all birds, reptiles, amphibians, lungfishes, coelacanths
Cys-Tyr-Ile-Gln-Ser-Cys-Pro-Ile-Gly-NH2	Seritocin	Frogs
Cys-Tyr-Ile-Ser-Asn-Cys-Pro-Ile-Gly-NH2	Isotocin	Bony fishes
Cys-Tyr-Ile-Ser-Asn-Cys-Pro-Gln-Gly-NH2	Glumitocin	Skates
Cys-Tyr-Ile-Asn/Gln-Asn-Cys-Pro-Leu/Val-Gly-NH2	Various tocins	Sharks
Invertebrate VP/OT Superfamily		
Cys-Leu-Ile-Thr-Asn-Cys-Pro-Arg-Gly-NH2	Diuretic Hormone	Locust
Cys-Phe-Val-Arg-Asn-Cys-Pro-Thr-Gly-NH2	Annetocin	Earthworm
Cys-Phe-Ile-Arg-Asn-Cys-Pro-Lys-Gly-NH2	Lys-Connopressin	Geography & imperial cone snail, pond snail, sea hare, leech
Cys-Ile-Ile-Arg-Asn-Cys-Pro-Arg-Gly-NH2	Arg-Connopressin	Striped cone snail
Cys-Tyr-Phe-Arg-Asn-Cys-Pro-Ile-Gly-NH2	Cephalotocin	Octopus
Cys-Phe-Trp-Thr-Ser-Cys-Pro-Ile-Gly-NH2	Octopressin	Octopus

*Vasotocin is the evolutionary progenitor of all the vertebrate neurohypophysial hormones.

Role in Disease

Decreased vasopressin release or decreased renal sensitivity to AVP leads to diabetes insipidus, a condition featuring hypernatremia (increased blood sodium concentration), polyuria (excess urine production), and polydipsia (thirst).

High levels of AVP secretion (syndrome of inappropriate antidiuretic hormone, SIADH) and resultant hyponatremia (low blood sodium levels) occurs in brain diseases and conditions of the lungs (Small cell lung carcinoma). In the perioperative period, the effects of surgical stress and some commonly used medications (e.g., opiates, syntocinon, anti-emetics) lead to a similar state of excess vasopressin secretion. This may cause mild hyponatremia for several days.

Hyponatremia can be treated pharmaceutically through the use of vasopressin receptor antagonists. These include the approved drug Vaprisol and the phase III drug lixivaptan.

PARATHYROID HORMONES

Parathyroid hormone-related protein (or **PTHrP**) is a protein member of the parathyroid hormone family. It is occasionally secreted by cancer cells (breast cancer, certain types of lung cancer including squamous cell carcinoma). However, it also has normal functions.

FUNCTION

The PTHrP acts as an endocrine, autocrine, paracrine, and intracrine hormone. It regulates endochondral bone development by maintaining the endochondral growth plate at a constant width. It also regulates epithelial-mesenchymal interactions during the formation of the mammary glands. PTHrP is critical in the intraosseous phase of tooth eruption where it acts as a signalling molecule to stimulate local bone resorption. Without PTHrP, the bony crypt surrounding the tooth follicle will not resorb, and therefore the tooth will not erupt. In the context of tooth eruption, PTHrP is secreted by the cells of the Reduced Enamel Epithelium.Writing a job description for parathyroid hormone is straightforward: if calcium ion concentrations in extracellular fluid fall below normal, bring them back within the normal range. In conjunction with increasing calcium concentration, the concentration of phosphate ion in blood is reduced. Parathyroid hormone accomplishes its job by stimulating at least three processes:

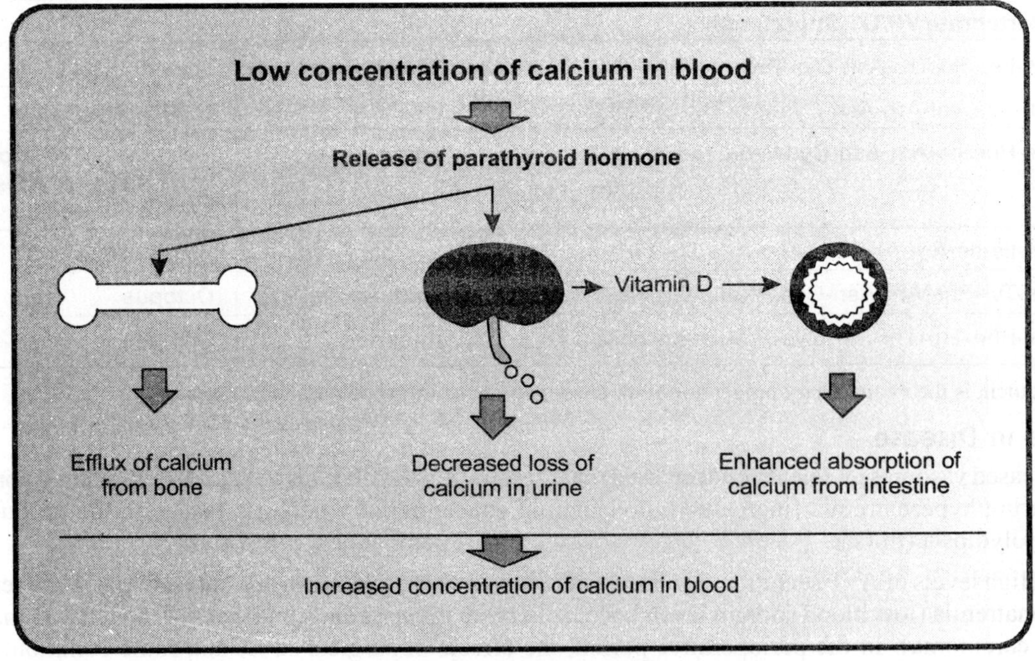

Fig. 12.27 : The Role of PTH in Bone Development

- **Mobilization of calcium from bone**: Although the mechanisms remain obscure, a well-documented effect of parathyroid hormone is to stimulate osteoclasts to reabsorb bone mineral, liberating calcium into blood.
- **Enhancing absorption of calcium from the small intestine**: Facilitating calcium absorption from the small intestine would clearly serve to elevate blood levels of calcium. Parathyroid hormone stimulates this process, but *indirectly* by stimulating production of the active form of vitamin D in the kidney. Vitamin D induces synthesis of a calcium-binding protein in intestinal epithelial cells that facilitates efficient absorption of calcium into blood.
- **Suppression of calcium loss in urine**: In addition to stimulating fluxes of calcium into blood from bone and intestine, parathyroid hormone puts a brake on excretion of calcium in urine, thus conserving calcium in blood. This effect is mediated by stimulating tubular reabsorption of calcium. Another effect of parathyroid hormone on the kidney is to stimulate loss of phosphate ions in urine.

CONTROL OF PARATHYROID HORMONE SECRETION

Parathyroid hormone is released in response to low extracellular concentrations of free calcium. Changes in blood phosphate concentration can be associated with changes in parathyroid hormone secretion, but this appears to be an indirect effect and phosphate *per se* is not a significant regulator of this hormone. When calcium concentrations fall below the normal range, there is a steep increase in secretion of parathyroid hormone. Low levels of the hormone are secreted even when blood calcium levels are high. Figure 12.28 depicts parathyroid hormone release from cells cultured *in vitro* in differing concentrations of calcium. The parathyroid cell monitors extracellular free calcium concentration via an integral membrane protein that functions as a calcium-sensing receptor.

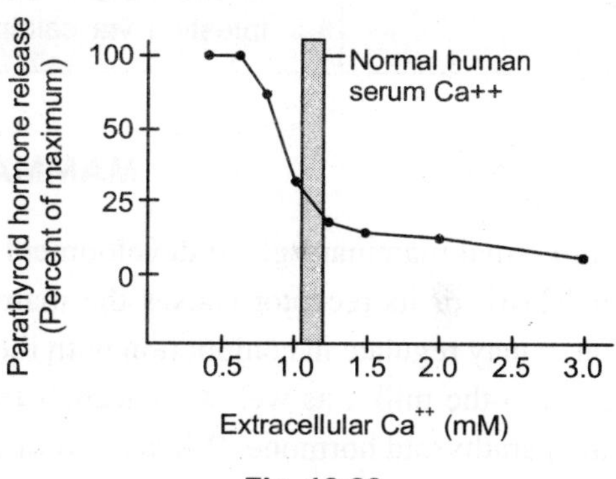

Fig. 12.28.

REGULATION OF SERUM CALCIUM

Parathyroid hormone regulates serum calcium through its effects on the following tissues:

Region	Effect
1	2
bone	It enhances the release of calcium from the large reservoir contained in the bones. Bone resorption is the normal destruction of bone by osteoclasts, which are indirectly stimulated by PTH. Stimulation is indirect since osteoclasts do not have a receptor for PTH; rather, PTH binds to

(Contd...)

1	2
	osteoblasts, the cells responsible for creating bone. Binding stimulates osteoblasts to increase their expression of RANKL, which can bind to osteoclast precursors containing RANK, a receptor for RANKL. The binding of RANKL to RANK stimulates these precursors to fuse, forming new osteoclasts which ultimately enhances bone resorption.
kidney	It enhances active reabsorption of calcium and magnesium from distal tubules and the thick ascending limb. As bone is degraded both calcium and phosphate are released. It also greatly increases the excretion of phosphate, with a net loss in plasma phosphate concentration. By increasing the calcium:phosphate ratio more calcium is therefore free in the circulation.
intestine via kidney	It enhances the absorption of calcium in the intestine by increasing the production of activated vitamin D. Vitamin D activation occurs in the kidney. PTH up-regulates 25-hydroxyvitamin D3 1-alpha-hydroxylase, the enzyme responsible for 1-alpha hydroxylation of 25-hydroxy vitamin D, converting vitamin D to its active form (1,25-dihydroxy vitamin D). This activated form of vitamin D increases the absorption of calcium (as Ca2+ ions) by the intestine via calbindin.

MAMMARY GLANDS

It aids in normal mammary gland development. It is necessary for maintenance of the mammary bud. Loss of PTHrP or its receptor causes the mammary bud cell fate to change back into epidermis. In lactation, it may regulate in conjunction with the calcium sensing receptor the mobilization and transfer of calcium to the milk , as well as placental transfer of calcium. PTHrP is related in function to the "normal" parathyroid hormone. When a tumor secretes PTHrP, this can lead to hypercalcemia. As this is sometimes the first sign of the malignancy, hypercalcemia caused by PTHrP is considered a paraneoplastic phenomenon. PTHR1 is responsible for most cases of humoral hypercalcemia of malignancy. PTHrP shares the same N-terminal end as parathyroid hormone and therefore it can bind to the same receptor, the Type I PTH receptor (PTHR1). PTHrP can simulate most of the actions of PTH including increases in bone resorption and distal tubular calcium reabsorption, and inhibition of proximal tubular phosphate transport.However, PTHrP is less likely than PTH to stimulate 1,25-dihydroxyvitamin D production. Therefore, PTHrP does not increase intestinal calcium absorption.

GONADOTROPIC HORMONES (SEX HORMNES)

The corpus luteum (Latin for "yellow body") (plural corpora lutea) is a temporary endocrine structure in mammals, involved in production of estrogen and progestogen, which is needed to maintain the endometrium. It is colored as a result of concentrating carotenoids from the diet.

DEVELOPMENT AND STRUCTURE

The corpus luteum develops from an ovarian follicle during the luteal phase of the menstrual cycle or estrous cycle, following the release of a secondary oocyte from the follicle during ovulation. The follicle first forms a corpus hemorrhagicum before it becomes a corpus luteum, but the term simply refers to the visible collection of blood left after rupture of the follicle and has no functional significance. While the oocyte (later the zygote) traverses the Fallopian tube into the uterus, the corpus luteum remains in the ovary. Corpus luteum is typically very large relative to the size of the ovary; in humans, the size of the structure ranges from under 2 cm to 5 cm in diameter. Its cells develop from the follicular cells surrounding the ovarian follicle.

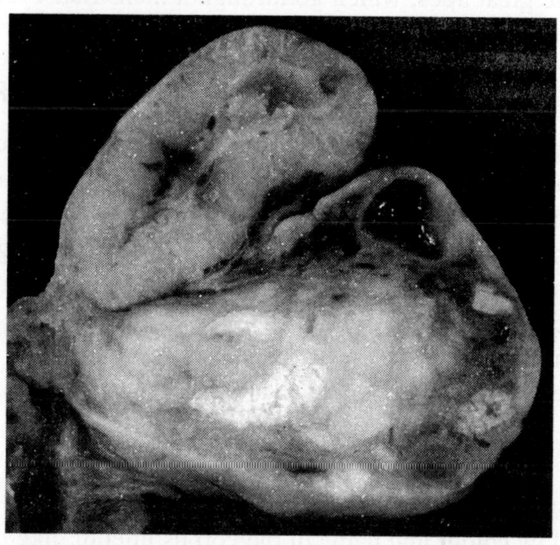

Fig. 12.29 : The Human Overy showing the Corpus Luteum

Follicular structure	Luteal structure	Secretion
Theca cells	Theca lutein cells	androgens and progesterone
Granulosa cells	Granulosa lutein cells	Progesterone, Estrogen

The previous theca cells lacked the aromatase enzyme that is necessary to produce estrogen, so the production of estrogen in theca lutein cells indicate presence of aromatase.

Once the corpus luteum regressed the remnants are known as the corpora albicantia.

FUNCTION

The corpus luteum is essential for establishing and maintaining pregnancy in females.In the ovary, the corpus luteum secretes estrogens and progesterone, which are steroid hormones responsible for the thickening of the endometrium and its development and maintenance, respectively. These hormones also inhibit Luteinizing hormone and Follicle-stimulating hormone.

WHEN EGG IS NOT FERTILIZED

If the egg is not fertilized, the corpus luteum stops secreting progesterone and decays (after approximately 14 days in humans). It then degenerates into a corpus albicans, which is a mass of fibrous scar tissue.

The uterine lining sloughs off without progesterone and is expelled through the vagina (in humans and some great apes, which go through a menstrual cycle). In an estrus cycle, the lining degenerates back to normal size.

WHEN EGG IS FERTILIZED

If the egg is fertilized and implantation occurs, the trophoblast cells of the blastocyst secrete the hormone human chorionic gonadotropin (hCG, or a similar hormone in other species). Human chorionic gonadotropin signals the corpus luteum to continue progesterone secretion, thereby maintaining the thick lining (endometrium) of the uterus and providing an area rich in blood vessels in which the zygote(s) can develop. From this point on, the corpus luteum is called the *corpus luteum graviditatis*.The introduction of prostaglandins at this point causes the degeneration of the corpus luteum and the abortion of the fetus. However, in placental animals such as humans, the placenta eventually takes over progesterone production and the corpus luteum degrades into a corpus albicans without embryo/foetus loss.

CONTENT OF CAROTENOIDS

The yellow colour and name of the corpus luteum, like that of the macula lutea of the retina, is due to its concentration of certain carotenoids. Following a 1968 report that beta-carotene was synthesized in laboratory conditions in slices of corpus luteum from cows, an organ known to concentrates beta-carotene (hence its color and name), attempts have been made to replicate these findings, but have not succeeded. The idea is not presently accepted by the scientific community. Rather, the corpus luteum concentrates carotenoids from the diet of the mammal.

THE IMPORTANT FEMALE SEX HORMONES

In this section we are going to discuss the most important female sex hormones the estrogen and Progestrogen with their activities in the female human body and synthesis.

THE ESTROGEN HORMONE

Estrogens (AmE), **oestrogens** (BE), or **œstrogens**, are a group of steroid compounds, named for their importance in the estrous cycle, and functioning as the primary female sex hormone, their name comes from estrus/oistros (period of fertility for female mammals) + gen/gonos = to generate.

Estrogens are synthesized in all vertebrates as well as some insects. The presence of these steroids in both vertebrate and insects suggests that estrogenic sex hormones have an ancient evolutionary history.

Estrogens are used as part of some oral contraceptives, in estrogen replacement therapy for postmenopausal women, and in hormone replacement therapy for trans women.

Like all steroid hormones, estrogens readily diffuse across the cell membrane. Once inside the cell, they bind to and activate estrogen receptors which in turn up-regulate the expression of many genes. Additionally, estrogens have been shown to activate a G protein-coupled receptor, GPR30.

HISTORY

In 1929 Adolf Butenandt and Edward Adelbert Doisy independently isolated and determined the structure of estrogen. Thereafter, the market for hormonal drug research opened up.

The "first orally effective estrogen", Emmenin, derived from the late-pregnancy urine of Canadian women, was introduced in 1930 by Collip and Ayerst Laboratories. Estrogens are not water-soluble and cannot be given orally, but the urine was found to contain estriol glucuronide which is water soluble and becomes active in the body after hydrolization.

Scientists continued to search for new sources of estrogen because of concerns associated with the practicality of introducing the drug into the market. At the same time, a German pharmaceutical drug company, formulated a similar product as Emmenin that was introduced to German women to treat menopausal symptoms.

In 1938, British scientists obtained a patent on a newly formulated nonsteroidal estrogen, diethylstilbestrol (DES), that was cheaper and more powerful than the previously manufactured estrogens. Soon after, concerns over the side effects of DES were raised in scientific journals while the drug manufacturers came together to lobby for governmental approval of DES. It was only until 1941 when estrogen therapy was finally approved by the Food and Drug Administration (FDA) for the treatment of menopausal symptoms

TYPES STEROIDAL

The three major naturally occurring estrogens in women are estrone (E1), estradiol (E2), and estriol (E3). Estradiol (E2) is the predominant form in nonpregnant females, estrone is produced during menopause, and estriol is the primary estrogen of pregnancy. In the body these are all produced from androgens through actions of enzymes.

- From menarche to menopause the primary estrogen is 17â-estradiol. In postmenopausal women more estrone is present than estradiol.
- Estradiol is produced from testosterone and estrone from androstenedione by aromatase.
- Estrone is weaker than estradiol.

Premarin, a commonly prescribed estrogenic drug, contains the steroidal estrogens equilin and equilenin, in addition to estrone sulfate but due to its health risk, more genetic estrogen named Progynova (estradiol valerate) are now more often prescribed.

NON-STEROIDAL

A range of synthetic and natural substances have been identified that also possess estrogenic activity.

- Synthetic substances of this kind are known as xenoestrogens.
- Plant products with estrogenic activity are called phytoestrogens.
- Those produced by fungi are known as mycoestrogens.

Unlike estrogens produced by mammals, these substances are not necessarily steroids.

BIOSYNTHESIS

Estrogens are produced primarily by developing follicles in the ovaries, the corpus luteum, and the placenta. Luteinizing hormone (LH) stimulates the production of estrogen in the ovaries. Some estrogens are also produced in smaller amounts by other tissues such as the liver, adrenal glands, and the breasts. These secondary sources of estrogens are especially important in postmenopausal women. Fat cells also produce estrogen, potentially being the reason why underweight or overweight are risk factors for infertility.

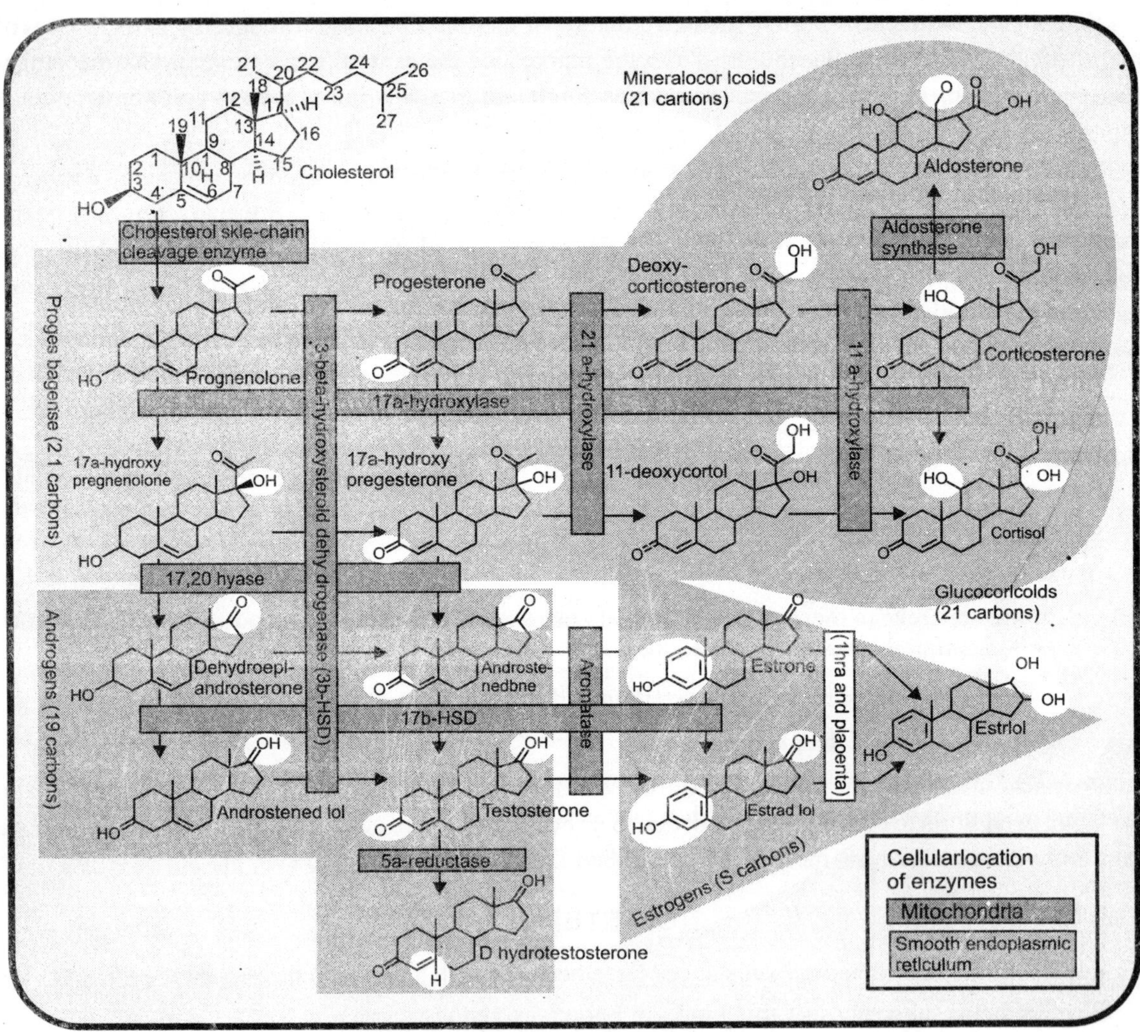

Fig. 12.30 : The Process of Steroidogenesis, showing Estrogens at Bottom Right as Triangle

In females, synthesis of estrogens starts in theca interna cells in the ovary, by the synthesis of androstenedione from cholesterol. Androstenedione is a substance of moderate androgenic activity. This compound crosses the basal membrane into the surrounding granulosa cells, where it is converted to estrone or estradiol, either immediately or through testosterone. The conversion of testosterone to estradiol, and of androstenedione to estrone, is catalyzed by the enzyme aromatase.

Estradiol levels vary through the menstrual cycle, with levels highest just before ovulation.

FUNCTION

While estrogens are present in both men and women, they are usually present at significantly higher levels in women of reproductive age. They promote the development of female secondary sexual characteristics, such as breasts, and are also involved in the thickening of the endometrium and other aspects of regulating the menstrual cycle. In males, estrogen regulates certain functions of the reproductive system important to the maturation of sperm and may be necessary for a healthy libido.[11][12] Furthermore, there are several other structural changes induced by estrogen in addition to other functions. In dentistry, it reduces hyperkeratinization of the gingiva and increase vascular permeability, exudation, and edema.

- Structural
  - promote formation of female secondary sex characteristics
  - accelerate metabolism (burn fat)
  - reduce muscle mass
  - stimulate endometrial growth
  - increase uterine growth
  - increase vaginal lubrication
  - thicken the vaginal wall
  - maintenance of vessel and skin
  - reduce bone resorption, increase bone formation
  - morphic change (endomorphic -> mesomorphic -> ectomorphic)
- protein synthesis
  - increase hepatic production of binding proteins
- coagulation
  - increase circulating level of factors 2, 7, 9, 10, plasminogen
  - decrease antithrombin III
  - increase platelet adhesiveness
- Lipid
  - increase HDL, triglyceride
  - decrease LDL, fat deposition

- Fluid balance
  - salt (sodium) and water retention
  - increase cortisol, SHBG
- Gastrointestinal tract
  - reduce bowel motility
  - increase cholesterol in bile
- Melanin
  - increase pheomelanin, reduce eumelanin
- Cancer
  - support hormone-sensitive breast cancers (see section below)
- Lung function
  - promotes lung function by supporting alveoli (in rodents but probably in humans).

Sexual desire is dependent on androgen levels rather than estrogen levels. In mice, estrogens (which are locally aromatized from androgens in the brain) play an important role in psychosexual differentiation, for example, by masculinizing territorial behavior, same is not true in human In humans, the masculinizing effects of prenatal androgens on behavior (and other tissues, with the possible exception of effects on bone) appear to act exclusively through the androgen receptor. As a result, the utility of rodent models for studying human psychosexual differentiation has been questioned. Estrogen is considered to play a significant role in women's mental health. Sudden estrogen withdrawal, fluctuating estrogen, and periods of sustained estrogen low levels correlates with significant mood lowering. Clinical recovery from postpartum, perimenopause, and postmenopause depression has been shown to be effective after levels of estrogen were stabilized and/or restored.

THE ESTROGEN DEFICIENCY

Low estrogen levels in male lab mice may be one cause of obsessive–compulsive disorder (OCD). When estrogen levels were raised through the increased activity of the enzyme aromatase in male lab mice, OCD rituals were dramatically decreased. Hypothalamic protein levels in the gene COMT are enhanced by increasing estrogen levels which is believed to return mice that displayed OCD rituals to normal activity. Aromatase deficiency is ultimately suspected which is involved in the synthesis of estrogen in humans and has therapeutic implications in humans having obsessive-compulsive disorder.

PROGESTERONE

Willard Myron Allen co-discovered progesterone with his anatomy professor George Washington Corner at the University of Rochester Medical School in 1933. Allen first determined its melting point, molecular weight, and partial molecular structure. He also gave it the name **Progesterone** derived from **Prog**estational **Ster**oidal k**etone**.

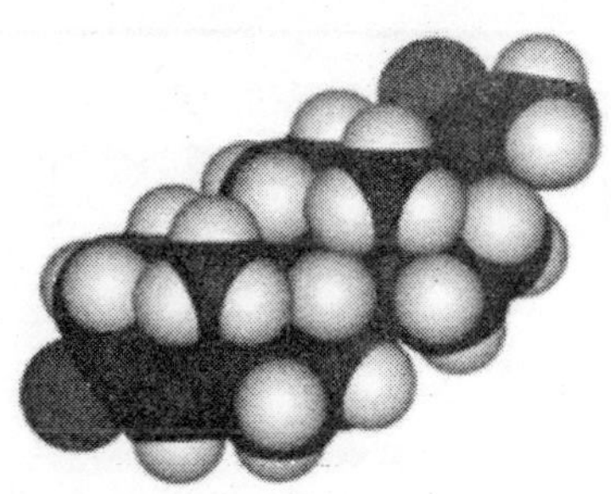

Progesterone also known as **P4** (pregn-**4**-ene-3,20-dione) is a C-21 steroid hormone involved in the female menstrual cycle, pregnancy (supports *gestation*) and embryogenesis of humans and other species. Progesterone belongs to a class of hormones called progestogens, and is the major naturally occurring human progestogen.Progesterone is commonly manufactured from the yam family, Dioscorea. Dioscorea produces large amounts of a steroid called diosgenin, which can be converted into progesterone in the laboratory.

CHEMISTRY OF PROGESTERONE

Like other steroids, progesterone consists of four interconnected cyclic hydrocarbons. Progesterone contains ketone and oxygenated functional groups, as well as two methyl branches. Like all steroid hormones, it is hydrophobic

SOURCCES OF PROGESTERONE

Progesterone is produced in the ovaries (to be specific, after ovulation in the corpus luteum), the adrenal glands (near the kidney), and, during pregnancy, in the placenta. Progesterone is also stored in adipose (fat) tissue.

In humans, increasing amounts of progesterone are produced during pregnancy:

- At first, the source is the corpus luteum that has been "rescued" by the presence of human chorionic gonadotropins (hCG) from the conceptus.
- However, after the 8th week, production of progesterone shifts to the placenta. The placenta utilizes maternal cholesterol as the initial substrate, and most of the produced progesterone enters the maternal circulation, but some is picked up by the fetal circulation and used as substrate for fetal corticosteroids. At term the placenta produces about 250 mg progesterone per day.
- An additional source of progesterone is milk products. They contain much progesterone because on dairy farms cows are milked during pregnancy, when the progesterone content of the milk is high. After consumption of milk products the level of bioavailable progesterone goes up.

BIOSYNTHESIS

Top: Conversion of cholesterol (**1**) into pregnenolone (**3**) to progesterone (**6**).

Bottom: Progesterone is important for aldosterone (mineralocorticoid) synthesis, as 17-hydroxyprogesterone is for cortisol (glucocorticoid), and androstenedione for sex steroids.

In mammals, progesterone (**6**), like all other steroid hormones, is synthesized from pregnenolone (**3**), which in turn is derived from cholesterol (**1**).

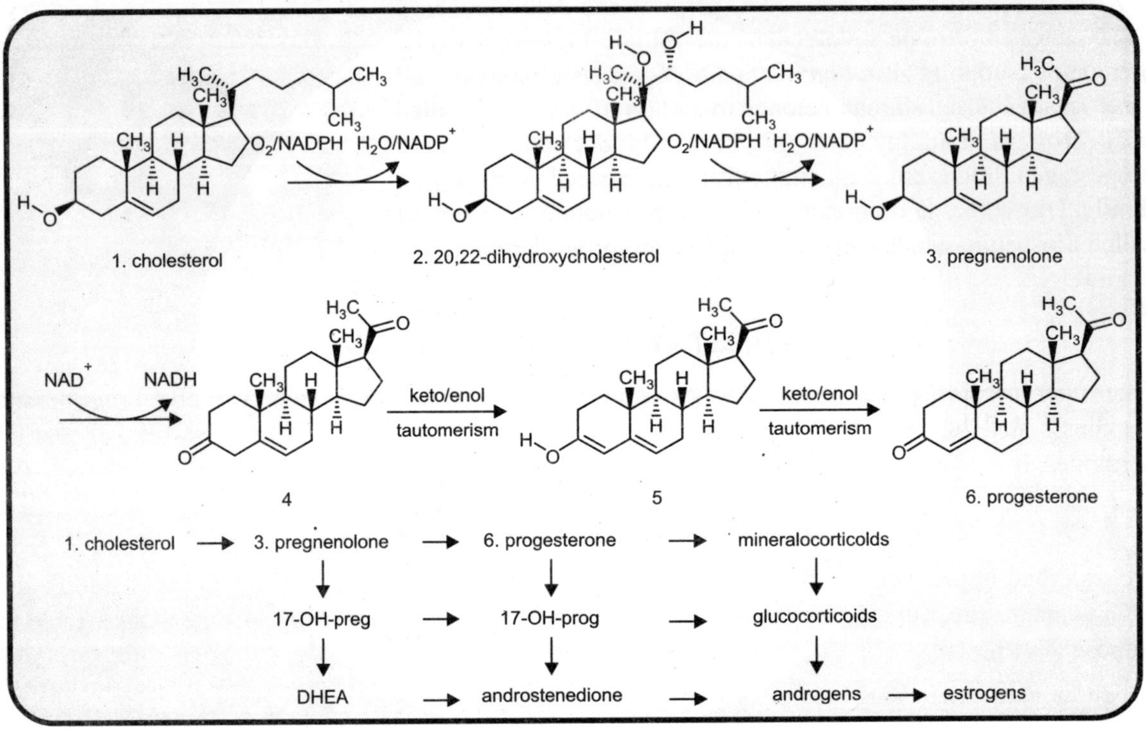

1. cholesterol → 2. 20,22-dihydroxycholesterol → 3. pregnenolone (O_2/NADPH, H_2O/$NADP^+$)

NAD^+ → NADH; 4 → keto/enol tautomerism → 5 → keto/enol tautomerism → 6. progesterone

1. cholesterol → 3. pregnenolone → 6. progesterone → mineralocorticolds

17-OH-preg → 17-OH-prog → glucocorticoids

DHEA → androstenedione → androgens → estrogens

Fig. 12.31 : The Biosynthesis of Progesterone

Cholesterol (**1**) undergoes double oxidation to produce 20,22-dihydroxycholesterol (**2**). This vicinal diol is then further oxidized with loss of the side chain starting at position C-22 to produce pregnenolone (**3**). This reaction is catalyzed by cytochrome P450scc. The conversion of pregnenolone to progesterone takes place in two steps. First, the 3-hydroxyl group is oxidized to a keto group (**4**) and second, the double bond is moved to C-4, from C-5 through a keto/enol tautomerization reaction. This reaction is catalyzed by 3beta-hydroxysteroid dehydrogenase/delta(5)-delta(4)isomerase.Progesterone in turn (see lower half of figure to the right) is the precursor of the mineralocorticoid aldosterone, and after conversion to 17-hydroxyprogesterone (another natural progestogen) of cortisol and androstenedione. Androstenedione can be converted to testosterone, estrone and estradiol.Pregenolone and progesterone can also be synthesized by yeast.

An economical semisynthesis of progesterone from the plant steroid diosgenin isolated from yams was developed by Russell Marker in 1940 for the Parke-Davis pharmaceutical company (Fig. 12.32). This synthesis is known as the Marker degradation. Additional semisyntheses of progesterone have also been reported starting from a variety of steroids. For the example, cortisone can be simultaneously deoxygenated at the C-17 and C-21 position by treatment with iodotrimethylsilane in chloroform to produce 11-keto-progesterone (ketogestin), which in turn can be reduced at position-11 to yield progesterone.

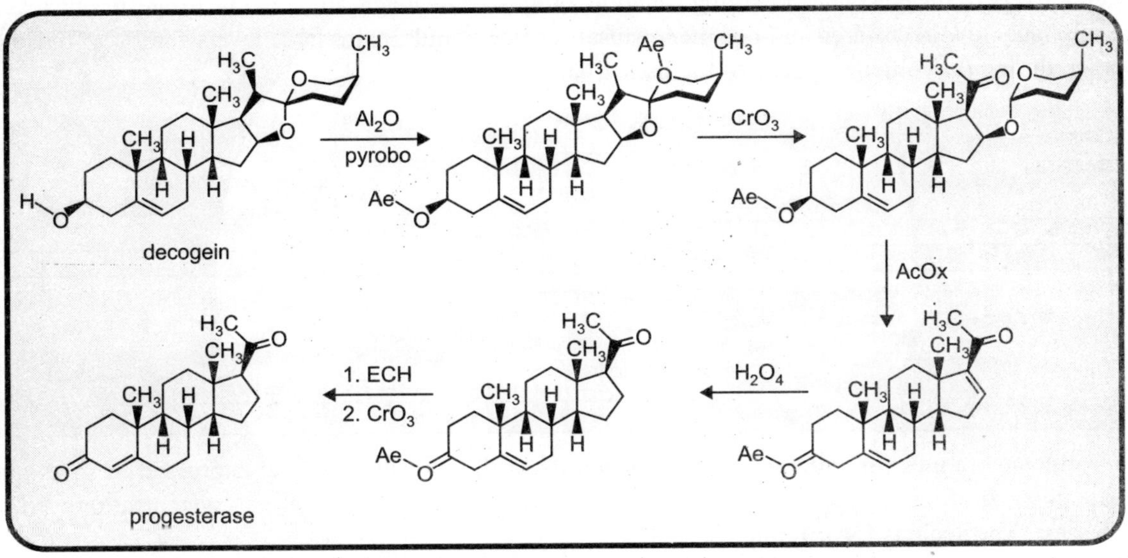

Fig. 12.32 : The Marker Semi-synthesis of Progesterone from Diosgenin

SUB-SECTION 12.24 L—LEVELS

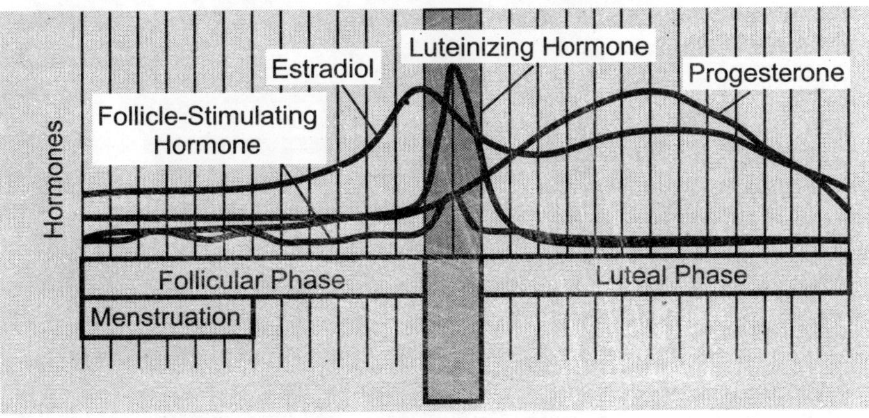

Fig. 12.33 : Progesterone Levels (black line) during the Menstrual Cycle

In women, progesterone levels are relatively low during the preovulatory phase of the menstrual cycle, rise after ovulation, and are elevated during the luteal phase. Progesterone levels tend to be < 2 ng/ml prior to ovulation, and > 5 ng/ml after ovulation. If pregnancy occurs, progesterone levels are initially maintained at luteal levels. With the onset of the luteal-placental shift in progesterone support of the pregnancy, levels start to rise further and may reach 100-200 ng/ml. at term. Whether a decrease in progesterone levels is critical for the initiation of labor has been argued and may be species-specific.

After delivery of the placenta and during lactation, progesterone levels are very low.Progesterone levels are relatively low in children and postmenopausal women. Adult males have levels similar to those in women during the follicular phase of the menstrual cycle.

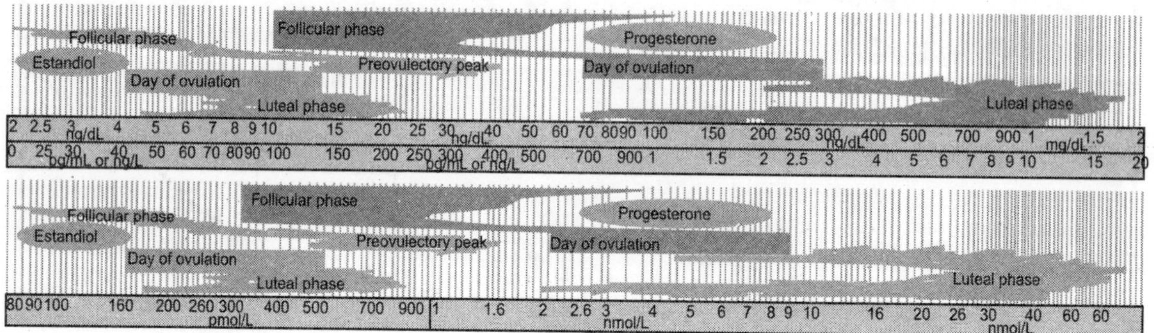

Reference ranges for estradiol and *progesterone* in the menstrual cycle, expressed in mass and molar concentration. Because of slightly different molar mass, the relative concentrations differ somewhat. The scale is logarithmic.

FUNCTION

Progesterone exerts its primary action through the intracellular progesterone receptor although a distinct, membrane bound progesterone receptor has also been postulated. In addition, progesterone is a highly potent antagonist of the mineralocorticoid receptor (MR, the receptor for aldosterone and other mineralo-corticosteroids). It prevents MR activation by binding to this receptor with an affinity exceeding even those of aldosterone and other corticosteroids such as cortisol and corticosterone. Progesterone has a number of physiological effects that are amplified in the presence of estrogen. Estrogen through estrogen receptors upregulates the expression of progesterone receptors. Also, elevated levels of progesterone potently reduce the sodium-retaining activity of aldosterone, resulting in natriuresis and a reduction in extracellular fluid volume. Progesterone withdrawal, on the other hand, is associated with a temporary increase in sodium retention

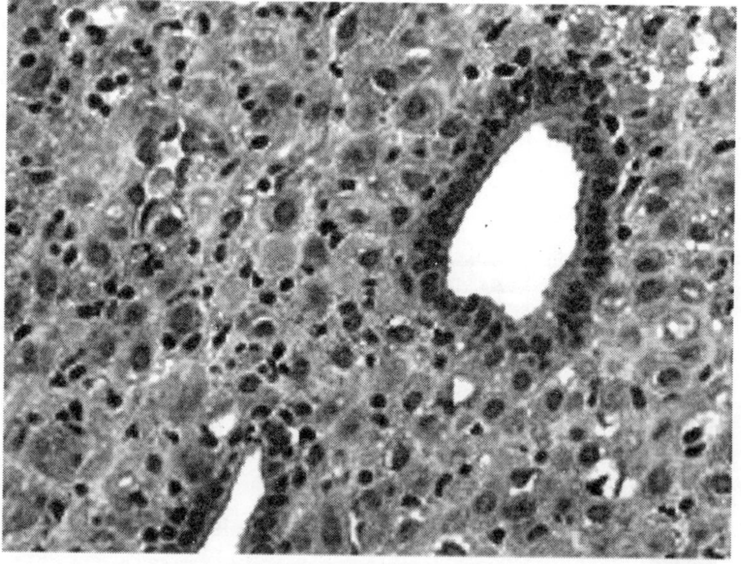

Fig. 12.34 : Micrograph showing Changes to the Endometrium due to Progesterone (Decidualization) H&E stain (1500x)

(reduced natriuresis, with an increase in extracellular fluid volume) due to the compensatory increase in aldosterone production, which combats the blockade of the mineralocorticoid receptor by the previously elevated level of progesterone.Progesterone has key effects via non-genomic signalling on human sperm as they migrate through the female tract before fertilisation occurs, though the receptor(s) as yet remain unidentified. Detailed characterisation of the events occurring in sperm in response to progesterone has elucidated certain events including intracellular calcium transients and maintained changes, slow calcium oscillations, now thought to possibly regulate motility. Interstingly progesterone has also been shown to demonstrate effects on octopus spermatozoa.

Progesterone is sometimes called the "hormone of pregnancy", and it has many roles relating to the development of the fetus:

- Progesterone converts the endometrium to its secretory stage to prepare the uterus for implantation. At the same time progesterone affects the vaginal epithelium and cervical mucus, making it thick and impermeable to sperm. If pregnancy does not occur, progesterone levels will decrease, leading, in the human, to menstruation. Normal menstrual bleeding is progesterone-withdrawal bleeding.
- During implantation and gestation, progesterone appears to decrease the maternal immune response to allow for the acceptance of the pregnancy.
- Progesterone decreases contractility of the uterine smooth muscle.
- In addition progesterone inhibits lactation during pregnancy. The fall in progesterone levels following delivery is one of the triggers for milk production.
- A drop in progesterone levels is possibly one step that facilitates the onset of labor.

The fetus metabolizes placental progesterone in the production of adrenal steroids.

Progesterone, like pregnenolone and dehydroepiandrosterone, belongs to the group of neurosteroids. It can be synthesized within the central nervous system and also serves as a precursor to another major neurosteroid, allopregnanolone.

Neurosteroids affect synaptic functioning, are neuroprotective, and affect myelination. They are investigated for their potential to improve memory and cognitive ability. Progesterone affects regulation of apoptotic genes.

Its effect as a neurosteroid works predominantly through the GSK-3 beta pathway, as an inhibitor. (Other GSK-3 beta inhibitors include bipolar mood stabilizers, lithium and valproic acid.)

- It raises epidermal growth factor-1 levels, a factor often used to induce proliferation, and used to sustain cultures, of stem cells.
- It increases core temperature (thermogenic function) during ovulation.
- It reduces spasm and relaxes smooth muscle. Bronchi are widened and mucus regulated. (Progesterone receptors are widely present in submucosal tissue.)
- It acts as an antiinflammatory agent and regulates the immune response.
- It reduces gall-bladder activity.

- It normalizes blood clotting and vascular tone, zinc and copper levels, cell oxygen levels, and use of fat stores for energy.
- It may affect gum health, increasing risk of gingivitis (gum inflammation) and tooth decay.
- It appears to prevent endometrial cancer (involving the uterine lining) by regulating the effects of estrogen.

Pill form of progesterone (actually a synthetic version such as Progestogen) taken at 400 mg as cited by the following patent can cause increased fluid retention, which may result in epilepsy, migraine, asthma, cardiac or renal dysfunction. Blood clots that can result in strokes and heart attacks, which may lead to death or long-term disability, may develop; pulmonary embolus or breast cancer can also develop as a result of progesterone therapy. Progesterone is associated with an increased risk of thrombotic disorders such as thrombophlebitis, cerebrovascular disorders, pulmonary embolism, and retinal thrombosis.

Common adverse effects include cramps, abdominal pain, skeletal pain, perineal pain, headache, arthralgia, constipation, dyspareunia, nocturia, diarrhea, nausea, vomiting, breast enlargement, joint pain, flatulence, hot flushes, decreased libido, thirst, increased appetite, nervousness, drowsiness, excessive urination at night. Psychiatric effects including depression, mood swings, emotional instability, aggression, abnormal crying, insomnia, forgetfulness, sleep disorders.

Less frequent adverse effects that may occur include allergy, anemia, bloating, fatigue, tremor, urticaria, pain, conjunctivitis, dizziness, vomiting, myalgia, back pain, breast pain, genital itching, genital yeast infection, upper respiratory tract infection, cystitis, dysuria, asthenia, xerophthalmia, syncope, dysmenorrhea, premenstrual tension, gastritis, urinary tract infection, vaginal discharge, pharyngitis, sweating, hyperventilation, vaginal dryness, dyspnea, fever, edema, flu-like symptoms, dry mouth, rhinitis, leg pain, skin discoloration, skin disorders, seborrhea, sinusitis, acne. Current research suggests that progesterone plays an important role in the signaling of insulin release and pancreatic function, and may affect the susceptibility to diabetes. It has been shown that women with high levels of progesterone during pregnancy are more likely to develop glucose abnormalities.

A total synthesis of progesterone was reported in 1971 by W.S. Johnson (see Fig. 12.35 to the right). The synthesis begins with reacting the phosphonium salt 7 with phenyl lithium to produce the phosphonium ylide **8**. The ylide **8** is reacted with an aldehyde to produce the alkene **9**. The ketal protecting groups of **9** are hydrolyzed to produce the diketone **10**, which in turn is cyclized to from the cyclopentenone **11**. The ketone of **11** is reacted with methyl lithium to yield the tertiary alcohol **12**, which in turn is treated with acid to produce the tertiary cation **13**. The key step of the synthesis is the π-cation cyclization of **13** in which the B-, C-, and D-rings of the steroid are simultaneously formed to produce **14**. This step resembles the cationic cyclization reaction used in the biosynthesis of steroids and hence is referred to as *biomimetic*. In the next step the enol orthoester is hydrolyzed to produce the ketone **15**. The cyclopentene A-ring is then opened by oxidizing with ozone to produce **16**. Finally, the diketone **17** undergoes an intramolecular aldol condensation by treating with aqueous potassium hydroxide to produce progesterone.

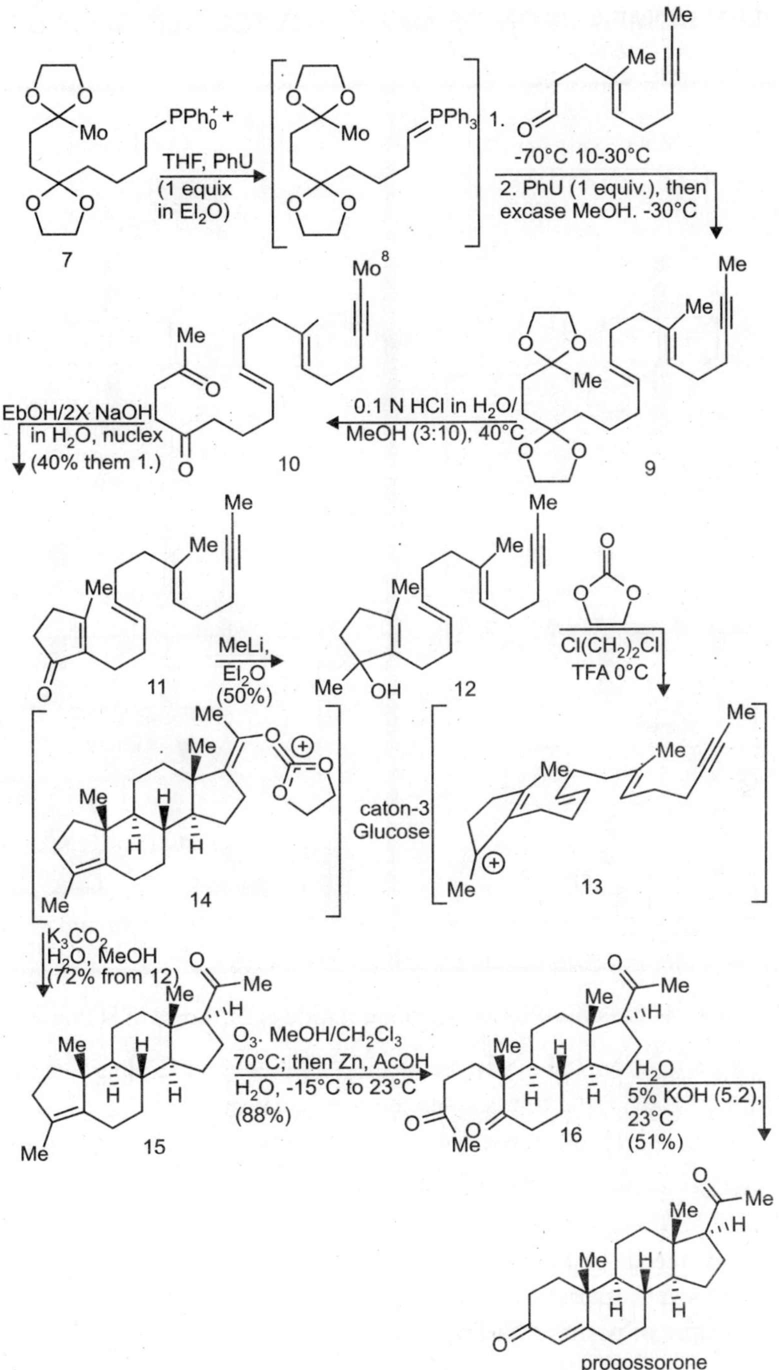

Fig. 12.35 : The Johnson Total Synthesis of Progesterone

THE COMPARISON OF MALE AND FEMALE GONADS

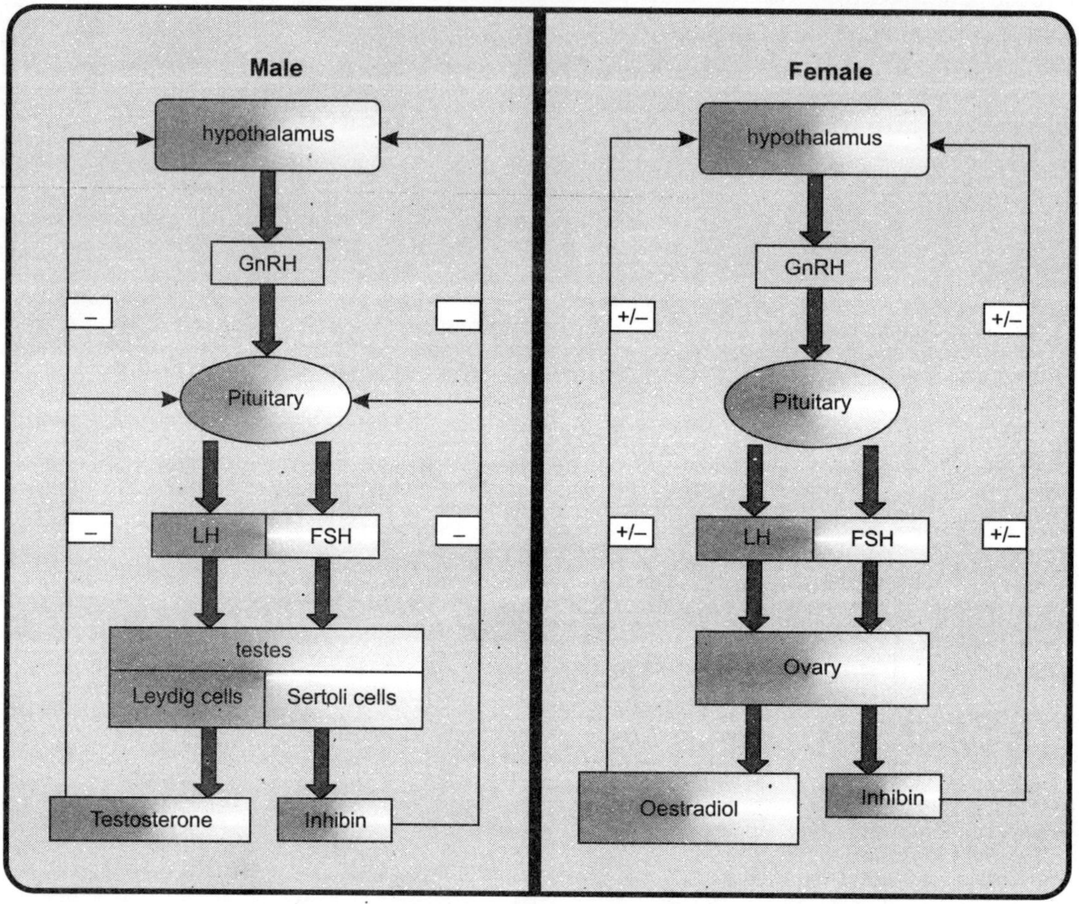

Fig. 12.36 : The Hypothalamo-Pituitary-Gonadal Axis is Different in Males and Females

In females GnRH is secreted from the hypothalamus in a cyclical way leading to a cyclical secretion of LH and FSH from the pituitary, which maintains the menstrual cycle.

- LH acts on the ovarian follicle and it induces ovulation and maintains the corpus luteum.
- FSH causes development of the ovarian follicle and stimulates secretion of oestradiol and progesterone.
- The sex steroids feed back to inhibit release of GnRH and therefore LH and FSH. At sustained high levels however, oestradiol causes a sharp increase in LH secretion linked to ovulation. This is an example of positive feedback.

In males GnRH causes the release of LH and FSH from the anterior pituitary, as in females.

- LH acts on the Leydig cells of the testes to produce testosterone.

- FSH acts on the Sertoli cells of the testes to maintain spermatogenesis as well as production of sex-hormone binding globulin.

In males and females FSH stimulates the production of inhibin, which has a negative feedback effect on the hypothalamus and pituitary.

What can go Wrong

Insufficient GnRH causes a fall in gonadotrophin production, which leads to amenorrhoea, due to lack of ovulation, in women and impotence and infertility in men. This is sometimes seen when nutrient intake is too low, as in anorexia nervosa, or when excessive exercise is undertaken. It may also be due to a prolactin secreting tumour because prolactin has effects on the hypothalamo-pituitary-gonadal axis. The problems caused by lack of GnRH may be treated by administration of GnRH, gonadotrophins such as recombinant FSH, or sex steroids.

REGULATION OF ESTROGEN AND PROGESTERONE

The synthesis and secretion of **estrogens** is stimulated by follicle-stimulating hormone (FSH), which is, in turn, controlled by the hypothalamic gonadotropin releasing hormone (GnRH).

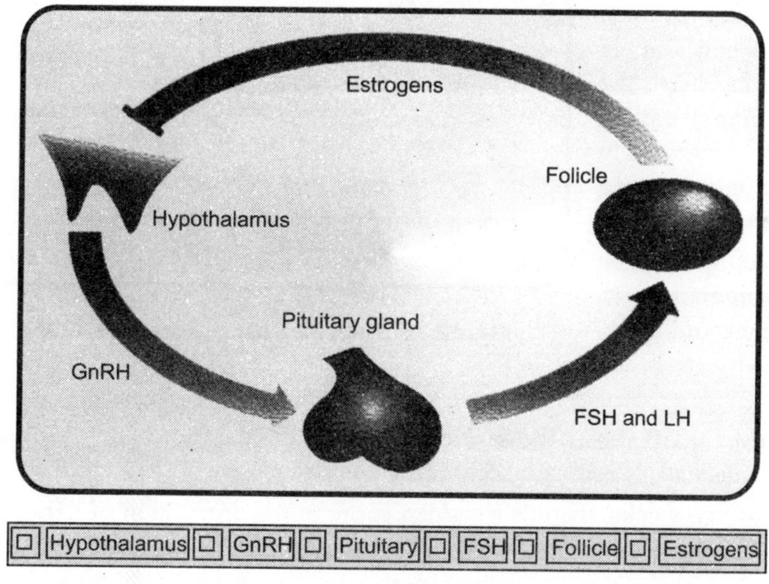

Fig. 12.37 : The Regulation of Estrogen Progesterone Cycle

High levels of estrogens **suppress** the release of GnRH (bar) providing a negative-feedback control of hormone levels.

It works like this: Secretion of GnRH depends on certain neurons in the hypothalamus which express a gene (*KiSS 1*) encoding a protein of 145 amino acids. From this are cut several short peptides collectively called **kisspeptin**. These are secreted and bind to G-protein-coupled receptors on the surface of the GnRH neurons stimulating them to release GnRH. However, high levels of estrogen (or progesterone or testosterone) inhibit the secretion of kisspeptin and suppress further production of those hormones.

Progesterone production is stimulated by luteinizing hormone (LH), which is also stimulated by GnRH.

Elevated levels of progesterone control themselves by the same negative feedback loop used by estrogen (and testosterone).

THE MENSTRUAL CYCLE

About every 28 days, some blood and other products of the disintegration of the inner lining of the uterus (the **endometrium**) are discharged from the uterus, a process called **menstruation**. During this time a new follicle begins to develop in one of the ovaries. After menstruation ceases, the follicle continues to develop, secreting an increasing amount of **estrogen** as it does so.

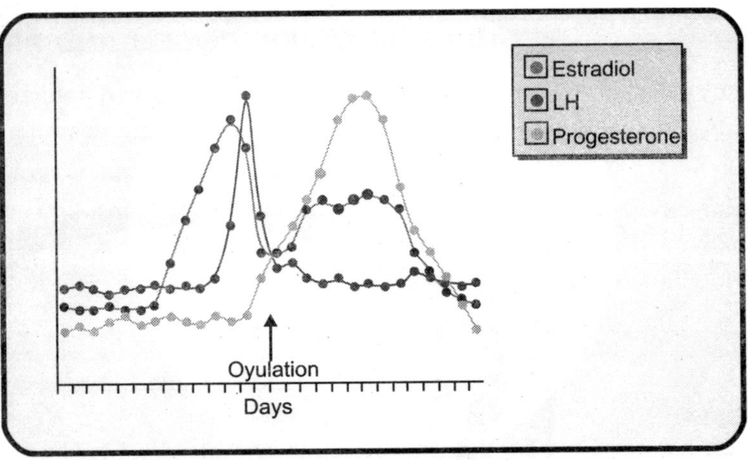

Fig. 12.38: The Level of Pogesterone during Ovulation

The rising level of estrogen causes the endometrium to become thicker and more richly supplied with blood vessels and glands.

- A rising level of **LH** causes the developing egg within the follicle to complete the first meiotic division (meiosis I), forming a **secondary oocyte**.
- After about two weeks, there is a sudden surge in the production of LH.
- This surge in LH triggers **ovulation**: the release of the secondary oocyte into the fallopian tube.
- Under the continued influence of LH, the now-empty follicle develops into a **corpus luteum** (hence the name luteinizing hormone for LH).
- Stimulated by LH, the corpus luteum secretes **progesterone** which
  - continues the preparation of the endometrium for a possible pregnancy
  - inhibits the contraction of the uterus
  - inhibits the development of a new follicle

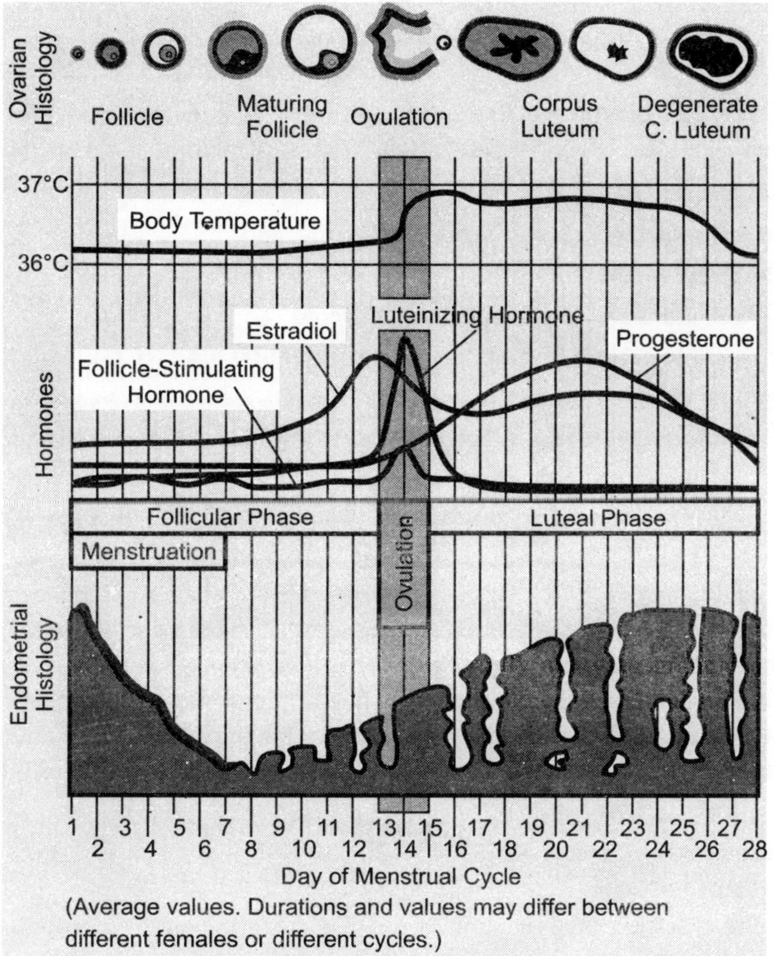

Fig. 12.39: The Menstrural Cycle

- If fertilization does **not** occur (which is usually the case),
  - the rising level of progesterone inhibits the release of GnRH which, in turn,
  - inhibits further production of progesterone.
- As the progesterone level drops,
  - the corpus luteum begins to degenerate;
  - the endometrium begins to break down, its cells committing programmed cell death (apoptosis);
  - the inhibition of uterine contraction is lifted, and
  - the bleeding and cramps of menstruation begin.

PREGNANCY

Fertilization of the egg takes place within the fallopian tube. As the fertilized egg passes down the tube, it undergoes its first mitotic divisions. By the end of the week, the developing embryo has become a hollow ball of cells called a **blastocyst**. At this time, the blastocyst reaches the uterus and embeds itself in the endometrium, a process called **implantation**. With implantation, pregnancy is established.

The blastocyst has two parts:

- the **inner cell mass**, which will become the baby, and
- the **trophoblast**, which will
  - develop into the **placenta** and umbilical cord
  - and begin to secrete **human chorionic gonadotropin** (**HCG**).

HCG is a glycoprotein. It is a dimer of

- the same alpha subunit (of 92 amino acids) used by TSH, FSH, and LH and
- a unique beta subunit (of 145 amino acids).

HCG behaves much like FSH and LH with one crucial exception: it is **NOT** inhibited by a rising level of progesterone. Thus HCG prevents the deterioration of the corpus luteum at the end of the fourth week and enables pregnancy to continue beyond the end of the normal menstrual cycle. Because only the implanted trophoblast makes HCG, its early appearance in the urine of pregnant women provides the basis for the most widely used test for pregnancy (which can provide a positive signal even before menstruation would have otherwise begun). As pregnancy continues, the placenta becomes a major source of **progesterone**, and its presence is essential to maintain pregnancy. Mothers at risk of giving birth too soon can be given a synthetic progestin to help them retain the fetus until it is full-term.

BIRTH

Toward the end of pregnancy,

- The **placenta** releases large amounts of CRH which stimulates the **pituitary glands** of both mother and her fetus to secrete
- ACTH, which acts on their adrenal glands causing them to release the estrogen precursor **dehydroepiandrosterone sulfate** (**DHEAS**).
- This is converted into **estrogen** by the placenta.
- The rising level of estrogen causes the smooth muscle cells of the uterus to
  - synthesize **connexins** and form gap junctions. Gap junctions connect the cells electrically so that they contract together as labor begins.
  - express receptors for **oxytocin**.
- **Oxytocin** is secreted by the posterior lobe of the pituitary as well as by the uterus.
- **Prostaglandins** are synthesized in the placenta and uterus.
- The normal inhibition of uterine contraction by progesterone is turned off by several mechanisms while.

- both oxytocin and prostaglandins cause the uterus to contract and labor begins.

Three or four days after the baby is born, the breasts begin to secrete milk.

- Milk synthesis is stimulated by the pituitary hormone **prolactin (PRL)**, and
- its release from the breast is stimulated by **oxytocin**.
- Milk contains an **inhibitory peptide**. If the breasts are not fully emptied, the peptide accumulates and inhibits milk production. This autocrine action thus matches supply with demand.

OTHER HORMONES

- **Relaxin**

  As the time of birth approaches in some animals (e.g., pigs, rats) , this polypeptide has been found to:

  - relax the pubic ligaments
  - soften and enlarge the opening to the cervix.

  Relaxin is found in pregnant humans but at higher levels early in pregnancy than close to the time of birth. Relaxin promotes angiogenesis, and in humans it probably plays a more important role in the development of the interface between the uterus and the placenta that it does in the birth process.

- **Activins, Inhibins, Follistatin**.

  These proteins are synthesized within the follicle. Activins and inhibins bind to follistatin. Activins increase the action of FSH; inhibins, as their name suggests, inhibit it. How important they are in humans remains to be seen. However the important role that activin and follistatin play in the embryonic development of vertebrates justifies mentioning them here.

RU-486

RU-486 (also known as mifepristone) is a synthetic steroid related to **progesterone**. Unlike the synthetic progestins used in oral contraceptives that mimic the actions of progesterone, RU-486 is a progesterone **antagonist**; that is, it blocks the action of progesterone. It does this by binding more tightly to the **progesterone receptor** than progesterone itself but without the normal biological effects:

- The RU-486/receptor complex is not active as a transcription factor.
- Thus genes that are turned on by progesterone are turned off by RU-486.
- The proteins needed to establish and maintain pregnancy are no longer synthesized.
- The endometrium breaks down.
- The embryo detaches from it and can no longer make chorionic gonadotropin (**HCG**).
- Consequently the corpus luteum ceases its production of progesterone.
- The inhibition on uterine contraction is lifted.
- Soon the embryo and the breakdown products of the endometrium are expelled.

These properties of RU-486 have caused it to be used to **induce abortion** of an unwanted fetus. In practice, the physician assists the process by giving a synthetic prostaglandin (e.g., **misoprostol**

[Cytotec®]) 36-48 hours after giving the dose of RU-486. Use of RU-486 is generally limited to the first seven weeks of pregnancy. RU-486 has been used for many years in some countries. However, the controversies surrounding abortion in the United States kept it from being authorized for use here until September 2000.

MENOPAUSE

The menstrual cycle continues for many years. But eventually, usually between 42 and 52 years of age, the follicles become less responsive to FSH and LH. They begin to secrete less estrogen. Ovulation and menstruation become irregular and finally cease. This cessation is called menopause. With levels of estrogen now running one-tenth or less of what they had been, the hypothalamus is released from their inhibitory influence (bar). As a result it now stimulates the pituitary to increased activity. The concentrations of FSH and LH in the blood rise to ten or more times their former values. These elevated levels may cause a variety of unpleasant physical and emotional symptoms.

HORMONE REPLACEMENT THERAPY (HRT)

Many menopausal women elect to take a combination of estrogen and progesterone after they cease to make their own. The benefits are:

- reduction in the unpleasant symptoms of the menopause
- a reduction in the loss of calcium from bones and thus a reduction in osteoporosis and the fractures that accompany it.

It was also believed that HRT reduced the risk of cardiovascular disease. However, a recent study of 16,000 menopausal women was stopped 3 years early when it was found that, in fact, HRT increased (albeit only slightly) not decreased the incidence of cardiovascular disease. Perhaps synthetic **selective estrogen response modulators** or SERMs (raloxifene is an example) will provide the protective effects without the harmful ones. Stay tuned.

Environmental Estrogens

Some substances that find their way into the environment, such as

- **DDE**, a breakdown product of the once widely-used insecticide DDT,
- **DDT** itself (still used in some countries (e.g., Mexico), and
- **PCBs**, chemicals once used in a wide variety of industrial applications

can bind to the estrogen (and androgen) receptors and mimic (weakly) the effects of the hormone. This has created anxiety that they may be responsible for harmful effects such as cancer and low sperm counts.

However, there is as yet little evidence to support these worries.

- No epidemiological relationship has been found between the incidence of breast cancer and the levels of these compounds in the body.
- As for laboratory studies that found a synergistic effect of two of these substances on receptor binding (findings that created the great alarm), these have not been replicated in other laboratories, and the authors of the original report have since withdrawn it as invalid.

REPRODUCTIVE HORMONES OF MALE

Androgen, also called androgenic hormones or testoids, is the generic term for any natural or synthetic compound, usually a steroid hormone, that stimulates or controls the development and maintenance of male characteristics in vertebrates by binding to androgen receptors. This includes the activity of the accessory male sex organs and development of male secondary sex characteristics. Androgens were first discovered in 1936. Androgens are also the original anabolic steroids and the precursor of all estrogens, the female sex hormones. The primary and most well-known androgen is testosterone. Androgen ablation can be used as an effective therapy in prostate cancer.

TYPES

A subset of androgens, **adrenal androgens**, includes any of the 19-carbon steroids synthesized by the adrenal cortex, the outer portion of the adrenal gland (zonula reticularis—innermost region of the adrenal cortex), that function as weak steroids or steroid precursors, including dehydroepiandrosterone (DHEA), dehydroepiandrosterone sulfate (DHEA-S), and androstenedione.

Besides testosterone, other androgens include:

- **Dehydroepiandrosterone (DHEA):** a steroid hormone produced in the adrenal cortex from cholesterol. It is the primary precursor of natural estrogens. DHEA is also called dehydroisoandrosterone or dehydroandrosterone.
- **Androstenedione (Andro):** an androgenic steroid produced by the testes, adrenal cortex, and ovaries. While androstenediones are converted metabolically to testosterone and other **androgens**, they are also the parent structure of estrone. Use of androstenedione as an athletic or body building supplement has been banned by the International Olympic Committee as well as other sporting organizations.
- **Androstenediol:** the steroid metabolite that is thought to act as the main regulator of gonadotropin secretion.
- **Androsterone:** a chemical by-product created during the breakdown of androgens, or derived from progesterone, that also exerts minor masculinising effects, but with one-seventh the intensity of testosterone. It is found in approximately equal amounts in the plasma and urine of both males and females.
- **Dihydrotestosterone (DHT):** a metabolite of testosterone, and a more potent androgen than testosterone in that it binds more strongly to androgen receptors. It is produced in the adrenal cortex.

FUNCTIONS

During mammalian development, the gonads are at first capable of becoming either ovaries or testes. In humans, starting at about week 4 the gonadal rudiments are present within the intermediate mesoderm adjacent to the developing kidneys. At about week 6, epithelial sex cords develop within the forming testes and incorporate the germ cells as they migrate into the gonads. In males, certain Y chromosome genes, particularly SRY, control development of the male phenotype, including conversion of the early bipotential gonad into testes. In males, the sex cords fully invade the developing gonads.The mesoderm-derived epithelial cells of the sex cords in developing testes become the Sertoli cells which will function to support sperm cell formation. A minor population of non-epithelial cells appear between the tubules by week 8 of human fetal development.

These are Leydig cells. Soon after they differentiate, Leydig cells begin to produce androgens.The androgens function as paracrine hormones required by the Sertoli cells in order to support sperm production. They are also required for masculinization of the developing male fetus (including penis and scrotum formation). Under the influence of androgens, remnants of the mesonephron, the Wolffian ducts, develop into the epididymis, vas deferens and seminal vesicles. This action of androgens is supported by a hormone from Sertoli cells,MIH (Mullerian Inhibitory Hormone), which prevents the embryonic Müllerian ducts from developing into fallopian tubes and other female reproductive tract tissues in male embryos.

MIH and androgens cooperate to allow for the normal movement of testes into the scrotum.Before the production of the pituitary hormone LH by the embryo starting at about weeks 11–12, human chorionic gonadotrophin (hCG) promotes the differentiation of Leydig cells and their production of androgens. Androgen action in target tissues often involves conversion of testosterone to 5α-dihydrotestosterone (DHT).

During puberty, androgen, LH and FSH production increase and the sex cords hollow out, forming the seminiferous tubules, and the germ cells start to differentiate into sperm. Throughout adulthood, androgens and FSH cooperatively act on Sertoli cells in the testes to support sperm production. Exogenous androgen supplements can be used as a male contraceptive. Elevated androgen levels caused by use of androgen supplements can inhibit production of LH and block production of endogenous androgens by Leydig cells. Without the locally high levels of androgens in testes due to androgen production by Leydig cells, the seminiferous tubules can degenerate resulting in infertility.

For this reason, many transdermal androgen patches are applied to the scrotum.Males typically have less adipose tissue than females. Recent results indicate that androgens inhibit the ability of some fat cells to store lipids by blocking a signal transduction pathway that normally supports adipocyte function. Also, androgens, but not estrogens, increase beta adrenergic receptors while decreasing alpha adrenergic receptors- which results in increased levels of epinephrine/ norepinephrine due to lack of alpha-2 receptor negative feedback and decreased fat accumulation due to epinephrine/ norepinephrine then acting on lipolysis-inducing beta receptors.

Males typically have more skeletal muscle mass than females. Androgens promote the enlargement of skeletal muscle cells and probably act in a coordinated manner to function by acting on several cell types in skeletal muscle tissue. One type of cell that conveys hormone signals to generating muscle is the myoblast. Higher androgen levels lead to increased expression of androgen receptor. Fusion of myoblasts generates myotubes, in a process that is linked to androgen receptor levels. Circulating levels of androgens can influence human behavior because some neurons are sensitive to steroid hormones. Androgen levels have been implicated in the regulation of human aggression and libido. Indeed, androgens are capable of altering the structure of the brain in several species, including mice, rats, and primates, producing sex differences. Although, their potential for conversion makes identifying which alterations in neuroanatomy stem from androgens or estrogens slightly difficult, numerous reports have outlined that androgens alone are capable of altering the structure of the brain. Reduced ability of a XY karyotype fetus to respond to androgens can result in one of several conditions, including infertility and several forms of intersex conditions.

TESTOSTERONE

Testosterone is a steroid hormone from the androgen group and is found in mammals, reptiles, birds, and other vertebrates. In mammals, testosterone is primarily secreted in the testes of males and the ovaries of females, although small amounts are also secreted by the adrenal glands. It is the principal male sex hormone and an anabolic steroid. In men, testosterone plays a key role in the development of male reproductive tissues such as the testis and prostate as well as promoting secondary sexual characteristics such as increased muscle and bone mass and hair growth. In addition, testosterone is essential for health and well-being as well as the prevention of osteoporosis. On average, an adult human male body produces about ten times more testosterone than an adult human female body, but females are, from a behavioral perspective (rather than from an anatomical or biological perspective), more sensitive to the hormone. However, the overall ranges for male and female are very wide, such that the ranges actually overlap at the low end and high end respectively.Testosterone is conserved through most vertebrates, although fish make a slightly different form called 11-ketotestosterone. Its counterpart in insects is ecdysone. These ubiquitous steroids suggest that sex hormones have an ancient evolutionary history.

BIOSYNTHESIS AND REGULATION

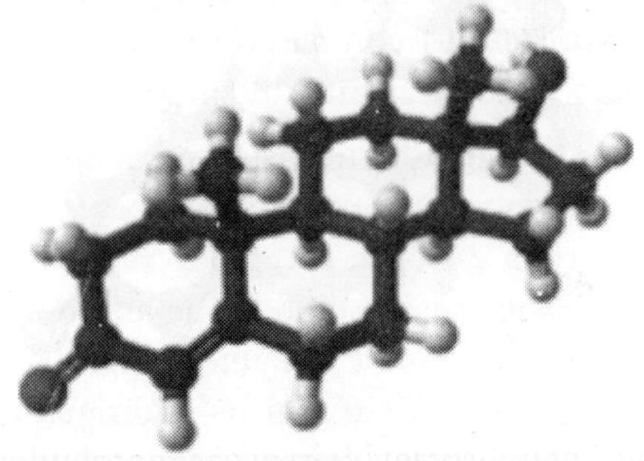

Like other steroid hormones, testosterone is derived from cholesterol (see figure to the right). The first step in the biosynthesis involves the cleavage of the sidechain of cholesterol by CYP11A, a mitochondrial cytochrome P450 oxidase with the loss of six carbon atoms to give pregnenolone. In the next step, two additional carbon atoms are removed by the CYP17A enzyme in the endoplasmic reticulum to yield a variety of C_{19} steroids. In addition, the 3-hydroxyl group is oxidized by 3-β-HSD to produce androstenedione. In the final and rate limiting step, the C-17 keto group androstenedione is reduced by 17-β hydroxysteroid dehydrogenase to yield testosterone.The largest amounts of testosterone (>95%) are produced by the testes in men.[3] It is also synthesized in far smaller quantities in women by the thecal cells of the ovaries, by the placenta, as well as by the zona reticularis of the adrenal cortex in both sexes. In the testes, testosterone is produced by the Leydig cells.[44] The male generative glands also contain Sertoli cells which require testosterone for spermatogenesis. Like most hormones, testosterone is supplied to target tissues in the blood where much of it is transported bound to a specific plasma protein, sex hormone binding globulin (SHBG).In males, testosterone is primarily synthesized in Leydig cells. The number of Leydig cells in turn is regulated by luteinizing hormone (LH) and follicle stimulating hormone (FSH). In addition, the amount of testosterone produced by existing Leydig cells is under the control of LH which regulates the expression of 17-β hydroxysteroid dehydrogenase.

Environmental factors affecting testosterone levels include:

- Loss of status or dominance in men may result in a decreased testosterone level.

- Implicit power motivation predicts an increased testosterone release in men.
- Aging reduces testosterone release.
- Hypogonadism
- Sleep (REM dream) increases nocturnal testosterone levels.
- Resistance training increases testosterone levels, however, in older men, that increase can be avoided by protein ingestion.
- Zinc deficiency lowers testosterone levels but over supplementation has no effect on serum testosterone.
- Licorice. The active ingredient in licorice root, glycyrrhizinic acid has been linked to small, clinically non-significant decreases in testosterone levels. In contrast, a more recent study found that licorice administration produced a substantial testosterone decrease in a small, female-only sample.

Approximately 7% of testosterone is reduced to 5α-dihydrotesto-sterone (DHT) by the cytochrome P_{450} enzyme 5α-reductase, an enzyme highly expressed in male accessory sex organs and hair follicles. Approximately 0.3% of testosterone is converted into estradiol by aromatase (CYP19A1) an enzyme expressed in the brain, liver, and adipose tissues.

DHT is a more potent form of testosterone while estradiol has completely different activities (feminization) compared to testosterone (masculinization). Finally testosterone and DHT may be deactivated or cleared by enzymes that hydroxylate at the 6, 7, 15 or 16 positions.

The effects of testosterone in humans and other vertebrates occur by way of two main mechanisms: by activation of the androgen receptor (directly or as DHT), and by conversion to estradiol and activation of certain estrogen receptors. Free testosterone (T) is transported into the cytoplasm of target tissue cells, where it can bind to the androgen receptor, or can be reduced to 5α-dihydrotestosterone (DHT) by the cytoplasmic enzyme 5-alpha reductase. DHT binds to the same androgen receptor even more strongly than T, so that its androgenic potency is about 5 times that of T. The T-receptor or DHT-receptor complex undergoes a structural change that allows it to move into the cell nucleus and bind directly to specific nucleotide sequences of the chromosomal DNA. The areas of binding are called hormone response elements (HREs), and influence transcriptional activity of certain genes, producing the androgen effects. It is important to note that if there is a 5-alpha reductase deficiency, the body (of a human) will *continue* growing into a female with testicles.Androgen receptors occur in many different vertebrate body system tissues, and both males and females respond similarly to similar levels. Greatly differing amounts of testosterone prenatally, at puberty, and throughout life account for a share of biological differences between males and females.The bones and the brain are two important tissues in humans where the primary effect of testosterone is by way of aromatization to estradiol. In the bones, estradiol accelerates maturation of cartilage into bone, leading to closure of the epiphyses and conclusion of growth. In the central nervous system, testosterone is aromatized to estradiol. Estradiol rather than testosterone serves as the most important feedback signal to the hypothalamus (especially affecting LH secretion). In many mammals, prenatal or perinatal "masculinization" of the sexually dimorphic areas of the brain by estradiol derived from testosterone programs later male sexual behavior. The human hormone testosterone is produced in greater amounts by males, and less by females. The human hormone estrogen is produced in greater amounts by females, and less by males. Testosterone causes the appearance of masculine traits (i.e., deepening voice, pubic and facial hairs, muscular build, etc.) Like men, women rely on testosterone to maintain libido, bone density and muscle mass throughout their lives. In men, inappropriately high levels of estrogens lower testosterone, decrease muscle mass, stunt growth in teenagers, introduce gynecomastia, increase feminine characteristics, and decrease susceptibility to prostate cancer, reduces libido and causes erectile dysfunction and can cause excessive sweating and hot flushes. However, an appropriate amount of estrogens is required in the male in order to ensure well-being, bone density, libido, erectile function, etc

Sex Hormone-binding Globulin (SHBG)

It is a glycoprotein that binds to sex hormones, specifically testosterone and estradiol. Other steroid hormones such as progesterone, cortisol, and other corticosteroids are bound by transcortin

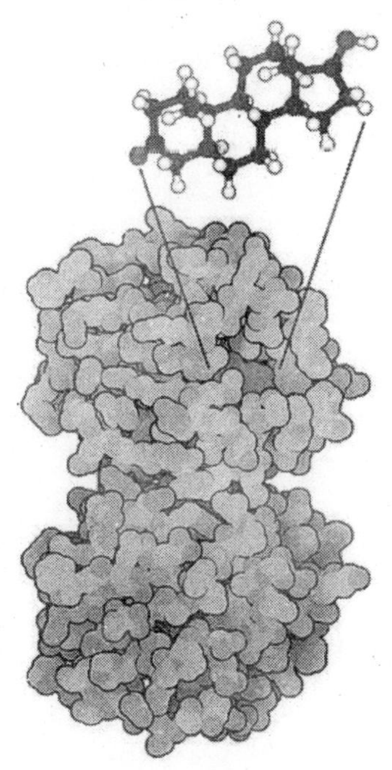

Testosterone and estradiol circulate in the bloodstream, bound mostly to SHBG and to some degree bound to serum albumin. Only a small fraction is unbound, or "free," and thus biologically active and able to enter a cell and activate its receptor. The SHBG inhibits the function of these hormones. Thus bioavailability of sex hormones is influenced by the level of SHBG. SHBG has a higher affinity for Dihydrotestosterone than for Testosterone or for Estradiol making it essential in women for regulating the bio-availability of Dihydrotestosterone.

SHBG is produced most by the liver and is released into the bloodstream. Other sites that produce SHBG include the brain, uterus, testes, and placenta. Testes-produced SHBG is called androgen-binding protein. The gene for SHBG is located on chromosome 17SHBG levels appear to be controlled by a delicate balance of enhancing and inhibiting factors. Its level is decreased by high levels of insulin and insulin-like growth factor 1 (IGF-1). Also, high androgen and transcortin levels decrease SHBG, while high growth hormone, estrogen and thyroxine levels increase it.However, recent evidence suggests that it is the liver's production of fats that reduces SHBG levels, not any direct effect of insulin and specific genetic mechanisms have been found that do this.Conditions with low SHBG include polycystic ovary syndrome, diabetes, and hypothyroidism. Conditions with high SHBG include pregnancy, hyperthyroidism, and anorexia nervosa. There has recently been research to link high SHBG levels with breast and testicular cancer as well.

APPENDIX

BIOLOGICAL STAINS

Stain	Use	Specimen Colour
1	2	3
Acid Fuchsin	Stains collagen fibers red and smooth muscle in contrast to collagen.	Red/magenta/ violet
Congo Red	Used to stain amyloid. (A starchlike substance. A hard, waxy deposit consisting of protein and polysaccharides that results from the degeneration of tissue.)	Red
Crystal Violet	The (positive) dye of choice in Grams stain. Bacteria	Blue violet
Eosin B	Interchangeable with Eosin Y. Ionization – acid (Color is red)	Bluish cast
Eosin Y	Stains alkaline cell parts (like cytoplasm). Use on plants, animals and blood.	Pink
Eosin Y ws	Common counterstain to alum hematoxylin and Eosin method	Red
Gram Stain	Positive – Staphylococcus, Streptococcus, Bacillus - subtilis (crystal violet)Negative – Neisseria, *E. coli* (Eosin Y ws)	Dark blue or purple Red
Hematoxylin	Stains nuclear chromatin using aluminum mordant. Acid resistant nuclear staining, muscle striations and some glial fibers (The delicate network of branched cells and fibers that supports the tissue of the central nervous system) with ferric salts mordant.	Yellow/brown
Iodine	Carbohydrates in plant and animal specimens Glycogen (Animal starch.. A polysaccharide, $(C_6H_{10}O_5)_n$, that is the main form of carbohydrate storage in animals.	Brown or Blue BlackRed
Mercurochrome	Marks tissue margins and faces for orientation. (Antiseptic) Similar to Eosin Y ws.	Red

(Contd...)

1	2	3
Methyl blue	Stains collagen and connective tissue.	Blue
Methylene blue	Acidic cell parts (Use on animal, bacteria and blood specimens.	Blue
Saffron	Colors connective tissue yellow in contrast to the pink cytoplasm given by phyloxine.	Yellow orange
Safranin O	Commonly used for counterstaining nuclei red. Cartilage is stained yellowish.	Red
Toluidene blue	Stains acidic cell parts (like nucleus). Good to show mitosis in plant cells. MITOSIS = The process in cell division by which the nucleus divides, typically consisting of four stages, prophase, metaphase, anaphase, and telophase, and normally resulting in two new nuclei, each of which contains a complete copy of the parental chromosomes. Also called karyokinesis.	Dark Blue

AMYLOID

Alkaline Congo Red for Amyloid

Amyloidosis is a disease characterised by the deposition of amorphous eosinophilic, predominantly extracellular, material (amyloid) in various body tissues forming confluent masses and progressively replacing the parenchymatous cells of vital organs, resulting in gradual loss of function. Amyloid is composed of 95% protein + 5% carbohydrate.

Principle

Congo Red dye forms non-polar hydrogen bonds with amyloid and red to green birefringence occurs when viewed by polarised light due to parallel alignment of the dye molecules on the linearly arranged amyloid fibrils. The Alkaline Congo Red Technique uses high concentrations of sodium chloride which act as ionic competitors for the dye thus eliminating background electrochemical (polar) staining while enhancing the hydrogen (non-polar) binding of Congo Red and amyloid.

Specimen

Standard paraffin section Frozen section (gives greater intensity of staining).

Control

Tissue positive for amyloid.

Note: Unstained control sections tend to lose reactivity if stored for long periods Tissue containing massive deposits, gives less intense reaction than small newly formed deposits.

Reagents

Stock Solutions

(A) Stock alcoholic salt solution Saturated sodium chloride in 80% alcohol (keeps well)

(B) 1% aqueous sodium hydroxide Dissolve 1g sodium hydroxide in 100 mL distilled water

(C) Stock Congo Red solution.

Caution: Possible teratogen - Minimise contact - Wear gloves and mask - Wash thoroughly after handling Make a saturated solution of Congo Red C.I. 22120 in stock alcoholic salt solution (solution A). Approx 0.2g of dye required to saturate 100 mL. Keeps well. Store in fridge. Working solutions Alcoholic sodium chloride-hydroxide Solution Place in a coplin jar

Stock Alcoholic Salt Solution

(A) 50 mL 1% aqueous Sodium hydroxide.

(B) 0.5 mL Use within 20 minutes of preparation. Congo Red Working Solution Place in a coplin jar.

Stock Congo Red Solution

(C) 50 mL 1% aqueous sodium hydroxide (B) 0.5 mL Mix and filter, use within 20 minutes of preparation.

N.B. Takes approximately 20 minutes to filter.

Procedure

1. Deparaffinise test and control sections and hydrate to water.
2. Stain nuclei with an alum haematoxylin (use linear stainer).
3. Wash in water.
4. Differentiate in 0.15% acid-alcohol approximately 10 seconds (in linear stainer).
5. Wash in water.
6. "Blue" in Lithium carbonate solution pH 7.8 to 8.6 (in linear stainer).
7. Wash in water.
8. Treat in working Alcoholic Sodium chloride-hydroxide solution for 20 minutes.
9. Drain.
10. Treat with working Congo red solution for 20 minutes.
11. Drain well.
12. Dehydrate rapidly in alcohols.
13. Clear and mount.

Results

Amyloid (eosinophil granules, elastin) Dull red Nuclei Blue Background Clear Polarised light: Amyloid Green birefringence:

2.1 Verhoeff-van Gieson

2.2 Masson's Trichrome

2.3 Gomori's One-step Trichrome

2.4 Van Gieson

2.5 Reticulin

2.6 Gordon & Sweets

2.7 Mallory's Phosphotungstic Acid Haematoxylin

2.8 Sirius Red

CONNECTIVE TISSUES

Verhoeff-van Gieson Stain For Elastic Fibres

Verhoeff's Technique (Verhoeff 1908) Elastic Fibres are composed of the protein, elastin. They are found widely distributed throughout the body but are especially associated with the respiratory system, circulatory system and the skin.

Principle

The rationale of the technique is obscure; both iodine and ferric chloride with which the haematoxylin is combined, are oxidising agents and it is this property which probably accounts for the production of a black dye with cationic properties rather than a simple dye- mordant-tissue mechanism.

Specimen

Standard paraffin section. This technique works after any fixative.

Control

Dermis of skin, medium-sized arteries.

Notes: Bulk staining is not possible due to the differentiation involved in this technique. Careful differentiation of each section is required, to ensure that the fine fibres are stained and at the same time a well-differentiated background is obtained. Elastic staining is extracted by the Van Gieson counterstain, therefore the section should be under-differentiated with ferric chloride. Pre-treatment with 1% potassium permanganate for 5 minutes, followed by oxalic acid, improves sharpness and intensity of staining.

Reagents

(A) 5 % Alcoholic Haematoxylin

Haematoxylin 5.0g Absolute alcohol 100mL Dissolve the haematoxylin in approx 20mL of warm alcohol. Make up volume to 100 mL then filter. Leave to ripen naturally for one month before use.

(B) 10% Aqueous Ferric Chloride

Take 16.5 mL commercial aqueous ferric chloride solution and make up to 100mL with distilled water.

(C) Verhoeff Iodine Solution

Dissolve 4 g of Potassium iodide in approx 10mL of distilled water. Add and dissolve 2 g of iodine in this solution and finally make up to 200mL with distilled water.

(D) Verhoeff's Mixture

Working Solution To 5.0 mL of A add 2.0 mL of B, mix and add 2.0 mL of C, mix. May be used up to 48 hours after mixing.

(E) 2% Ferric Chloride

Take 3.3 mL commercial 60% ferric chloride solution and make up to 100 mL with distilled water.

(F) Van Gieson Stain

Saturated Picric Acid 100mL 1 % aqueous Acid Fuchsin 10mL Boil for 3 mins, cool and filter, then add 0.25 mL concentrated Hydrochloric Acid.

Procedure

1. Deparaffinise sections and hydrate to distilled water.
2. Stain in freshly prepared Verhoeff's mixture in a closed container for 15 mins.
3. Wash in water.
4. Differentiate in 2% Ferric chloride solution, checking each section microscopically until the nuclei and fine elastic fibres are black, but the background is still weakly stained.
5. Wash in water then in 95 % alcohol momentarily to remove iodine colouration.
6. Wash in water for 5 mins.
7. Counter stain with Van Gieson stain for 3 mins. Drain.
8. Dehydrate rapidly in alcohols - (Van Gieson stain will be removed if dehydration is prolonged).
9. Clear and mount.

Results

Nuclei and elastic fibres Black

Collagen fibres Red

Background Yellow

Masson's Trichrome Stain for Collagen

Masson's Trichrome Method (1929) - Connective Tissue. The term "trichrome stain" is a general name for a number of techniques for the selective demonstration of muscle, collagen fibres, fibrin and erythrocytes. Three dyes are used, one of which may be a nuclear stain.

Principle

Whilst the firm attachment of dye to tissue is probably electrostatic in nature, the means by which selective sequential staining with a number of anionic dyes is achieved is almost certainly bound up with physical structure of dye and tissue. The less porous tissues are coloured by the smallest dye molecule; whenever a dye of large molecular size is able to penetrate, it will always do so at the expense of the smaller molecule. Dyes, which are low in pH, tend to decolourise the nuclear stain, therefore iron haematoxylins which are more resistant to acid are used in preference to alum haematoxylins. Phosphomolybdic acid acts as a differentiator. Masson's method is a sensitive technique for displaying fine collagen fibres, basement membranes, fibrin and hyaline. Muscle is well shown in contrast and striations of voluntary muscle can be seen.

Specimen

Mercuric chloride or picric acid fixation gives best results. If tissue was fixed in neutral buffered formalin, staining is enhanced by using Bouin's fluid as a mordant. Standard paraffin sections.

Control

Large artery: tunica media should stain red and the tunica adventitia blue or green. Section containing connective tissue, eg skin

Reagents

(A) Bouin's Fixative

Pre-heated to 56 - 60°C. Picric acid, saturated aqueous solution 750 mL 37% - 40% formalin 250 mL Glacial Acetic acid 50 mL

(B) Weigert's Haematoxylin (IRON)

These solutions keep for approximately six months. (a) Dissolve 1 g haematoxylin in 100 mL of 95% alcohol. Filter. Allow to ripen naturally for four weeks before use. (b) Dilute 2 mL 60% commercial aqueous ferric chloride in 97 mL of distilled water. Add 1 mL concentrated hydrochloric acid. Working Solution Mix equal volumes of (A) and (B) prior to use. This mixture should be a violet black colour and must be discarded if it is brown.

(C) Acid Fuchsin Solution (kept in fridge)

Acid fuchsin (CI 42685) 0.5 g Glacial acetic acid 5 mL Distilled water 100 mL

(D) Phosphomolybdic Acid Solution

Phosphomolybdic acid 1.0 g Distilled water 100 mL

(E) Methyl Blue Solution (kept in fridge)

Methyl blue (CI 42780) 2.0 g Glacial acetic acid 2.5 mL Distilled water 100 mL

Procedure

1. Deparaffinise sections and hydrate to distilled water.
2. Mordant in Bouin's Fixative for one hour at 56°C in water bath (or overnight at room temperature).
3. Cool and wash in running water until yellow colour disappears.
4. Stain in Weigert's haematoxylin for ten minutes.
5. Wash well in tap water.
6. Stain in acid fuchsin solution for five minutes.
7. Rinse in distilled water.
8. Treat with phosphomolybdic acid solution for five minutes.
9. Drain.
10. Stain with methyl blue solution for five minutes.
11. Rinse in distilled water.
12. Treat with one percent acetic acid for two minutes.
13. Dehydrate through alcohols.
14. Clear and mount.

Results

Nuclei Black

Cytoplasm, muscle and erythrocytes Red

Collagen Blue

Note: Light green counterstain may be substituted for methyl blue.

GOMORI'S ONE-STEP TRICHROME STAIN

Specimen

Standard paraffin section of well-fixed tissue or alcohol-fixed smear.

Reagents

(A) Bouin's Fixative Solution

Preheated to 56 -60°C. Prepare as for method 2.2.

(B) Weigert's Iron Haematoxylin Working Solution

Prepare as for method Masson's Trichrome Stain for Collagen.

(C) Gomori's Trichrome Stain

Chromotrope 2R 0.6 g

Light green CI 42095 0.3 g

Glacial acetic acid 1.0 mL

Phosphotingstic acid 0.8 g

Distilled water 100 mL

(D) 1% Glacial Acetic Acid Solution

Alternatively, use acid-alcohol step on linear stainer.

Procedure

1. Deparaffinise and hydrate to distilled water.
2. Place in Bouin's fixative in oven at 56°C for one hour.
3. Wash well in running water until sections are clear.
4. Stain in Weigert's iron haematoxylin working solution for 10 minutes.
5. Rinse in tap water.
6. Stain in Gomori's trichrome stain for 15 to 20 minutes.
7. Rinse in one percent acetic acid (or use acid-alcohol in linear stainer).
8. Rinse in distilled water.
9. Dehydrate and clear in 95% ethyl alcohol, absolute ethyl alcohol and histolene, two changes each
10. Two minutes each.
11. Mount with safety-mount.

Results

Muscle fibres red

Collagen green

Nuclei blue to green

Notes

1. Gomori's chrómium haematoxylin may be substituted for the Weigert's iron haematoxylin at step 4.
2. Aniline blue can be substituted for the light green.
3. Slides may be left in Bouin's solution overnight at room temperature.
4. If sections are too red, differentiate in 100 mL of 1% acetic acid to which 0.7 g of phosphotungstic acid has been added.

Van Gieson Technique for Collagen

The Van Gieson Technique is a "trichrome" stain for the demonstration of connective tissue. It is suitable for showing coarse collagen fibres, but not recommended for fine collagen fibres or muscle collagen comparisons.

Principle

The mixture of the two acid dyes in acid solution compete for available linkages. Acid fuchsin being a larger molecular size is restricted to more permeable collagen fibres whereas the diffusible picric acid penetrates the compact muscle, cytoplasm and red cells.

References

p 4 H C Cook, Manual of Histological Demonstration Techniques p 127 Bancroft JD & Stevens, Theory and Practice of Histological Techniques 4th ed.

Specimen

Standard paraffin sections Fixation - 10% neutral buffered formalin.

Control

Section containing connective tissue, e.g. skin.

Reagents

(A) Weigerts Iron Haematoxylin

These solutions keep for approximately 6 months.

1. Dissolve one gram haematoxylin in 100 mL of 95% alcohol. Filter. Allow to ripen naturally for four weeks before use.
2. Dilute 2 mL 60% commercial aqueous ferric chloride in 97 mL of distilled water. Add 1 mL concentrated hydrochloric acid. Working Solution Mix equal volumes of (1) and (2) prior to use. This mixture should be a violet black colour and must be discarded if it is brown.

(B) Van Gieson's Stain

100mLs saturated aqueous Picric Acid 10 mLs 1% aqueous acid fuchsin (C.I. 42685) Boil for three minutes, cool and filter then add 0.25mL concentrated hydrochloric acid.

Procedure

1. Deparaffinize and hydrate to distilled water.
2. Stain nuclei with Weigert's Iron Haematoxylin for 15 to 30 mins.
3. Wash in water until nuclei are dark blue-black and the background a paler blue-black.
4. Stain with Van Gieson's solution for 3 mins.
5. Dehydrate rapidly in alcohol.
6. Clear and mount.

Notes

Acid fuchsin is extracted by water and picric acid by alcohol, hence rapid washing, blotting and brief dehydration are advised. Iron haematoxylin is used in preference to alum haematoxylin as it is more resistant to extraction by picric acid.

Results

Nuclei Brown - black Collagen Red Muscle, RBCs, cytoplasm Yellow.

RETICULIN FIBRES

Principle

Reticulin fibres have little affinity for silver solutions and must be treated with a suitable solution, (2.5% iron alum), to produce sensitised sites on the fibres where silver deposition may be initiated. A separate reducing agent is required to cause deposition of silver in the form of the metal. Any unprecipitated, excess silver is removed by treatment with sodium thiosulphate solution.

Clinical Significance

This method demonstrates the presence of reticulin in tissue sections.

Specimen

A 5-micron paraffin section of tissue.

Control

A section of normal liver tissue provides a good positive control.

Reagents

(A) Acidified Potassium Permanganate 0.5% Potassium permanganate 0.5g Distilled water 100mL 3% Sulphuric acid 5mL Mix immediately before use.

(B) 1% Oxalic acid

(C) 2.5% Iron alum (ferric ammonium sulphate).

(D) Ammoniacal silver 10% silver nitrate. 6mL Absolute alcohol 3mL Add strong ammonia dropwise, until precipitate that first forms, is just dissolved. (approx. 17 drops).

Note

Use a conical flask to make up the silver solution and swirl it well in between each additional drop. This is important because the addition of too much ammonia renders the silver solution less effective. The silver solution should be made fresh each day it is required.

(E) Developer Formalin (40%) 10mL Absolute alcohol 20mL Distilled water 30mL

Procedure

1. De-paraffinise and rehydrate to distilled water
2. Acidified potassium permanganate- 10 minutes..
3. Rinse off in distilled water
4. Oxalic acid - three minutes.
5. Rinse in distilled water
6. Iron alum - 15 minutes
7. Rinse with distilled water 3-4 times
8. This cuts down background staining.

9. Ammoniacal silver - 70 seconds.
10. For this step, best results are achieved by staining each slide separately. When applying the silver solution, pour on, tip off and re-apply all in one action. This ensures even staining of the section.
11. Wash in distilled water - each wash 10 seconds.
12. Developer - about three minutes when it turns golden.
13. Wash well in water.
14. Fix in sodium thiosulphate (hypo). - three minutes
15. Wash well in water
16. Counterstain in 0.1% neutral red - one minute
17. Dehydrate, clear and mount.

Results

Reticulin fibres - black

Tissue - red.

GORDON AND SWEETS' METHOD FOR RETICULIN FIBRES

(Gordon & Sweets 1936) Reticulin fibres are fine delicate fibres, which are normally found connected to stronger and coarser collagenic fibres. They make up the bulk of the supporting framework of the liver, spleen and lymph nodes.

Principle

Silver from silver oxides is selectively deposited on the reticulin fibres, which appear black after conversion to reduced silver, by the reducing agent (formalin). Gold chloride is used as a toner to give a clearer background and unreduced silver is removed by treatment with Sodium thiosulphate.

Specimen

Standard paraffin sections Fixation 10% neutral buffered formalin

Control

Liver/spleen

Reagents

(A) SILVER SOLUTION to 5mL of 10% Silver nitrate solution add strong ammonia until the precipitate formed is just dissolved. Avoid excess of ammonia. Add 5mL of 3% Sodium hydroxide solution. Re-dissolve the precipitate by adding strong ammonia drop by drop until the resultant solution retains a faint opalescence. If at this stage any excess ammonia is present, indicated by the absence of opalescence, add a few drops of 10% silver nitrate solution to produce a light precipitate. Make up to 50 mL with distilled water. Store in a dark bottle at 4°C - keeps three months. Filter before use (approximately 2 mL per slide).

Note: All glassware should be rinsed in distilled water and dried before use.

(B) ACIDIFIED POTASSIUM PERMANGANATE
0.5% Potassium permanganate 190 mL 3% Sulphuric acid 10 mL

(C) 1% OXALIC ACID

(D) 4% IRON ALUM (Ammonium Ferric Sulphate)

(E) 10% FORMALIN (not buffered) - dilute in tap water not distilled

(F) 0.2% GOLD CHLORIDE (Sodium Chloro-aurate)

(G) 5% SODIUM THIOSULPHATE

(H) 1% NEUTRAL RED (C.I. 50040)

Procedure

(use PLASTIC forceps - not metal)

1. Deparaffinize and hydrate to distilled water
2. Acidified potassium permanganate two min
3. Wash in water
4. 1 % Oxalic acid two min
5. Wash well in distilled water
6. 4 % Iron alum 10 min
7. Wash in distilled water
8. Silver solution (filter 2mLs per slide) 1 min
9. Rinse in distilled water
10. 10 % Formalin two min
11. Wash in water followed by distilled water
12. 0.2 % Gold chloride two min
13. Rinse in distilled water
14. 5 % (Sodium thiosulphate) two min
15. Wash in water
16. Counterstain with 1 % Neutral Red two min
17. Dehydrate rapidly, clear and mount

Note: Dehydrate rapidly as the neutral Red counterstain is removed in alcohol.

Results

Reticulin fibres Black

Nuclei Black

Background Red

MALLORY'S PHOSPHOTUNGSTIC ACID HAEMATOXYLIN (PTAH) FOR MUSCLE STRIATIONS

The PTAH stain demonstrates many tissue structures, particularly fibrin, muscle striations, cilia and glial fibres, plus many CNS structures.

Principle

The mechanism by which two-colour staining is achieved from a mixture of haematein and phosphotungstic acid is obscure. Garland and Gaer (1964) concluded that the blue colour produced was a metachromatic-type staining effect.

Control

Skeletal (striated) muscle

Reagents

(A) PTAH SOLUTION

1. PTAH solution naturally oxidised .

   Haematoxylin (C.I. 75290) 0.5 g, Phosphotungstic acid 5.0 g, Distilled water 500 mL. The haematoxylin is dissolved in 100mL of the distilled water, and the phosphotungstic acid in the remaining 400mL. The two solutions are then mixed. The stain is allowed to ripen naturally in a loosely stoppered bottle at room temperature in the light for several months. This solution is stable and useable for many years, or
2. PTAH Solution using Haematein Haematein 0.5 g, Phosphotungstic acid 5.0 g, Distilled water 500 mL. The solids are dissolved in separate portions of the distilled water as above, then mixed. Store in airtight bottle. The stain is ready for use in 24 hours and does not need to be chemically ripened due to the use of Haematein, or
3. PTAH Solution chemically oxidised Haematoxylin (C.I. 75290) 0.5 g, Phosphotungstic acid 10.0 g, Distilled water 500 mL 0.25%. Aqueous potassium permanganate 25 mL The solids are dissolved in separate portions of distilled water as above, then mixed. The potassium permanganate is then added to chemically ripen the stain, which can be used in 24 hours, but peak staining activity is reached in 7 days. Continuing oxidation means that this stain has a comparatively short life.

(B) 0.25% Aqueous Potassium Permanganate

(C) 5% Aqueous Oxalic Acid

Procedure

1. Deparaffinize and hydrate to distilled water
2. Oxidise in potassium permanganate for 5 min
3. Wash in water
4. Bleach in Oxalic acid until white

5. Wash in water
6. Filter and stain in PTAH overnight at room temperature
7. Drain off excess stain
8. Dehydrate rapidly
9. Clear and mount

Note

Dehydration should be rapid since water and alcohol may remove the stain. Dehydration may be commenced in 95% alcohol, and CNS sections may require thorough washing in 95% alcohol for several minutes to remove excess red stain.

Results

Muscle striations, neuroglia fibres Dark blue

Fibrin, amoebae Dark blue

Nuclei, cilia, red blood cells Blue

Myelin Lighter blue

Collagen, osteoid, cartilage, elastic fibres Deep brownish/red

Cytoplasm Pale pinkish brown

SIRIUS RED FOR COLLAGEN AND BILE PIGMENTS IN LIVER TISSUE

The Sirius Red technique is a useful stain for collagen and bile pigment in liver biopsies.

Principle

Collagen forms the ground substance of connective tissue. It is composed of three amino- acids and stains strongly with acid red dyes due to the affinity of the cationic groups of the proteins for the anionic reactive groups of the acid dyes.

Specimen

Standard paraffin section. Neutral buffered formalin fixation.

Control

Normal liver tissue.

Reagents

(A) Fouchet's Reagent:

Trichloracetic acid 25 g
Distilled water 100 mL
10% Ferric chloride 10 mL

(B) Sirius Red (working solution):

Saturated aqueous Picric acid 98 mL

1% Aqueous Sirius Red 2 mL (Direct Red 80 C.I. 35780)

(C) Celestine Blue:

Celestine Blue (C.I. 51050) 0.5 g

Iron Alum (Ammonium Ferric Sulphate) 5 g

Distilled water 100 mL

Procedure

1. Deparaffinize and hydrate test and control sections to distilled water.
2. Fouchet's reagent for 5 minutes.
3. Rinse in distilled water.
4. Celestine blue for 2 minutes.
5. Rinse in distilled water.
6. Alum Haematoxylin for 3 minutes - use linear stainer.
7. Wash, - do NOT differentiate- blue in Lithium Carbonate, wash.
8. Sirius Red WORKING solution for 20 minutes.
9. Rinse in distilled water.
10. Dehydrate, clear and mount.

Results

Bile products Emerald green

Red blood cells Light/pale green

Collagen Bright red

Hepatocytes Dull yellow

Nuclei Black

FIBRIN

Martius Yellow-Brilliant Crystal Scarlet Blue (Msb) Technique—For Fibrin

Fibrin is an insoluble fibrillar protein formed by polymerisation of fibrinogen, which is one of the plasma proteins and is seen in tissues where there has been tissue damage.

Principle

Martius yellow, a small molecule dye, together with phosphotungstic acid in alcoholic solution stains red cells. Early fibrin deposits may be coloured, but the phosphotungstic acid blocks the staining of muscle, collagen and connective tissue fibres. Brilliant crystal scarlet, a medium sized molecule, stains muscle and mature fibrin. Phosphotungstic acid removes any red stain from the collagen. The large molecule dye aniline blue stains the collagen and old fibrin.

References

p 101 Cook H C, Manual of Histological Demonstration Techniques. p 134 Bancroft JD & Stevens, Theory and Practice of Histological Techniques, 2ndEd.

Specimen

Standard paraffin section - coated slide may be used to aid adhesion Fixation -10% neutral buffered formalin Post fixation in Bouin's fluid enhances staining

Control

Use control positive for fibrin e.g. inflammatory appendix due to peritonitis.

Reagents

(A) Weigert's Iron Haematoxylin

(B) Martius Yellow Solution
95% alcohol 100.mL
Phosphotungstic acid 2.0 g
Martius yellow C.I. 10315 (Acid yellow 24) 0.5 g

(C) 1% Brilliant Crystal Scarlet
C.I.16250 in 2.5% Acetic Acid (Synonym: Crystal Ponceau Acid Red 44)

(D) 1% Phosphotungstic Acid
(Synonym: Dodecatungstophosphoric Acid)

(E) 1% Aniline (Soluble) Blue C.I. 42755 in 1% Acetic Acid
Or 0.5% Methyl Blue (Acid Blue 93) C.I. 42780 in 1% Acetic Acid

Procedure

1. Deparaffinize and hydrate to distilled water.
2. Post-fix in Bouin's fluid at 60°C for one hour.
3. Wash in water
4. Stain with Weigert's Haematoxylin - 15-30 mins.
5. Wash in water
6. Stain in Martius Yellow in a coplin jar for 2-3 mins and rinse in distilled water.
7. Stain with Brilliant Crystal Scarlet for 10mins and rinse in distilled water.
8. Treat with Phosphotungstic acid for 5 mins until no red remains in the collagen and rinse in distilled water.
9. Stain with Aniline Blue for 3-5 mins (control microscopically)
10. Rinse in water, dehydrate, clear and mount.

Results

Fibrin, Red (early fibrin may colour yellow and very old fibrin blue)

Red blood cells Yellow

Collagen Blue

Nuclei Blue/black

Muscle Red (paler than fibrin)

LENDRUM'S PICRO-MALLORY METHOD FOR FIBRIN

Fibrin is present in tissue where there has been tissue damage and is one of the plasma proteins.

Principle

This stain is very similar in principle to the trichrome techniques where small molecule dyes are used to stain less porous tissue followed by larger molecule dyes to stain looser textured structures such as collagen.

Specimen

Standard paraffin section - coated slide may be used to enhance adhesion Fixation -10% neutral buffered formalin

Control

Fibrin positive control eg inflamed appendix (peritonitis)

Reagents

(A) Weigert's Iron Haematoxylin (see Van Gieson's Technique) 2.4

(B) Picro-Orange
Dissolve 0.2g of Orange G [CI 16230] in 100 mL of 80% alcohol saturated with picric acid.

(C) Acid Fuchsin (CI 42685)
1% Acid Fuchsin in 3% trichloroacetic acid.

(D) 1% Phosphomolybdic Acid.

(E) 2% Aniline (Soluble) Blue C.I. 42755 in 2% Acetic Acid.

Procedure

1. Deparaffinize and hydrate to distilled water
2. Post-fix in Bouin's fluid at 60°C for one hour (to enhance staining).
3. Wash in water to remove yellow colour.
4. Stain with Weigert's Haematoxylin - 15-30 mins.
5. Wash in water-10 mins
6. Rinse in 95% alcohol

7. Stain in Picro-orange for 2 mins
8. Wash in water 5-10 secs.
9. Stain with acid fuchsin for 5 mins, then rinse in water.
10. Dip sections in equal parts Picro-Orange and 80% alcohol for a few seconds.
11. Differentiate in one percent Phosphomolybdic acid until colours are clear (5-10 mins), then rinse in water.
12. Stain with soluble Aniline Blue for 2-10 mins, then rinse in water.
13. Dehydrate, clear and mount.

Results

Fibrin Clear red

Red blood cells Orange/yellow

Collagen Blue

Nuclei Blue/black

Muscle Paler red

LIPIDS

Oil Red O for the Demonstration of Lipids

This method is suitable for tissue containing lipids in tiny droplets. Lipids dissolve less rapidly in this staining solution than in the Sudan Technique.

Principle

Lipids are soluble in the solvents used in standard histological tissue processing, therefore cryostat sections on fresh or fixed tissue are used. Oil-soluble dyes stain lipids by being more soluble in the lipids than in their solvents.

Control

Liver with fatty changes fixed in 10% neutral buffered formalin.

Reagents

(A) Oil Red O Stock Solution Oil Red O (C.I. 26125) 0.5 g Isopropanol 500 mL Warm the dye and alcohol in a long necked flask over a water bath at 56°C for 30-60 mins. Cool.

Oil Red O Working Solution Stock solution 30 mL or 12 mL Distilled water 20 mL 8 mL Mix well and stand for 5 to 10 mins. Filter. Use same day.

(B) Formal Calcium 40% formaldehyde 100 mL Distilled water 900 mL 10% Calcium chloride 100 mL

(C) Harris or Mayer's Haematoxylin

Procedure

1. Stain in working solution—5-10 mins in a closed container.
2. Wash well in water but gently (sections tend to float).
3. Harris Haematoxylin—$^1/_2$-1 min or Mayer's Haematoxylin three minutes.
4. Blue in tap water.
5. Rinse in distilled water.
6. Mount in aqueous mounting medium.

Note: Use a closed container when staining in working solution to avoid precipitation on the section due to evaporation of the solvent.

Results

Triglycerides (neutral fats) Deep orange red

Nuclei Blue

NILE BLUE SULPHATE METHOD FOR ACIDIC AND NEUTRAL LIPIDS

Principle

Nile blue sulphate serves as a preliminary test to indicate the distribution of neutral versus acid lipids. Nile blue sulphate comprises two components; a red oxazone which dissolves in neutral lipids and a blue oxazine which is basic and reacts with phospholipids and free fatty acids.

Specimen

1. 6 - 8 æ frozen sections on fixed tissue post fixed for one hour in formal-calcium.
2. 6 - 8 æ Cryostat sections on tissue fixed in 10 % neutral buffered formalin

Control

Phospholipids myelin Neutral fat subcutaneous tissue or fatty liver.

Reagents

(A) Nile Blue Sulphate Solution

1% aqueous Nile Blue Sulphate solution should be tested for the presence of oxazone (Nile Red) by shaking a little of it with a little amount of xylol. The xylol, after 20-30 seconds, should be a definite red. A poor oxazone content may be improved by the following method. Add 10mLs of 1% Sulphuric Acid to 200mLs of 1% Nile Blue Sulphate (C.I. 51180) in water. Boil under reflux for 4 hours. The solution should be at pH2 so that non-lipid reaction is minimal.

(B) Formal Calcium

40% formaldehyde 100 mL Distilled water 900 mL 10 % Calcium chloride 100 mL

(C) 1% Methyl Green, Zinc Chloride Salt

(Crystal Violet Free) Aldrich # 32,382 - 9. or 1% Chloroform-washed Methyl Green (C.I. 42585) 1% Aqueous methyl green which has been washed in a separating funnel with chloroform until all trace of methyl violet is removed and chloroform remains colourless.

Procedure

1. Cut frozen sections and fix for 1 hour in formal calcium.
2. Stain in Nile Blue Sulphate at 60°C for 30 mins.
3. Differentiate in 1% acetic acid for 1-2 mins.
4. Wash well and counterstain nuclei with 1% chloroform-washed methyl green 5 mins.
5. Wash well and mount in glycerine jelly.

Results

Unsaturated hydrophobic lipids Pink

Free fatty acids Pink to blue Phospholipids Blue

MICRO-ORGANISMS

Gram Stain

Bacteria Acid Fast Bacilli Donovan Bodies Viral Inclusions Fungi and Pneumocystis carinii

Control for Gram Stain

A method for making Gram Positive and Negative Controls in Tissue

Principle

The probability of obtaining a well-fixed piece of tissue containing both Gram-positive and Gram-negative bacteria in the exponential growth phase, uncompromised by antimicrobial therapy and therefore, displaying typical Gram staining, is remote:

1. Gram controls can be made using placental tissue which has been incubated with a heavy suspension of bacteria in a nutrient broth.
2. Lung may also be used - obtain from autopsy. Culture 24 hours then process.

Specimen

Fresh unfixed placenta or lung - for use on 2nd day of procedure. Gram - positive cocci (Staphylococcus)eg. Staph. aureus Gram - negative bacillus eg. E. coli, Proteus or Klebsiella

Reagents

(A) Blood Agar Plates (Microbiology Dept)

(B) Brain Heart Infusion Broth (Microbiology Dept)

Procedure

Day 1: Blood agar plates are heavily inoculated with pure cultures of a Gram-positive cocci (Staphylococcus) and a Gram-negative bacillus (Proteus or Klebsiella) and incubated overnight at 37°C.

Day 2: The next day the entire inoculum from each plate is suspended in 15 mL of Brain Heart Infusion broth. Small 0.5cm clippings of placenta are suspended in the inoculated broth and incubated at 37°C for three hours. Placenta is chosen because it is porous and is available, fresh and unfixed, from the labour and delivery rooms of most hospitals. After incubation, the clippings of placenta are fixed in 10% neutral buffered formalin overnight and processed in the usual manner.

Results

The stained slides show extracellular Gram-positive and Gram-negative bacteria throughout the tissue.

Note that the control may exhibit better Gram staining than the test slide as bacteria in the control are from an exponential growth phase population and fixed under ideal conditions.

GRAM STAIN FOR BACTERIA

A method for Gram Positive and Gram Negative Bacteria (Brown and Brenn)

Principle

Crystal Violet stains the nucleic acids of the bacteria (and background tissue) and after treatment with iodine, the sections are differentiated in acetone and counterstained with basic fuchsin. The tissue background and Gram-negative bacteria lose their blue staining and are subsequently stained with basic fuchsin. Gram-positive bacteria retain the crystal violet/iodine blue staining.

Specimen

Standard paraffin section Fixative 10% neutral buffered formalin

Control

Use control section containing both Gram positive and Gram negative bacteria. See method of preparing a Gram control-method 5 1.1

Reagents

(A) 1% Crystal Violet Solution
Crystal violet (C.I 42555) 1.0 g Distilled water 100 mL

(B) 5% Sodium Bicarbonate Solution
Sodium bicarbonate 5.0 g Distilled water 100 mL

(C) Gram's Iodine Solution
Iodine 1.0 g Potassium iodide 2.0 g Distilled water 300 mL

(D) Acetone
Solution 1 & 3) Flammable liquid. Store in refrigerator.

(E) 0.25% Basic Fuchsin Solution (Stock)
Basic Fuchsin (C.I. 42510) 0.25 g Distilled water 100 mL

(F) Basic Fuchsin Solution (Working)
Basic fuchsin solution (Stock) 10 mL Distilled water 100 mL

(G) Picric - Acid Acetone (Solution 2) (stored in fridge) Picric acid 0.1 g Acetone 100 mL

(H) Acetone - Histolene Solution (Solution 4) Acetone 50 mL Histolene 50 mL

Procedure

1. Deparaffinize and hydrate to distilled water.
2. Place slides on a staining rack. Pour on approx. 1 mL (or 20 drops) of crystal violet solution and add 5 drops of sodium bicarbonate solution for 1 minute. Agitate gently.
3. Rinse in tap water.
4. Flood slides with Gram's iodine solution for 1 minute.
5. Rinse in water and blot with filter paper to complete dryness.
6. Differentiate briefly with acetone (Solution 1 in coplin jar) until no more colour runs off. Do not overdifferentiate as Gram-positive bacteria will become negative. Rinse in water.
7. Repeat Steps 5 and 6 for further differentiation if section background is too blue.
8. Working basic fuchsin solution for 3 minutes. Wash in water.
9. Blot gently, but do not allow sections to dry completely.
10. Dip in acetone to start differentiation reaction.
11. Differentiate immediately with picric acid-acetone (Solution 2) until sections are yellowish pink.
12. Rinse quickly in acetone (Solution 3), then in acetone-histolene (Solution 4).
13. Clear in histolene, several changes.
14. Mount with Depex or Safe-t-mount.

Notes:

1. This method is invaluable in the demonstration of the filaments of Nocardia and Actinomyces. It must be realised however, that these filaments are not completely or strongly Gram positive.
2. It is possible to obtain either Gram positive or Gram negative results depending on the degree of differentiation. It is suggested therefore, that at least two slides be run with varying degrees of differentiation in order to lessen the possibility of over-differentiation at Step 9 of the staining procedure.

Results

Gram positive bacteria Blue

Gram negative bacteria Red

Filaments of Nocardia & Actinomyces Blue

Nuclei Red Other tissue elements Yellow

DIFF-QUIK STAIN FOR CAMPYLOBACTER-LIKE ORGANISMS

Diff-Quik Stain. Modified for the demonstration of Campylobacter-like organisms in gastric biopsies.

Principle

The Diff-Quik stain is a modification of Wright's stain which is used mainly for blood films and bone marrow preparations. The original Diff-Quik method included a fixative stage for use with unfixed blood and bone marrow films. For use with paraffin sections we omitted this step.

Specimen

Fixation: Neutral buffered formalin Sections: Standard paraffin section

Control Sections

Tissue containing Helicobacter pylori.

Reagents

1. Commercial Diff-Quik Solution (Red) in a staining jar.
2. Commercial Diff-Quik Solution 3 (Blue) in a staining jar.

Procedure

1. Deparaffinise sections and hydrate to water.
2. Dip slides into Diff-Quik Solution 2 (Red) for 3 seconds. (3×1 second dips) Allow excess stain to drain.
3. Dip slides into Diff-Quik Solution 3 (blue) for 6 seconds (6×1 second dips) Allow excess stain to drain.
4. Rinse with distilled water.
5. Blot sections carefully and air-dry using gentle warmth.
6. Dip in histolene and mount in Safety-Mount.

Result

CLOs Dark blue

Tissue elements Shades of blue, pink and red

ZIEHL-NEELSEN STAIN FOR TUBERCLE BACILLI (A.F.B.)

Principle

Mycobacteria tuberculosis (tubercle bacilli) have a lipid-rich cell wall which is capable of taking up strong phenol dye solutions (e.g. carbol fuchsin solution) such that the dye is retained upon subsequent differentiation in acid or alcohol ie. they are said to be acid and alcohol fast (AAFB = acid and alcohol fast bacilli).

Specimen

Standard paraffin section Fixative 10% neutral buffered formalin

Control

Use a control positive for tubercle bacilli.

Reagents

(A) Carbol Fuchsin Solution
Dissolve 1 g basic fuchsin (CI 42510) in 10 mL absolute alcohol Dissolve 5 g of phenol in 100 mL distilled water Mix the two solutions and filter before use Warning: Phenol is highly toxic/ corrosive Wear gloves. Re-seal bottle with paraffin wax

(B) 1% Acid Alcohol 70% alcohol 99.0 mL Conc. HCl 1.0 mL

(C) Stock Ethylene Blue Solution
Dissolve 1.4g methylene blue (CI 52015) in 100 mL 95 % alcohol

Working Methylene Blue Solution

Stock methylene blue 10 mL Tap water 90 mL

Procedure

1. Deparaffinise sections and hydrate to distilled water.
2. Pre-heat filtered carbol fuchsin to 56°C in Coplin jar in water bath/oven for 15 mins, then add sections and heat in water bath/oven for 30 mins.
3. Wash well in tap water.
4. Differentiate in acid alcohol until faint pink (do NOT over differentiate).
5. Wash in water.
6. Counterstain lightly with Working methylene blue solution for 2 - 3 seconds.
7. Dehydrate in alcohol.
8. Clear and mount.

Note: Blue counterstain should be pale to facilitate detection of the red staining organisms.

Results

AFB (acid-fast bacilli) Magenta

Background Weak blue

WADE - FITE FOR DEMONSTRATION OF MYCOBACTERIUM LEPRAE

Principle

The Wade-Fite technique is recommended for the demonstration of Mycobacterium leprae. In this technique the more gentle dewaxing of the section, and the minimal exposure to organic solvents, protects the rather precarious acid-and alcohol fastness of the organism.

Specimen

Standard paraffin section Fixative 10% neutral buffered formalin

Control

Always use a control that is positive for Mycobacterium leprae.

Reagents

(A) Deparaffinising Solution
Prepare two coplin jars of the following mixture. Turpentine Oil 40 mL Xylene substitute (Histolene) 20 mL

(B) Carbol New Fuchsin Solution
Dissolve 0.5 g new fuchsin (CI 42520) in 10 mL absolute alcohol Dissolve 5 g phenol in 90 mL distilled water Mix the two solutions and filter. Caution: Phenol is highly toxic/corrosive Wear gloves Re-seal bottle with paraffin wax.

(C) 10% Aqueous Sulphuric Acid

(D) Working Methylene Blue Solution Stock methylene blue 10 mL Tap water 90 mL

Procedure

1. Take section straight from hot-plate and de-paraffinise in two changes of Turpentine Oil/ Histolene mixture - approximately 5 minutes each change.
2. Blot dry on paper hand towel.
3. Boil Carbol Fuchsin Solution and filter onto section - leave on for 10 minutes.
4. Wash in water 10 minutes.
5. Differentiate in 10% aq sulphuric acid approximately 3 seconds depending on tissue. Sections turn blue/black.
6. Wash in running water 15 minutes.
7. Counterstain with 2-3 dips in working Methylene Blue Solution.
8. Dehydrate rapidly.

Results

Leprosy bacilli Magenta

Background Pale blue

WARTHIN STARRY METHOD FOR SPIROCHAETES AND DONOVAN BODIES

Treponema pallidum is the causative organism of syphilis and can be demonstrated with this silver technique. Granuloma inguinale is a venereal disease caused by a gram-negative bacteria, Calymmatobacterium granulomatosis. These bean-shaped bacilli, are also called Donovan bodies and can be demonstrated with the Warthin-Starry silver technique.

Principle

Organisms are demonstrated by silver impregnation technique.

Specimen

Standard paraffin section - three per block Fixative 10% neutral buffered formalin

Control

Use a Control that is positive for Treponema pallidum or Donovan bodies. 3 Control slides, developed from 5 to 10 minutes are used.

Equipment

Plastic forceps, 2 x water baths, one at 43°C and one at 54°C

Reagents

All glassware should be clean, rinsed in distilled water and dried.

(A) Acidulated Water 1000 mLs (Distilled water brought to pH4 with a few granules of Citric Acid) All solutions are prepared using acidulated water. Prepare fresh solutions in acidulated water of the following:

(B) 2% Silver Nitrate 100 mL

(C) 5% Gelatin 100 mL

(D) 0.15% Hydroquinone 100 mL

(E) Developer Solution 2% silver nitrate (in acidulated water) 15 mL 5% gelatin (in acidulated water) 37.5 mL 0.15% hydroquinone (in acidulated water) 20.0 mL Heat separately to 54°C Mix well just before use.

Procedure

1. Rinse all glassware with acidulated water.
2. Deparaffinise test sections × 3 + Control Sections × 3, hydrate to distilled water and then rinse in acidulated water.
3. Dilute 25 mLs of the 2% solution silver nitrate with 25 mLs of acidulated water. Heat the 1% silver solution in a water bath to 43°C for 15 mins. Impregnate the sections with 1% silver nitrate at 43°C for 30 mins.
4. Measure solutions for developer and keep separately at 54øC. Place a coplin jar of acidulated water into 54°C water bath for rinsing after development.
5. Mix developer in a coplin jar. Place slides in developer solution at 54°C (approx. 5 - 10 minutes).

   Notes:

   (a) Development is critical. Sections develop from pale yellow (under-developed) to golden brown (over-developed).

(b) Process 3 control sections and remove the patient's slide and a control slide at 3 different time intervals e.g. 3, 5 and 8 minutes.

(c) Developing time varies from approximately 3 to 10 mins and even up to 15 minutes.

(d) Donovan bodies require longer development than spirochaetes.

(e) Correct temperatures and impregnation times are very important.

(f) Rinse in acidulated water at 54°C

(g) Rinse in distilled water.

(h) Dehydrate, clear and mount.

Results

Spirochaetes Black

Donovan bodies Black

Background Yellow - brown

GIEMSA STAIN FOR DONOVAN BODIES AND PARASITES

Principle

This method is a modified version of the original Giemsa technique used for haematological smears and gives good results for sections. Giemsa is a Romanowsky stain which contains azure B and eosin Y and is capable of making subtle distinction in shades of staining. The acidic groupings of the nucleic acids and proteins of the cell nuclei determine their uptake of the basic dye azure B and the presence of basic groupings result in an affinity for acidic dyes and their staining by eosin.

Specimen

Standard paraffin section fixed in 10% neutral buffered formalin. Air dried smears, post fixed in methanol or 95% alcohol - 20 mins. CONTROL Use a positive control section to demonstrate Donovan bodies.

Reagents

(A) Stock Commercial Giemsa (Gurr's Improved R66 # 35086)

(B) Phosphate Buffer The pH 6.8 buffer can be prepared from buffer tablets. Dissolve one tablet in distilled water and make up to one litre. Alternative method. Stock A 0.2 M Sodium dihydrogen orthophosphate (M.W156) 3.12 g Sodium dihydrogen orthophosphate in 100 mLs distilled water Stock B 0.2 M disodium hydrogen orthophosphate (M.W142) 2.83 g disodium hydrogen orthophosphate in 100 mLs distilled water Composition of Buffer pH 6.8 25.5 mL Stock A + 24.5 mL Stock B made up to 100 mLs with distilled water.

(C) 1% Aqueous Acetic Acid. Overnight Staining Dilute 1.0 mL of (A) with 50 mL of (B) 40 Minute Staining Dilute 5 mL of (A) with 50 mL (B)

Procedure

1. Deparaffinise and hydrate sections to water.
2. Hydrate post-fixed smears (see specimen requirements) in water for one minute.
3. Stain in Giemsa overnight or 40 mins in stronger solution. Wash in water.
4. Differentiate with acetic acid until sections are pale purple in colour (approximately 4 secs).
5. Wash in water.
6. Air dry sections.
7. Mount in Safety Mount

Notes

To achieve a good colour balance it is necessary to initially overstain with the Giemsa stain, and then differentiate with 1% acetic acid.

Results

Nuclei Blue - purple
Azurophilic granules Blue
Collagen Pale pink
Donovan bodies Pale blue
Protozoons Dark blue
Eosinophil granules, Red blood cells & Pink Other acidiphil structures

PHLOXINE-TARTRAZINE TECHNIQUE FOR VIRAL INCLUSIONS

Principle

All tissues are stained red with phloxine, then the red stain is differentiated out with tartrazine in cellosolve which also acts as a yellow counterstain. This stain is used to demonstrate many structures according to the degree of differentiation.

Specimen

Standard paraffin section fixed in 10% neutral buffered formalin.

Control

Viral inclusion bodies - viral warts Molluscum contagiosum, cytomegalic disease Paneth cell granules - small intestine

Notes

1. The red colour is removed first from muscle and other connective tissues; the last structures to lose their red colour are viral inclusions, Paneth cell granules and red blood cells.
2. With less extensive differentiation this method can also be used to display pancreatic beta cells, Russell bodies (within plasma cell cytoplasm), keratin (e.g. keratin pearls and individual cell

keratinisation in squamous carcinoma, and inhaled squames in the lungs of still-births and neonatal deaths) and fibrin.

3. The variable staining of keratin can be a problem when attempting to demonstrate viral inclusions in skin lesions.

Reagents

(A) Phloxine Solution
Phloxine (Phloxin B CI 45410) 0.5 g Calcium chloride 0.5 g Distilled water 100 mL

(B) Tartrazine Solution (CI 19140)
Saturated solution of tartrazine in 2-ethoxy-ethanol ('cellosolve') - store in flammables cupboard.

Procedure

1. Dewax in histolene, then hydrate through graded alcohols to water.
2. Stain nuclei in alum haematoxylin for two minutes.
3. Wash in running tap water for five minutes.
4. Stain in phloxine solution for 20 minutes.
5. Rinse in tap water, then blot dry.
6. Stain in tartrazine solution under microscopic control until only the viral inclusions remain strongly red. An average time is 5-10 minutes.
7. Rinse in 95 % alcohol.
8. Dehydrate in absolute alcohol, clear in histoclear and mount in Safe-t-mount.

Results

Viral inclusions Bright red

Red blood cells Variably orange-red

Nuclei Blue - grey

Background Yellow

ORCEIN STAIN FOR HEPATITIS B ANTIGEN (HBSAG) IN LIVER TISSUE

(Antigen) The Hepatitis B virus causes an acute inflammation of the liver. Hepatic cell necrosis is associated with leucocytic and histiocytic reaction and infiltration. HBSAg may be stained by orcein in liver of carriers and chronic hepatitis, not acute.

Specimen

Standard paraffin section Fixation Neutral buffered Formalin.

Control

HBAg in liver tissue.

Reagents

(A) Acidified Potassium Permanganate Solution
Potassium permanganate 0.3g, Distilled water 100mL, Concentrated Sulphuric acid 0.3mL

(B) 1.5% Aqueous Oxalic Acid

(C) Orcein Stain
Dissolve 1.0 g of synthetic orcein in small amount of 70 % alcohol with aid of gentle heat e.g. 56øC water bath. Make up final volume to 100 mL of 70% alcohol, then add 1.0 mL concentrated hydrochloric acid.

(D) 1% Acid Alcohol

Procedure

1. Deparaffinize and hydrate to distilled water.
2. Oxidise in permanganate for five mins.
3. Wash in water
4. Bleach in oxalic acid until white
5. Wash in 70% alcohol.
6. Stain in orcein for one and a half hours.
7. Wash in water.
8. Differentiate in acid alcohol until no colour runs off the slides.
9. Wash in water.
10. Dehydrate, clear and mount.

Results

Australia antigen purplish (The distribution of Australian antigen varies: fine diffuse granules in the cytoplasm in sub- lobar clumps of hepatocytes to round or oval cytoplasmic clumps in single scattered hepatocytes.) Cytoplasm much lighter brown

VICTORIA BLUE—NUCLEAR FAST RED FOR HBSAG

Principle

To demonstrate copper associated protein, elastic and Hepatitis B surface antigen (HBSAg) in paraffin wax sections.

Specimen

10 per cent neutral buffered formalin. Standard paraffin wax sections.

Control

Known case of primary biliary cirrhosis.

Reagents

(A) Victoria Blue

Distilled water 200 mL, Dextrine 0.5 g, Victoria Blue 2 g, Resorcinol 4 g, Gradually warm up the mixed solution of all the above until it boils. Remove form heat then gradually add 25mL of boiling 29% ferric chloride solution (12mL commercial 60% ferric chloride solution plus 13mL distilled water) and boil the solution for a further 3 minutes. Cool. After cooling filter it. Dry the filtrate on the filter paper in a 56°C oven until completely dry. Dissolve the dried filtrate in 400mL of 70% alcohol. Finally add 4mL concentrated hydrochloric acid and 6g phenol. It is best to leave this solution to mature for at least two weeks prior to use.

(B) Nuclear Fast Red

Dissolve 0.5g nuclear fast red in 500mL of warmed 5% aluminium sulphate. Filter when cool.

(C) 4% Aqueous Sodium Metabisulphite

Procedure

1. Deparaffinize and hydrate to distilled water.
2. Treat with acidified potassium permanganate (as for Gordon and Sweet's Reticular fibres method 2.4) for 5 minutes.
3. Treat with 4% sodium bisulphite for 1 minute.
4. Wash in running tap water.
5. Wash well with 70% alcohol.
6. Stain in the Victoria Blue solution, in a Coplin jar, for a minimum of four hours (preferably overnight).
7. Wash well with 70% alcohol for approx one minute. This is the differentiation step - ensure the background of the section is clear.
8. Wash in running tap water for one minute.
9. Stain with nuclear fast red solution for five minutes.
10. Wash in running water for two minutes.
11. Dehydrate, clear and mount.

Results

HBSAg, elastic fibres, copper associated protein blue

All other tissue components pink/red

Health and Safety

Potassium permanganate Harmful by ingestion. Irritating to eyes and skin Ferric chloride Irritant. Hydrochloric acid Vesicant. Causes burns. Phenol Harmful by ingestion. Irritating to eyes and skin.

GROCOTT-GOMORI METHENAMINE SILVER METHOD FOR FUNGI

Principle

This method depends upon the reduction of the silver by the aldehyde groups produced after oxidation of fungal wall components with chromic acid.

Specimen

Standard Paraffin section fixation Neutral buffered Formalin.

Control

Use a control positive for Fungi.

Reagents

(A) 5% Aqueous Chromic Acid (Chromium trioxide)

(B) 1% Aqueous Sodium Metabisulphite

(C) Stock Hexamine-Silver Solution

To 100 mL 3% hexamine (methenamine) add 5 mL of 5% aqueous silver nitrate. (Highly Toxic-Oxidiser)

A white precipitate will form which dissolves on shaking. This solution will keep for a few months at 4°C.

Working Hexamine - Silver Solution Dilute 2 mL of 5% aqueous sodium tetraborate (borax) with 25 mL distilled water, then add 25 mL stock solution. Filter and warm in 56°C oven.

(D) Stock Light Green (CI 42095) Solution

Dissolve 0.2 g of light green in 100 mL distilled water. Add 0.2 mL glacial acetic acid.

Working Light Green Solution

Dilute 10 mL stock solution with 50 mL distilled water. Add 1 drop glacial acetic acid.

(E) 5% Sodium Thiosulphate

(F) 0.2% Gold Chloride (Sodium Chloro-aurate) Irritant

Kept with reticulin fibre stains method 2.5.

Procedure

1. Deparaffinize and hydrate to distilled water.
2. Oxidise in chromic acid for 1 hour.
3. Wash in water
4. Bleach in metabisulphite for 1 minute.
5. Wash in distilled water.
6. Treat sections in pre-heated Working hexamine-silver solution at 56°C water bath/oven until sections are yellowish/brown in colour ($^1/_2$-1 hour).

7. Wash in distilled water.
8. Tone in gold chloride for two minutes.
9. Wash in water.
10. Fix in 5 % thiosulphate for five minutes.
11. Wash in water.
12. Counterstain lightly with Working light green solution 5-10 seconds.
13. Dehydrate in alcohol.
14. Clear and mount.

Results

Fungal elements Black

Background Pale green

Uric acid & Urates Black

Pneumocystis carinii Black

GRAM WEIGERT STAIN FOR PNEUMOCYSTIS CARINII AND GRAM- POSITIVE ORGANISMS

Pneumocystis carinii is a protozoon which mainly infects the lungs and is often a fatal pneumonia in patients on immunosuppressive therapy or patients with AIDS.

Principle

This is a modified Gram technique. The Gram stain is superimposed on a haematoxylin and eosin stain. Only Gram-positive organisms are demonstrated by this technique and not Gram-negative organisms. Pneumocystis carinii can also be demonstrated.

Specimen

Standard paraffin section Fixation Neutral buffered Formalin.

Control

Use a control positive for *Pneumocystis carinii.*

Reagents

(A) 2.5% Aqueous Eosin

(B) Aniline Crystal Violet
Crystal violet CI 42555 5 g, Absolute alcohol 10 mL, Aniline 2 mL, Distilled water 88 mL. This solution keeps well

(C) Lugol's Iodine
Iodine 1 g, Potassium iodide 2 g, Distilled water to 100 mL. Dissolve the potassium iodide in 4-5 mL of water; dissolve the iodine in this. Dilute to 100 mL to make Lugol's solution.

(D) 50/50 Aniline/Histolene

Procedure

1. Deparaffinize and hydrate to distilled water.
2. Stain nuclei lightly (approx 1 minute) with alum haematoxylin.
3. Blue in tap water
4. Stain in 2.5 % aqueous phloxine or eosin for 10 minutes at 56°C
5. Wash in water.
6. Stain in aniline crystal violet for $^1/_2$-1 hour.
7. Rinse in water.
8. Treat with Lugol's iodine for 1 minute.
9. Blot with fine filter paper.
10. Differentiate with equal part of histoclear and aniline until only bacteria and fibrin are blue-black (it is almost impossible to over-differentiate).
11. Rinse in xylol to remove aniline, and mount in Canada balsam or DPX.

Results

Gram-positive organisms Blue -black

Nuclei Blue

Other tissue constituents Shades of red

Pneumocystis carinii Blue with one side denser Blue stain.

TISSUE CARBOHYDRATES—GLYCOGEN AND MUCINS

Periodic Acid Schiff (Pas) for the Demonstration of Glycogen

Glycogen is a simple intracytoplasmic polysaccharide found in greatest amount in liver, cardiac and skeletal muscle, with significant quantities also present in hair follicles, endometrial glands, vaginal and cervical epithelium, umbilical cord, neutrophil leucocytes and megakaryocytes.

Principle

Tissue containing vicinal glycol groups or their amino or alkylamino derivatives are oxidized by periodic acid to form dialdehydes, which combine with Schiff's reagent to form an insoluble magenta compound. The chromphoric groups of basic fuchsin in Schiff's reagent are broken by sulphuration to form a colourless solution. In the presence of free aldehyde groups an insoluble coloured compound similar to the original dye is formed.

Specimen

Standard paraffin section fixed in 10% neutral buffered formalin.

Control Normal liver

Reagents

(A) 1% Aqueous Periodic Acid Schiff Reagent

Commercial Schiff reagent Australian Biostain Product AB1200.0600 (kept in fridge) If unavailable, use the following: Basic fuchsin (CI 42500) Sigma cat # P1528 2.0 g Distilled water 400 mL Potassium or sodium metabisulphite 4.0 g Conc. hydrochloric acid 4 mL Decolourising activated charcoal 0.4 g

1. Bring water to boil, remove flame and add basic fuchsin slowly and stir until dissolved.

   Warning: Use a conical flask and always remove from heat after boiling, then add basic fuchsin slowly. If added too fast solution will bubble explosively out of the flask.

2. Cool to 50°C, add metabisulphite. Dissolve and cool to room temperature.
3. Add conc. HCl and mix.
4. Leave overnight in the dark at room temperature. Add charcoal and shake for 1 - 2 minutes. Filter and store at 4°C in dark container. Bring solution to room temperature before use. With use it will eventually be found that the originally colourless to pale yellow solution will turn pink due to loss of sulphur dioxide causing restoration of the basic fuchsin colour. When this happens, the solution should be discarded. If a white precipitate forms it should be filtered out.

   Notes: Schiff Reagent - test of activity: The Schiff reagent may be tested by adding a few drops to 3 to 5 mL of 40 formaldehyde on a petri dish: active Schiff reagent will lead to the rapid development of a reddish purple colour; delayed development of a deep bluish purple indicates that the reagent is going off.

Procedure

1. Deparaffinize and hydrate to distilled water.
2. Oxidise with periodic acid for 5 minutes.
3. Rinse in distilled water.
4. Treat with Schiff's reagent for 5 minutes.
5. Wash in running water for 10 minutes; this intensifies the colour reaction.
6. Stain the nuclei with Mayer's haematoxylin for one minute (no need to blue).
7. Wash in water.
8. Dehydrate, clear and mount.

Results

Nuclei Blue

PAS positive material Magenta

Glycogen Cellulose Chitin Neutral mucin Sialomucin: enzyme-labile Starch Sulphated sialomucins Amoebae Fungi Basement membranes.

PERIODIC ACID SCHIFF (PAS) WITH DIASTASE FOR THE ENZYME EXTRACTION OF GLYCOGEN

Principle

Glycogen is digested with certain forms of amylase. Commercially available diastase, which contains both α and β amylase or salivary amylase from saliva can be used to digest glycogen in tissue sections.

Specimen

Standard paraffin section fixed in 10% neutral buffered formalin.

Control

Tissue containing glycogen e.g. liver. Use two positive controls and two test sections Treat one test section and one positive control with diastase; then stain all sections with PAS.

Reagents

(A) Amylase Solution

Amylase Type VI-B from porcine pancreas (Sigma A 3176) 0.25g (stored in fridge). Distilled water 10mL. If larger quantity required, use 1g Amylase in 40mL distilled water. Prepare fresh just before use. Warm water for 5 sec in microwave oven then add amylase.

Note : AQIS requirements for disposal. Place waste amylase into container and place into yellow biohazard bag for disposal by incineration..

Procedure

1. Deparaffinise two positive controls and two test sections and hydrate to distilled water.
2. Treat one control and one test section with Amylase for 25 minutes to digest Glycogen.
3. Wash in running water for 20 minutes.
4. Proceed with P.A.S. technique as follows.
5. Oxidise with periodic acid for 5 minutes.
6. Rinse in distilled water.
7. Treat with Schiff's reagent for 5 minutes.
8. Wash in running water for 10 minutes, this intensifies the colour reaction.
9. Stain the nuclei with Mayer's haematoxylin 1 minute (no need to blue).
10. Wash in water.
11. Dehydrate, clear and mount.

Results

Presence of glycogen will be evidenced by loss of staining after enzyme treatment when compared to the untreated sections.

ALCIAN BLUE TECHNIQUE FOR ACID MUCINS

Acid mucin - strongly sulphated connective tissue mucins react at low pH values with suitable cationic dyes and are usually PAS-negative.

Principle

Alcian blue is a copper phthalcyanin dye and contains positively charged groups capable of salt linkage with certain polyanions. These polyanions consist of the sulphate and carbocyl radicals of the acid mucins - the phosphate radicals of the nucleic acids do not react. Consequently, only the acid mucins are stained. By varying the pH of the solution of Alcian blue more information can be gained concerning the types of acid mucin present.

Specimen

Standard paraffin section fixed in neutral buffered formalin.

Control

Large bowel

Reagents

(A) Alcian Blue Solutions (CI 74240)
pH 3.1 - 1% Alcian blue in 0.5% acetic acid Routine: pH 2.5 - 1% Alcian blue in 3.0% acetic acid Use pH 2.5 for routine staining of acid mucins. Solutions keep for 6 months and should be filtered before use. pH 1.0 - 1% Alcian blue in N/10 hydrochloric acid. pH 0.2 - 1% Alcian blue in 10% sulphuric acid.

(B) 3% Aqueous Acetic Acid (C) 0.5% Neutral Red

Procedure

1. Deparaffinize and hydrate to distilled water.
2. Treat in 3% acetic acid for two minutes.
3. Drain off.
4. Stain in 1% alcian blue (pH 2.5) for 30 mins.
5. Wash in water.
6. Counterstain with neutral red lightly (three mins).
7. Dehydrate rapidly in alcohol.
8. Clear and mount.

Results

Alcian blue solution at: pH 3.1 and 2.5 most acid mucins (except some of the strongly sulphated group) Blue

pH 1.0 only weakly and strongly sulphated acid mucins Blue

pH 0.2 only the strongly sulphated acid mucins Blue

Nuclei Red

Background Very pale red or colourless

COMBINED ALCIAN BLUE PAS TECHNIQUE FOR ACID AND NEUTRAL MUCINS

Principle

This stain demonstrates the presence of mucins and clearly distinguishes between acid and neutral mucins. By first staining all acid mucins with alcian blue, those acid mucins which are also PAS positive will not react with subsequent PAS reaction, only the neutral mucins.

Control

Large bowel.

Reagents

(A) Alcian Blue Solutions (CI 74240) 1 g 3% Acetic Acid 100mL Store in fridge.

(B) Schiffs Reagent See P.A.S. technique

(C) 1% Periodic Acid

(D) 3% Acetic Acid

Procedure

1. Deparaffinize and hydrate to distilled water.
2. Acetic acid for two mins.
3. Drain.
4. Alcian Blue Solution 10 minutes.
5. Wash in water, then distilled water.
6. Treat with 1% periodic acid for five mins.
7. Rinse in distilled water.
8. Treat with Schiffs reagent five mins.
9. Wash in running tap water 5-10 minutes.
10. Stain nuclei with Mayer's haematoxylin for 1 minute 30 seconds. Nuclear overstaining causes masking of the Alcian Blue staining.
11. Wash in water.
12. Dehydrate, clear and mount.

Results

Acid mucins Blue

Neutral mucins Magenta

Nuclei Pale blue

COMBINED ALCIAN BLUE/DIASTASE/PAS FOR ACID AND NEUTRAL MUCINS

Principle

This stain is useful for the demonstration of acid and neutral mucins. Alcian blue stains acid mucins, subsequent treatment with diastase digests glycogen from the tissue and neutral mucins can then be stained with PAS.

Specimen

Standard paraffin section fixed in 10% neutral buffered formalin.

Control

Two positive (for glycogen) controls + two test sections. Treat one control + one test section with diastase. One control section + one test section are NOT treated.

Reagents

(A) 3% Acetic Acid

(B) Alcian Blue Solution

(C) Amylase Solution
0.25g in 10mL distilled water. Pre-warm water for 5 sec in microwave oven then add amylase.
Note: AQIS requirements for disposal.
Place waste amylase into container, cap and place into yellow biohazard bag for disposal by incineration.

(D) 1% Periodic Acid.

(E) Schiffs Reagent.

Procedure

1. Deparaffinize and hydrate to distilled water.
2. Treat one control and one test section with amylase for 25 minutes.
3. Wash in running water for 20 minutes.
4. Acetic acid two minutes.
5. Drain.
6. Alcian blue 10 minutes.
7. Wash in tap water then distilled water.
8. Treat with 1% periodic acid for 5 minutes.
9. Rinse in distilled water.
10. Treat with Schiff's reagent 5 minutes.
11. Wash in running tap water 5-10 minutes.
12. Stain nuclei with Mayer's Haematoxylin 1 minute 30 seconds (or 1 container of Harris haematoxylin in linear stainer).
13. Wash in water.
14. Dehydrate, clear and mount.

Results

Acid mucins Blue

Neutral mucins Magenta

BEST'S CARMINE TECHNIQUE FOR GLYCOGEN

Principle

This staining technique demonstrates glycogen by hydrogen bond formation between OH groups on the glycogen, and H atoms of the carminic acid. Fibrin and neutral mucin stain weakly with this method.

Control

Liver.

Reagents

(A) Carmine Stock

60 mL H20 + 2g carmine (CI 75470) 1 g Potassium carbonate 5 g Potassium chloride Boil gently in a large conical flask. 5 mins. Cool and add 20mLs concentrated ammonia Filter and store in a dark container at 4°C. This will keep 1 - 2 months.

(B) Working Solution

Stock solution 15 mL Ammonia concentrated solution 12.5 mL Methanol 12.5 mL Staining time to be increased as stock solution ages.

(C) Best's Differentiator

Methanol 40 mL Ethanol 80 mL Distilled water 100 mL PROCEDURE

1. Dewax test + positive control sections and hydrate to water.
2. Duplicate sections may be treated with diastase if desired.
3. Stain nuclei of all sections well using one of the iron haematoxylin solutions e.g. Weigerts iron haematoxylin.
4. Treat with carmine solution for 10 mins.
5. Transfer slides quickly to a Coplin jar of Differentiating solution.
6. Wash in alcohol (NOT water), clear in histoclear and mount.

Results

Glycogen Bright red Neutral mucin, mast cells, fibrin Paler red Nuclei Blue

SOUTHGATE'S MUCICARMINE

For Epithelial Mucin This method is useful to stain encapsulated fungi of *Cryptococcus neoformans*, but the combined Alcian Blue - PAS technique is more informative to establish presence or absence of tissue mucins.

Principle

Mucicarmine has large molecular size which allows the dye to penetrate and bind to acidic substrates of low density, i.e. mucins. Neutral mucins and some strongly acidic sulphated mucins do not show appreciable staining.

Specimen

Standard paraffin section fixed in 10% neutral buffered formalin.

Reagents

(A) Mucicarmine Solution

Carmine 2g, Aluminium hydroxide 2g, 50% alcohol, 100 mL Aluminium chloride, anhydrous 1g. Mix well, heat rapidly to boiling point and boil for 2 minutes. Cool and filter. Add the carmine and aluminium hydroxide to the alcohol in a long-necked 500 mL flask. Ground the aluminium chloride in a mortar and add it to the mixture. Note the warning. Warning: Aluminium chloride gives off HCl fumes from the bottle. Prepare in biochemistry fume hood. Wear gloves, safety glasses and mask when removing lid from and using aluminium chloride.

(B) 0.5% Acid Alcohol (0.5mL HCl in 100mL 70% alcohol).

Procedure

1. Deparaffinise and hydrate to distilled water.
2. Stain with haematoxylin in linear stainer.
3. Rinse in running water.
4. Differentiate quickly in 0.5% acid alcohol. All background should be clear with only nuclei stained.
5. Rinse in running water 5 minutes.
6. Stain in Mucicarmine solution for 2 hours.
7. Rinse in distilled water.
8. Dehydrate, clear and mount.

Results

Epithelial mucin Red

Nuclei Blue

NERVOUS TISSUE

BIELSCHOWSKY'S SILVER STAIN FOR AXONS (MODIFIED)

Principle

This is a silver technique. Sections are first impregnated with silver nitrate and the deposit is intensified by means of a solution containing silver nitrate and a reducing substance - formalin.

Control

Nervous tissue Spinal cord

Reagents

(A) 20% Silver Nitrate Solution
(Caution: Silver nitrate is highly toxic and stains clothing/skin-wear gloves)

(B) 0.1% Ammonia Solution

(C) Developer
10% formalin 20 mL Distilled water 100 mL Concentrated nitric acid 1 drop (Caution: Highly toxic - burns skin) Citric acid 0.5 g

(D) 0.2% Gold Chloride

(E) 5% Sodium Thiosulphate

All solutions should be freshly prepared using clean glassware and distilled water.

Equipment

37°C water bath/oven

Procedure

1. Deparaffinize and hydrate to distilled water. Wash well in distilled water.
2. Place sections in 50 mL of 20% silver nitrate in the dark at 37°C for 15 mins. Keep this silver solution for stage 4.
3. Wash well in distilled water.
4. Add strong ammonia to the 50 mL of silver nitrate (from stage 2) until the initial precipitate disappears. Leave section in this solution for 10 minutes at 37°C.
5. Wash section in 0.1% ammonia.
6. Add 6 - 10 drops of developer solution to the silver hydroxide solution (from stage 4). Stain section for 3 - 5 minutes or until sections turn almost black.
7. Wash in ammoniated water and then distilled water.
8. Tone in 0.2% gold chloride for 1-2 minutes.
9. Fix in 5% sodium thiosulphate for 1 minute.
10. Wash, dehydrate, clear and mount.

Results

Axons Black

LUXOL FAST BLUE TECHNIQUE FOR MYELIN

Principle

Luxol fast blue is one of the copper phthalocyanin dyes with a strong affinity for phospholipids and strong choline bases; hence presumably its affinity for myelin.

Specimen

Standard paraffin sections 8-10 frozen sections.

Control

Nervous tissue - cerebellum

Reagents

(A) Luxol Fast Blue Solution
Luxol fast blue 0.1 g 95% alcohol 100 mL 10% acetic acid 0.5 mL

(B) Cresyl Fast Violet Solution
0.1% (aq) NOTE: No CI number. Manufacture of Cresyl Fast Violet discontinued. Substitute Cresyl Violet Acetate (no CI number) Sigma C1791 or C5042.

(C) Lithium Carbonate Solution
0.05% (aq)

(D) Cresyl Violet Differentiator
95% alcohol 90 mL Chloroform 10 mL Glacial acetic acid 3 drops

Equipment

37°C water bath/oven

Procedure

1. Deparaffinize and hydrate to 95 % alcohol.
2. Stain in Luxol Fast Blue overnight at 37°C.
3. Wash in 95% alcohol, then in distilled water.
4. Commence differentiation by immersing sections in lithium carbonate solution for not more than 20 seconds.
5. Differentiate in 70% alcohol until grey and white matter is clearly distinguished (30-60 seconds).
6. Rinse in distilled water and examine under microscope.
7. If differentiation not complete repeat steps 4 to 6 with reduced times.
8. Wash well in distilled water.
9. Stain in cresyl violet for 10 minutes.
10. Wash in distilled water.
11. Wash in 70% alcohol.
12. Differentiate in cresyl violet differentiator 1-2 seconds.
13. Rinse in 95% alcohol.
14. Rinse in absolute alcohol and clear in histolene.
15. Check differentiation under microscope to ensure only nuclei (and Nissl substance) are stained. Repeat steps 11 - 14 if necessary.
16. Mount.

Results

Myelin, RBCs Blue Nuclei & Nissl substance Purple

PAS-ORANGE G METHOD FOR ANTERIOR PITUITARY CELLS

Principle

The pituitary gland has two main components, the adenohypophysis (anterior pituitary) and the neurohypophysis (posterior pituitary). The adenohypophysis is composed of solid masses of lightly packed epithelial cells which are divided into three groups on the basis of their staining characteristics:

1. Chromoplobe cells (approx. 50%) which have a scanty non-granular cytoplasm.
2. Eosinophil or acidophil cells (approx. 40%) found in clumps in the lateral wings of the anterior pituitary.
3. Basophil cells, which have blue staining (with H&E or aniline blue) granules in their cytoplasm. Numerous hormones are produced by the adenohypophysis, A.C.T.H., T.S.H., F.S.H. and L.H.

Specimen

Standard formalin.

Control

Normal pituitary gland

Reagents

(A) 1% Aqueous Periodic Acid

(B) Schiff Reagent
(Use commercial product or refer to 6.1 for method to make Schiffs reagent)

(C) 2% Orange G in 5% Phosphotungstic Acid

Procedure

1. Deparaffinize and hydrate to distilled water.
2. Oxidise with periodic acid for 8 minutes.
3. Rinse in water.
4. Treat with Schiff's reagent for 15 minutes.
5. Wash in running water for 10 minutes, this intensifies the colour reaction.
6. Stain the nuclei with an alum haematoxylin (no need to blue).
7. Stain in 2 % orange G in 5 % phosphotungstic acid 20 seconds.
8. Differentiate in tap water until macroscopically the section is pale yellow. Check microscopically that only the RBC's and acidophil cells are yellow.
9. Dehydrate, clear and mount in Safety Mount

Notes: (*a*) Other structures, e.g. collagen, will be PAS positive. If this obscures the delicate staining, the oxidation and Schiff's staining times should be reduced. (*b*) This method is sometimes referred to as the Tri-PAS technique.

Results

Basophil cells Magenta
Acidophil cells Yellow
Red blood cells Yellow
Nuclei Black
Chromophobes Pale blue - grey.

PIGMENTS AND MINERALS

REMOVAL OF FORMALIN PIGMENT

Principle

Formalin pigment may be present in tissue after long-term fixation in formalin with an acid pH. It is brown and usually extracellular, but may be present in macrophages. Short term fixation in neutral buffered formalin may prevent this artefact. It is easily removed with alcohol - ammonia solution.

Reagents

(A) Alcohol-Ammonia Solution
95% alcohol 50 mL Concentrated ammonia (0.880) 15 mL

Procedure

1. Deparaffinize and hydrate to distilled water.
2. Place in Alcohol-ammonia solution for 1 (one) hour.
3. Wash in water.
4. Stain using H&E or other technique.

REMOVAL OF MERCURY PIGMENT

Principle

Mercuric chloride is found to a variable extent after fixation in mercuric chloride - containing fixatives, especially after long fixation. The granular black deposit can easily be removed by treatment in Lugols iodine and subsequent treatment with sodium thiosulphate.

Reagents

(A) Lugols Iodine
Iodine 1.0 g Potassium iodide 2.0 g Distilled water 100 mL Alternatively, Verhoeff's solution C from Elastic stain may be used for 5 minutes.

(B) 5% Aqueous Sodium Thiosulphate
Use from retic. stain.

Procedure

1. Bring sections to water.
2. Place in Lugol's iodine for 15 minutes.
3. Wash in water.
4. Place in thiosulphate for 3 mins.
5. Wash in water.
6. Stain with H&E or other technique.

DEMONSTRATION OF BILE

DEMONSTRATION OF BILE AND HAEMATOIDIN

Bile pigment may be demonstrated by the following methods currently in use in this laboratory.

Haematoxylin and Eosin

Liver bile Golden brown

Gall bladder bile Gold

Haematoidin Gold

Van Gieson Method

Liver bile Bright green

Gall bladder bile Yellow

Haematoidin Yellow The green colour of liver bile by this method may be due to its high biliverdin content.

Sirius Red Method

Bile products Emerald green

FOUCHET'S REACTION FOR BILE PIGMENTS

Van Gieson method preferred.

Principle

In the presence of trichloracetic acid the bile pigments are oxidized by ferric chloride to form a mixture of green bilirubin and blue cholecyanin.

Specimen

Standard paraffin sections 10% neutral buffered formalin fixation. Avoid mercury salts.

Control

Gall bladder wall containing Bile pigment.

Note: Controls required:

1. Section not treated with Fouchet's reagent but counterstained with Van Gieson, i.e. Van Gieson only.
2. Section treated with Fouchet's but not counterstained with Van Gieson, i.e. Fouchet's Reagent only

Reagents

(A) Fouchets Reagent
25% trichloroacetic acid (aq) 25 mL 10% ferric chloride (aq) [corrosive] 25 mL

(B) Van Gieson's Solution
(see Van Gieson's method or Sirius Red Stain method).

Procedure

1. Deparaffinize and hydrate to distilled water.
2. Treat with Fouchet's reagent for 5 minutes.
3. Rinse well in distilled water.
4. Counterstain with Van Gieson's solution for 2 minutes or Sirius Red Stain 30 minutes.
5. Rinse rapidly in water.
6. Dehydrate rapidly, Clear and mount.

Note: Steps 5 and 6 - the delicate red and yellow counterstaining is easily lost in water and alcoholic solutions.

Results

Bile pigments Blue/green

Muscle Yellow

Collagen Red

DEMONSTRATION OF FERRIC IRON

PERL'S PRUSSIAN BLUE FOR FERRIC IRON

Principle

Ferric iron is released from any attachments to protein by treatment with dilute hydrochloric acid, which then reacts with a dilute solution of potassium ferrocyanide to produce an insoluble blue compound, ferric ferrocyanide (Prussian Blue).

Specimen

Standard paraffin sections 10% neutral buffered formalin fixation

Control

Tissue containing iron salts. See next page for modification and procedure currently in use.

Reagents

(A) 2% Aqueous Potassium Ferrocyanide
(B) 2% Aqueous Hydrochloric Acid
(C) 1% Aqueous Neutral Red

Procedure

1. Deparaffinize and hydrate to distilled water.
2. Treat with 2% potassium ferrocyanide for 5 min. While waiting, mix equal volumes of 2% HCl and 2% ferrocyanide solution.
3. Drain off.
4. Flood slides with the mixture for 15 mins.
5. Wash well in water.
6. Counterstain with neutral red lightly.
7. Wash in water.
8. Dehydrate rapidly in alcohols.
9. Clear and mount.

REMOVAL OF MELANIN PIGMENTS (BLEACH)

Melanins are brown-black pigments, which are bound to proteins and which are localised in cells in the cytoplasm. Melanins of various types are found in many sites including skin, eye and brain. Pathologically it is found in benign naevi and malignant melanoma.

Principle

Melanin can be bleached by strong oxidising agents such as potassium permanganate.

Specimen

Standard paraffin sections, formalin fixed

Control

H&E without bleaching procedure, and H&E with bleaching procedure for each patient.

Reagents

(A) 0.25% Aqueous Potassium Permanganate
(B) 5% Aqueous Oxalic Acid

Procedure

1. Deparaffinize and hydrate to distilled water.
2. Oxidise in permanganate for 30 minutes.

3. Water wash.
4. Bleach in oxalic acid until white.
5. Proceed staining as desired.

Results

Skin melanin is bleached within 30 minutes
Ocular melanin takes 2 - 4 hours.

MASSON-FONTANA METHOD FOR MELANIN

Principle

Melanin has the ability to reduce solutions of ammoniacal silver nitrate to metallic silver.

Specimen

Standard paraffin sections, formalin fixed

Control

Tissue containing melanin pigment

Reagents

(A) Preparation of Silver Solution
To 20 mL of 10% silver nitrate add concentrated ammonia drop by drop until the formed precipitate almost dissolves. A faint opalescence is evident when the end point is reached. If too much ammonia is added, 10% silver nitrate can be added drop by drop until the faint opalescence is obtained. To this silver solution add 20 mL of distilled water and filter. Store in a dark bottle. The solution will last for about four weeks in fridge 4°C.

(B) 5% Sodium Thiosulphate

(C) 1% Neutral Red

Procedure

1. Deparaffinize and hydrate to distilled water.
2. Treat with silver solution a) 45 mins at 56°C to demonstrate melanin. b) at room temperature over-night in the dark to demonstrate Argentaffin.
3. Wash well in several changes of distilled water.
4. Treat with 5% sodium thiosulphate for two minutes.
5. Wash in tap water.
6. Counterstain in 1% neutral red (aqueous) for three minutes.
7. Wash in tap water.
8. Dehydrate through graded alcohols to Histoclear.
9. Mount in Safety-mount.

Results

Melanin Black
Chromaffin Black
Some Lipofuscins Black
Nuclei Red
Argentaffin Black
Argyrophil cells do not stain

SCHMORL'S REACTION FOR MELANIN

(Masson-Fontana method Preferred)

Principle

Melanin has the ability to reduce ferricyanide to ferrocyanide, which in the presence of ferric ions forms Prussian Blue.

Specimen

Standard paraffin or frozen sections

Control

Melanin pigment

Reagents

(A) 1% Aqueous Ferric Chloride

(B) 1% Aqueous Potassium Ferricyanide (freshly prepared) Working Solution To 30 mL of 1% ferric chloride add 4 mL of 1% potassium ferricyanide, then add 6 mLs of distilled water. Use within 30 minutes.

(C) Van Gieson Stain to counterstain.

Procedure

1. Deparaffinize and hydrate to distilled water.
2. Treat with Working Solution for 10 minutes.
3. Wash in water.
4. Counterstain with Van Gieson Stain for two minutes.
5. Dehydrate rapidly in alcohols.
6. Clear and Mount.

Results

Melanin Dark blue
Argentaffin cells, Chromaffin cells, some Lipofuscins Blue
Nuclei Red

MODIFIED VON KOSSA'S METHOD FOR CALCIUM

A modification of the classic von Kossa method of 1901.

Principle

The classic method for the demonstration of calcium and certain other salts in tissues is that of von Kossa (1901). In the present context it is used as a negative stain for osteoid, the bone mineral being blackened by the deposition of silver and the osteoid remaining receptive to a contrasting stain.

Specimen

Standard paraffin section.

Control

Calcium in tissue.

Reagents

(A) 1% Aqueous Silver Nitrate
(B) 2.5% Sodium Thiosulphate
(C) 1% Neutral Red or Van Gieson's Picro-fuchsin

Procedure

1. Deparaffinize and hydrate to distilled water. Rinse well in distilled water
2. Place in Silver nitrate solution and expose to strong light - 10 to 60 minutes
3. Wash in three changes of distilled water.
4. Treat with Sodium thiosulphate for five minutes.
5. Wash well in distilled water.
6. Counterstain with neutral red three mins.
7. Dehydrate, clear and mount.

Results

Mineralised bone Black

Osteoid Red

TIMM'S COPPER STAIN—(MODIFIED)

Principle

Copper is a normal constituent of many tissues in such small amounts that it cannot be demonstrated, however in Wilson's disease it slowly accumulates in certain organs, mainly in the liver and the basal ganglia region of the brain. Timm's copper stain is a silver technique where copper sulfide is converted to silver sulfide.

Specimen

Formalin or fresh tissue.

Control

The following control procedures should be employed with contiguous sections:

1. Omit step 4 to stain all heavy metals.
2. Omit step 6.
3. Treat with 0.5 % potassium cyanide before developing to remove copper sulfide.
4. Stain with Perl's method for iron.

Reagents

(A) Solution A
5.0 % Silver nitrate aqueous solution.

(B) Solution B
Hydroquinone 2.0 g, Citric acid 5.0 g, Distilled water 100.0 mL. This solution can be stored at 4°C in the dark for one month.

(C) 0.1n HCl or 15% Trichloracetic Acid

Procedure

1. Deparaffinize in 2 changes of Xylene and hydrate sections in water through graded alcohols.
2. Place in 0.5 % ammonium sulfide solution (aqueous) for 5 minutes.
3. Wash well in distilled water.
4. Treat sections with 0.1 N HCl for 2 to 3 minutes to remove iron and zinc sulfides (15 % trichloroacetic acid may also be used).
5. Rinse in distilled water.
6. Develop sections in a freshly filtered solution of 1 part A and 5 parts B. This converts copper sulfide to silver sulfide.
7. Rinse slides in distilled water.
8. Dehydrate through alcohols, clear in Histoclear, and mount in synthetic resin.

Results

Copper deposits are stained black

DEMONSTRATION OF URIC ACID AND URATES

Principle

Uric acid and urates are breakdown products of purine metabolism which are further broken down and excreted in urine. In the disease gout, they are not broken down and accumulate in the body, becoming crystallised out into the tissues, particularly into the synovial membrane lining joint cavities. Uric acid and urates are soluble in water, so much of the material will be lost during normal aqueous fixation and subsequent processing. Staining with eosin only (no haematoxylin) provides limited contact with water.

Specimen

95% alcohol is recommended as the most suitable fixative for the specific demonstration of urates. These specimens should be processed on a day cycle starting at absolute alcohol (station 5). Cut section at standard thickness. Avoid prolonged floating-out of section on waterbath.

Reagents

Eosin as used in linear stainer for routine H&E.

Procedure

1. Floating-out of sections must not be prolonged.
2. Deparaffinise and hydrate section to 80 % alcohol (not to water).
3. Stain on lincar staincr starting with Eosin.

Results

Urates are birefringent and appear as pink needle-shaped crystals in Eosin only preparation.

Notes

If a specimen is received in formalin, processed routinely and stained by H&E, urate crystals may appear blue to purple if numerous urate crystals were originally present. An alternative method of staining is the Gomori methenamine silver method. See Bancroft and Stevens, Theory and Practice of Histological Techniques, 4th ed p260.

Results

Uric acid and urates black

INTRACELLULAR NEUROSECRETORY GRANULES

Argentaffin and Argyrophil Granules. Cells containing neurosecretory granules, arise from neuroectoderm and migrate to their final position in the body rather than differentiating *in situ*. In many sites, such as the adrenal medulla, pituitary and pancreatic islets, they become aggregated together to form a recognisable endocrine gland, but in other parts, particularly the G.I. tract, they are scattered throughout the mucosa, forming a diffuse system. Argentaffin cells containing 5 hydroxy-trytomine (5HT) and argyrophil cell containing 5- hydroxytrytophane are important in the investigation of carcinoid tumours These tumours are found in the appendix and ileum and are usually argentaffin positive and argyrophil negative and those of the stomach are argrophil positive. The argentaffin cells are potential reducers in that with formalin fixation beta carboline is formed from 5HT and will reducc silvcr nitratc. Argyrophil cells are not reducers and stain negatively with silver stains.

ALKALINE DIAZO TECHNIQUE FOR ARGENTAFFIN CELL GRANULES

Principle

The rationale of this technique is that certain diazonium salts are converted to diazo dyes by 5HT.

Specimen

Standard paraffin section of formalin fixed tissue.

Note: This technique is not successful on unfixed or alcohol fixed tissue.

Control

Distal small intestine.

Reagents

(A) 1% Fast red B (C.I 37125) 5 mL (Store at 4°C - diazonium salts deteriorate with time)

(B) Saturated Lithium Carbonate 2 mL Keep solutions at 4°C before mixing, then use immediately (allow 30 minutes for solutions to come to 4°C from room temperature).

Procedure

1. Deparaffinize and hydrate to distilled water.
2. Transfer sections to incubating solution at 4øC for 1 minute
3. Wash well in tap water
4. Lightly stain nuclei in haematoxylin
5. Wash well in tap water
6. Dehydrate through graded alcohols to histoclear.
7. Mount in Safety Mount
8. Haematoxylin staining must be light so that positive staining of granules is not masked.

Results

Argentaffin cell granules Orange

Nuclei Blue

Background Yellow

Notes : 1. Other diazonium salts may be used, *e.g.* Fast Garnet GBC

2. This technique is only successful when fresh fixed tissue is used; the method rarely works with necropsy material. It is perhaps unnecessary to stress that, by the nature of the reaction, which depends on amine-aldehyde concentration, the techniques cannot be applied to unfixed material, or to tissues fixed in a non-aldehyde solution.

GRIMELIUS SILVER METHOD FOR ARGYROPHIL CELLS

Principle

This is a silver technique where the silver salt first attaches to the tissue component and is then reduced by an externally added reducing reagent.

Specimen

Standard formalin fixed paraffin sections.

Control

Pancreas or stomach.

Reagents

(A) Acetate Buffer pH 5.6

Stock A: 0.2M Acetic Acid, 1.2 mL Glacial Acetic Acid in 100 mL distilled water

Stock B: 0.2M Sodium Acetate, 1.64 g Sodium Acetate (anhydrous) (MW 82), or 2.72 g Sodium Acetate Trihydrate (MW 136) in 100 mL distilled water

Mix 10 mL Stock A with 90 mL Stock B.

(B) Silver Solution

Acetate Buffer (pH 5.6) 10 mL Redistilled water 87 mL 1% silver nitrate (aqueous), fresh 3 mL

(C) Reducing Solution

Hydroquinone 1 g, Sodium sulphite (cryst) 5 g, Distilled water 100 mL Freshly prepared.

Procedure

1. Deparaffinize and hydrate to distilled water.
2. Transfer sections to silver solution at 60°C for 3 hours
3. Drain silver solution from slides
4. Place sections in freshly prepared reducing solution at 45°C for one minute
5. Rinse sections in distilled water
6. Counterstain if required
7. Wash in tap water
8. Dehydrate through graded alcohols
9. Clear in Histoclear and mount in Safety Mount

Notes :

1. If a weak positive reaction is obtained, the results can be improved by double impregnation. After the reducing stage, the sections are treated with 5 % sodium thiosulphate for 2 minutes, rinsed in distilled water and placed in freshly prepared silver solution at room temperature for 10 minutes. This is followed by fresh reducing solution for one minute at 45°C.
2. Counterstains are not normally used.

Results

Argyrophil islet cells (A or a2 cells) - Positive (brown-black)

B & D cells—Negative Many other argyrophil cells - Positive (brown-black)

HAEMATOLOGICAL CELLS ALSO REFER TO GIEMSA

METHYL GREEN—PYRONIN TECHNIQUE FOR PLASMA CELLS

Plasma cells are found in serous membranes, in the lining of the respiratory and intestinal tracts and in lymphoid tissue. They are plentiful in regions of chronic inflammation.

Principle

The methyl green-Pyronin technique can be used to demonstrate malignant plasma cell tumours such as plasmocytomas.

DNA has a differential affinity for methyl green and RNA has an affinity for pyronin, thus staining the nuclei green and the RNA (plasma cell cytoplasm) red.

Specimen

Standard formalin fixed Paraffin section

Control

Tissue with chronic inflammatory reaction.

Reagents

(A) pH 4.8 Acetate Buffer

Stock A 0.2M acetic acid (1.2 mL in 100 mL distilled water) Stock B 0.2M sodium acetate (1.6g in 100 mL distilled water) Mix A:B in ratio 2:3 (10 mL A with 15 mL B)

(B) Methyl Green—Pyronin Solution

Make a mixture of the following four reagents

2% aqueous methyl green (C.I.42585) 9mL (Purchase Methyl violet free reagent or extract with chloroform until the methyl violet contaminant is removed - difficult to remove)

2% aqueous pyronin Y or G (C.I. 45005) 4mL

Glycerol 14mL pH 4.8 acetate buffer 23mL

Stock A 0.2M acetic acid (1.2mL in 100mL distilled water)

Stock B 0.2M sodium acetate (16.4g/L or 1.6g/100mL)

Mix in ratio A:B = 2:3 Mix before use.

Procedure

1. Deparaffinize and hydrate to distilled water. Rinse with pH 4.8 buffer.
2. Stain with the methyl green-pyronin solution for 25 minutes.
3. Wash with buffer and blot dry.
4. Rinse in equal parts acetone-histoclear, then in histoclear and mount in Safety Mount.

Results

DNA (nuclei) Green, green-blue

RNA (plasma cell cytoplasm, Nissl substance, bacteria) Red

Some mucins Red

HITCHCOCK-EHRICH TECHNIQUE FOR PLASMA CELLS

Principle

This technique is similar in principle to the methyl green-pyronin but uses instead malachite green and acridine red with similar results.

DNA has a differential affinity for malachite green and RNA has an affinity for acridine red.

Specimen

Zenker-acetic fixation preferred or 10 % neutral buffered formalin. 4æ paraffin sections. Treat sections with 1% acetic acid if fixative other than Zenker-acetic fixative is used.

Control

Chronic inflammatory reaction in tissue.

Reagents

(A) Malachite Green Solution

Dissolve 3 g of malachite green in 150 mL of distilled water. Filter and label.

(B) Acridine Red Solution

Dissolve 9 g of acridine red in 450 mL of distilled water. Filter and label.

Working Solution

Mix one (1) volume of (A) and three (3) volumes of (B).

15 mL of (A) and 45 mL of (B) is sufficient for a coplin jar.

Procedure

1. Deparaffinize and hydrate to distilled water.
2. Stain in working solution for 3 minutes.
3. Rinse quickly in water
4. Dehydrate rapidly in alcohols.
5. Clear and mount.

Results

DNA (nuclei) Blue - green

RNA (plasma cell cytoplasm) Red.

UNNA'S METHOD FOR MAST CELLS

Mast cells contain heparin and histamine, usually in the form of granules which are difficult to see in an H&E preparation. They are found in connective tissue around blood vessels in many varied inflammatory conditions. Rapid fixation is essential as the granules deteriorate with time, hence they are better seen on surgical material than on postmortem material.

Principle

Mast cell granules stain metachromatically with methylene blue.

Specimen

Ten percent neutral buffered formalin or formol alcohol. Cut standard paraffin sections.

Control

Trachea.

Reagents

(A) Polychrome Methylene Blue Solution
Methylene blue 1 g Distilled water 100 mL Alcohol 95 % 20 mL Potassium carbonate 1 g

(B) Glycerin—Ether Solution
Glycerin 50 mL, Calcium Chloride anhydrous 10 g. Heat the above mixture until the calcium chloride dissolves completely. Cool to room temperature.

Procedure

1. Deparaffinize and hydrate to distilled water
2. Polychrome Methylene Blue solution for 10 minutes.
3. Rinse in distilled water.
4. Differentiate in Glycerin-ether solution, diluted 5 to 10 times with distilled water, for 30 seconds to 1 minute, or until section is a medium blue. Be careful not to differentiate too long.
5. Wash thoroughly in water for 2-5 minutes, then blot with filter paper.
6. Dehydrate rapidly in absolute alcohol.
7. Clear in Histoclear, two or three changes.
8. Mount with Safe-t-mount.

Results

Mast cells Red

Other cells Greenish-blue

Notes

This technique, for the differentiation of mast cells and plasma cells, is especially useful for highly cellular specimens, such as lymph node and spleen.

TOLUIDINE BLUE METHOD FOR MAST CELLS AND OTHER METACHROMATIC SUBSTANCES PRINCIPLE

Toluidine blue is a polychromatic stain which stains many tissue components.

References

p 164 Bancroft JD & Stevens A, Theory and Practice of Histological Techniques, 4th Ed. The exact origin of this method is unknown.

Specimen

Standard paraffin sections or 6æ frozen sections.

Control

Trachea.

Reagents

Toluidine blue (CI 52040) 0.25 g, Sodium acetate 2 g, Sodium barbiturate 3 g. Distilled water to 100 mL

Procedure

1. Paraffin or frozen sections may be used.
2. Bring sections to water.
3. Stain in toluidine blue solution, for 20 seconds .
4. Blot with filter paper; allow to dry and clear in histolene.
5. Mount in Safety Mount

Results

Metochromatic substances Red, pink or purple
Nuclei and other components
Blue Mast cell granules Purple

LUNA'S METHOD FOR ERYTHROCYTES AND EOSINOPHILS

Principle

The tissue is stained with a combined Haematoxylin-Biebrich Scarlet solution to demonstrate eosinophil granules against a blue background..

Specimen

Ten percent neutral buffered formalin. Standard paraffin sections.

Control

Known positive control containing eosinophils.

Reagents

(A) Weigerts Iron Haematoxylin

Same as reagent A in Van Gieson.

1. Dissolve 1g of haematoxylin in 100mLs of 95% ethanol (filter)
2. To 4mLs of 30% aq. ferric chloride add 95mLs of distilled water. Add 1mL of concentrated hydrochloric acid.

Working Solution

(A) Mix equal volumes of (1) and (2) prior to use.

(B) 1% Biebrich Scarlet Solution
Biebrich scarlet (CI 26905) 1 g Distilled water 100 mL

(C) Haematoxylin-biebrich Scarlet Solution (Working)
Weigert's iron haematoxylin solution (working) 45 mL Biebrich scarlet solution 5 mL

(D) 1% Acid Alcohol Solution 70 % alcohol 100 mL Concentrated HCl 1 mL

(E) 0.5% Lithium Carbonate Solution
Lithium carbonate 0.5 g Distilled water 100 mL

Procedure

1. Deparaffinize and hydrate to distilled water.
2. Stain in working haematoxylin-biebrich scarlet solution for 5 minutes.
3. Differentiate in 1% acid alcohol until desired nuclear detail is achieved (usually eight dips).
4. Rinse in tap water to remove acid alcohol. 5. Dip Lithium carbonate solution until section turns blue and erythrocytes are bright red (usually five dips).
6. Wash in running water for two minutes.
7. Dehydrate in 95% alcohol, absolute alcohol, and clear in Histolene, two changes each.
8. Mount with Safety-mount.

Results

Eosinophil granules Red

Erythrocytes Red

Charcott Leyden crystals

Red Background Blue

Notes

Especially useful for phagocytosis and sometimes useful for demonstrating Negri bodies.

DUNN THOMPSON METHOD FOR HAEMOGLOBIN

Haemoglobin is the physiological pigment which is responsible for the transport of oxygen and carbon dioxide to and from the tissues in the blood stream.

Principle

Haemoglobin stains orange-red by acid dyes such as eosin. The Dunn Thompson method is a modification of the Van Gieson stain and stains haemoglobin, red cell and haemoglobin urinary casts and phagocytosed haemoglobin an emerald green or greenish-black colour. Fixation in neutral buffered formalin is essential.

Specimen

Fixation -10% neutral buffered formalin - essential.

Standard paraffin sections.

Control

Haemoglobin is found normally within red blood cells.

Reagents

(A) Alum Haematoxylin
Commercial Harris haematoxylin as used in linear stainer.

(B) Van Gieson Solution
As used in Van Gieson's stain method 2.4 100mL saturated aqueous Picric Acid 10 mL 1% aqueous fuchsin (C.I. 42685)
Boil for 3 minutes, cool and filter then add 0.25mL concentrated hydrochloric acid.

(C) 4% Iron Alum
4 g ammonium ferric sulphate in 100 mL distilled water (as used in reticulin fibres stain.

Procedure

1. Deparaffinize and hydrate to distilled water.
2. Alum haematoxylin - 7 minutes (up to 15 minutes).
3. Wash in tap water.
4. Mordant in 4 % iron alum - 5 minutes (up to 10 minutes).
5. Rinse in tap water.
6. Stain in van Gieson solution - 15 minutes.
7. Differentiate in 95 % alcohol - 3 minutes.
8. Rinse in absolute alcohol
9. Clear in histolene.
10. Mount in Safety Mount.

Note

The double haematoxylin staining produces stronger haemoglobin colouration.

Results

RBCs Greenish / black

Haemoglobin casts Lighter green

OKAJIMA'S METHOD FOR HAEMOGLOBIN

Name: Okajima's Stain

Purpose: To stain hemoglobin

Techniques

Cut paraffin sections at 4 to 5 microns. Use control slide.

Solutions

10% Phosphomolybdic Acid - Stock (make fresh)

Phosphomolybdic acid 10.0 gm

Distilled water 100.0 ml

7.7% Alizarin Red S - Stock (shelf life: 1 year)

Alizarin Red S (C.I. #58005) 7.7 g

Distilled water 100.0 ml

Working Stain Solution: shelf life: 10 hours

10% Phosphomolybdic acid (stock) 10.0 ml

7.7% Alizarin red S (stock) 30.0 ml

Mix just before use. Discard after use.

Hematoxylin of choice

Procedure

1. Deparaffinize and hydrate to water.
2. Place in 10% phosphomolybdic acid solution for one minute.
3. Wash in water.
4. Preheat 20 ml of working stain for 10-15 seconds in microwave. Add slides and stain for 10 minutes. (or 1-2 hours in non-heated solution)
5. Wash in water.
6. Counterstain in hematoxylin for 30 seconds-1 minute (we use Gill's)
   (Follow whatever you normally do after your hematoxylin, but don't counterstain with Eosin!)
7. Dehydrate, clear, and coverslip with synthetic mounting medium.

Results

Hemoglobin.............orange to orange red

Nuclei..................................blue

Background.....................reddish brown

MISCELLANEOUS

ATTWOOD'S METHOD FOR AMNIOTIC FLUID EMBOLISM

Principle

All tissues are stained with phloxine, then the red stain is differentiated out with tartrazine in cellosolve. This method can be used to demonstrate many structures according to the degree of differentiation.

Specimen

Ten per cent neutral buffered formalin fixation. Standard paraffin sections CONTROL Known positive tissue.

Reagents

(A) 1% Alcian Green Solution
Alcian green 1 g Distilled water 100 mL Glacial acetic acid 2 mL

(B) Phloxine-calcium Chloride Solution
Phloxine 0.5 g Calcium chloride 0.5 g Distilled water 100 mL

(C) Tartrazine-cellosolve Solution
(Store in flammables cabinet) Tartrazine (CI 19140) 2.5 g Cellosolve (irritant to skin) 100 mL (also called 2-ethoxyethanol or ethylene glycol monoethyl ether)

(D) 0.5% Potassium Permanganate Solution
Potassium permanganate 0.5 g Distilled water 100 mL

(E) 0.5% Oxalic Acid Solution
Oxalic acid 0.5 g Distilled water 100 mL or Dilute 1% oxalic acid solution from reticulin fibre stain (method 2.5) 50/50 with distilled water.

Procedure

Use control slide.

1. Deparaffinise and hydrate to distilled water.
2. Alcian green solution for two minutes.
3. Rinse in distilled water.
4. Potassium permanganate solution for 2 minutes.
5. Rinse in distilled water.

6. Oxalic acid solution for 2 minutes.
7. Rinse in distilled water
8. Stain with Mayer's Haematoxylin 3 minutes (or alum haematoxylin in linear stainer).
9. Rinse and blue in lithium carbonate solution.
10. Phloxine solution for 30 minutes.
11. Rinse in distilled water, then in cellosolve.
12. Differentiate each slide with tartrazine-cellosolve solution, checking until the phloxine is removed from erythrocytes (approximately 45 seconds).
13. Dehydrate in cellosolve, clear in histolene, two changes each.
14. Mount with Permount or Safety-mount.

Results

Nuclei Blue/black

Mucin, ground substance of cartilage, granules of certain mast cells Green

Foetal squames, squamous epithelium, fibrin Red

Erythrocytes, plasma collagen Yellow

Inclusion bodies Red

Note : The permanganate-oxalic bleach is not strictly necessary but it assists in the subsequent differentiation of the phloxine and tends to remove any background staining by Alcian Green.

Any excess staining with tartrazine can be removed by rinsing in water until the desired intensity is obtained.

MALLORY BODIES (ALCOHOLIC HYALINE)

Alcoholic hyaline is a protein substance occurring within the cytoplasm of liver cells in certain patients who have liver disease due to excess alcohol consumption. It appears as a perinuclear pale-staining, often greyish zone in the liver cell. Alcoholic hyaline stains with the following stains but appears to be most satisfactorily stained with Luxol Fast Blue.

Refer to Method Luxol Fast Blue Technique for Myelin.

Specimen

Routine formalin fixation. Standard paraffin section.

Control

Use known positive section of liver.

Results

Stain Alcoholic hyaline H & E Weakly eosinophilic

Luxol Fast Blue Blue PTAH Blue Congo Red Red (weak)

GORDON AND SWEETS' METHOD FOR RETICULIN

- Dewax sections, rinse in alcohol, rinse in water.
- Oxidise in acid potassium permanganate - 5 minutes.
- Wash briefly in water.
- Bleach in 1% oxalic acid - 2 minutes.
- Rinse in distilled water followed by a thorough wash in tap water.
- Sensitise in 2.5% iron alum - 10 minutes.
- Wash in several changes of distilled water.
- Treat with silver solution in a coplin jar - 5 seconds.
- Rinse briefly in distilled water.
- Reduce in 10% 'Analar' formalin in distilled water - 1 minute. (In fume hood.)
- Wash in tap water followed by distilled. (In fume hood.)
- Tone in 0.2% gold chloride - 2 minutes.
- Wash briefly in distilled water.
- Fix in 5% sodium thiosulphate - 2 minutes.
- Wash well in tap water.
- Rinse in distilled water.
- Counterstain with nuclear fast red - 5 minutes.
- Wash in distilled water.
- Dehydrate, clear and mount.

Results

Reticulin - black.

Nuclei - red.

Collagen and cytoplasm - purple/pink.

Acid potassium permanganate

0.5% potassium permanganate in 3% sulphuric acid

Gold chloride = sodium tetrachloroaurate

Silver solution

To 5 mls of 10% silver nitrate add strong ammonia drop by drop until the formed precipitate just redissolves. Shake the solution for a few seconds after adding each drop. Add 5mls 3% sodium hydroxide and redissolve the precipitate with a few more drops of ammonia. Again, shake the solution after adding each drop. The finished solution should have a slight opalescence. If the solution goes completely clear add a drop or two of 10% silver nitrate until a slight opalescence remains. Make up to 50mls with distilled water. Kept in a sealed coplin jar at 4°C the solution can be used for up to one month.

P.T.A.H. (MALLORY'S PHOSPHOTUNGSTIC ACID HAEMATOXYLIN) FOR TISSUE OTHER THAN C.N.S.

- Dewax sections, rinse in alcohol, rinse in water.
- Treat in 1% potassium permanganate. - 5 minutes.
- Rinse in distilled water.
- Bleach in 1% oxalic acid - 2 minutes.
- Rinse well in distilled.
- Treat in 5% iron alum - 1 hour.
- Rinse in distilled.
- Stain in P.T.A.H. for 2-4 hours.
- Dehydrate, clear and mount.

Results

Nuclei, red cells, intracellular bridges of squamous epithelium, cilia, mitochondria - blue.

Muscle striations, fibrin - dark blue.

Collagen, cartilage, bone matrix, reticulin, elastic fibres - reddish brown.

Cytoplasm - pale brown to pink.

P.T.A.H. solution (naturally oxidised)

Haematoxylin - 1g

Phosphotungstic acid - 10g

Distilled water - 1 litre

Dissolve the components separately in portions of the distilled water then mix. This is the best method of preparation and although the solution will take up to a year to ripen it will keep for a very long time.

P.T.A.H. solution (Shum & Hon)

Take 0.08g of haematein and grind to a paste with 1ml of distilled water. The paste should appear chocolate brown in colour. Dissolve 0.9g phosphotungstic acid in 99ml of distilled water and mix with the haematein paste. Bring to the boil cool and filter. This is then ready for use and should keep for several months.

HISTOPATHOLOGY RECIPE FORMULARY

ACETIC LACMOID - *(La Cour)*

Lacmoid	1 g
Glacial acetic acid	45 ml
Water	55 ml

Place lacmoid and water in a flask, heat to dissolve, add acid and cool.

ACETIC ORCEIN - (*La Cour*)

Orcein	1 g
Glacial acetic acid	45 ml
Water	55 ml

Place orcein and acid in a flask and plug with cotton wool, bring to near boiling, cool and add water.

ACETO-CARMINE - *(Schneider)*

Carmine	0.5 g
Glacial acetic acid	45 ml
Water	55 ml

Place carmine and water in a 200 ml flask, bring to boil, add acetic acid, plug flask with cotton wool and boil again; cool and filter.

ACID ALCOHOL - *(Gurr)*

Ethanol 70%	100 ml
Hydrochloric acid	3 ml

ACID ALCOHOL DECOLORIZER - *(Lempert)*

Pure hydrochloric acid	0.5 ml
Sodium chloride	0.5 g
Methanol, pure	75 ml
Water	25 ml

ACID FUCHSIN - (*Mallory*)

Acid Fuchsin $^1/_2$% aqueous

ALBERT STAIN No. 1 - *(Gurr)*

Toluidine blue	0.15 g
Malachite green	0.2 g
Glacial acetic acid	1 ml
Ethanol (95%)	2 ml
Distilled Water	100 ml

ALBERT STAIN No. 2 - *(Gurr)*

Iodine	2 g
Potassium iodide	3 g
Distilled water	300 ml

Add iodine and iodine to about 25 ml. water, when solution comploete, add remainder of water.

ALCIAN BLUE - *(Gurr)*

Alcian blue 8GX	1 g
Water	100 ml

Dissolve and add a crystal of thymol or a few drops of chloroform to preserve.

ALUM CARMINE - *(Gurr)*

Carmine	10 g
Ammonia alum	25 g
Water	500 ml

Mix and bring to boil, cool, filter and make up to 500 ml: add 5 ml formaldehyde.

ANILINE BLUE - *(Masson)*

Aniline blue water soluble	2 g
Acetic acid 2%	100 ml

ANILINE BLUE-ORANGE G - (*Mallory*)

Aniline Blue water soluble	2.5g
Orange G	10 g
Phosphotungstic Acid	5 g
Water	500 ml

ANILINE BLUE-ORANGE G, ACETIC (for Azan Stain) - *(Gurr)*

Aniline blue water soluble	0.5 g
Orange G	2 g
Acetic acid 8%	100 ml

Warm to dissolve; filter when cold.

ANILINE GENTIAN VIOLET - *(Gurr)*

Gentian violet	10 g
Aniline	5 ml
Ethanol, absolute	20 ml
Water	175 ml

Warm water to about 40 degrees C., stir in genetian violet and aniline; add alcohol; filter

ANILINE XYLENE - *(Gurr)*

Aniline and Xylene	equal volumes

ARCHIBALD'S STAIN - *(Gurr)*

No. 1 Thionin	0.5 g
Formalin	1.0 ml
2.5%% Phenol in water	100 ml
No. 2 Methylene blue	0.5 g
Formalin	1.0 ml
2.5% Phenol in water	100 ml

Allow to stand 24 hours and mix equal parts before use.

AURAMINE PHENOL - *(Lempert)*

3% aqueous phenol 100 ml

Warm to about 30°C. and add

Auramine 0.3 g

Shake well and filter.

AZO BLACK (Chlorazol Black E) - *(Gurr)*

0.5% Azo black in water or 70% Alcohol.

AZOCARMINE (for Heidenhain Azan Stain) - *(Gurr)*

Azocarmine B (or G)	0.5 g
Acetic acid glacial	1 ml
Water	to 100 ml

AZURE A-ERIE GARNET - *(Gurr)*

1. Azure A 1% aqueous	100 ml
2. Erie garnet B$^{1}/_{2}$% aqueous	50 ml
3. Solution A	20 ml
Solution B	

Mix rapidly and filter.

Note: Solution 3 should be prepared by mixing the stock. Solutions (1 and 2) as and when required for use.

BEST'S CARMINE STOCK STOLUTION - *(Gurr)*

Carmine	2.5 g
Potass. carbonate	1.25 g
Potass. chloride	6.25 g
Water	75 ml

Place all ingredients in a beaker, stir and boil gently for a few minutes.

Cool and add liq. ammon. fort. 25 ml

Do not filter. Keep in refrigerator.

BIEBRICH SCARLET ACETIC - *(Gurr)*

Biebrich scarlet	0.2 g
1% acetic acid	100 ml

BIEBRICH SCARLET - (glycerol-alcohol) - *(Gurr)*

Biebrich scarlet 1% aqueous	25 ml
Glycerol	10 ml
Ethanol, absolute	40 ml
Water	425 ml

A low staining time (10-15 minutes) gives very good results

BORAX CARMINE - *(Grenacher)*

Carmine	2 g
Borax	8 g
Water	200 ml

Boil all together for half an hour; col and make up to 200 ml, then add - ethanol 70% 200 ml

BORAX METHYLENE BLUE *(Unna)*

Methylene blue	5 g
Borax	5 g
Water	500 ml

Heat water to 60°C and stir in solids; allow to cool slowly. Improves with age.

BORAX METHYLENE BLUE *(Manson)*

Methylene blue	10 g
Borax	25 g
Water	500 ml

Warm water to 60°C and stir in solids; allow to cool slowly. Improves with age.

BORREL'S METHYLENE BLUE *(Gurr)*

SOLUTION A

Methylene blue	1 g
Water	100 ml

SOLUTION B

Silver nitrate	0.5 g
Water	100 ml

Dissolve the silver nitrate in the water in a 500 ml beaker and add gradually 2% sodium hydroxide with constant stirring until no more precipitate is formed. Allow precipitate to settle and decent liquid. Wash precipitate with distilled water until washings are no longer alkaline to litmus. Add solution A to the washed silver hydroxide and boil for 5 minutes, cool and filter

BOUIN FIXATIVE (Picro-Formal) - *(Gurr)*

Picric acid saturated aqueous solution	75 ml
Formalin	25 ml

Glacial acetic acid	5 ml

BREED'S STAIN - Methylene Blue - *(Breed)*

Methylene blue	0.3 g
Ethanol, absolute	30 ml
Dissolve and add - phenol 2.5% aqueous	100 ml

BRUN'S GLUCOSE MOUNTANT - *(Gurr)*

Glucose	40 g
Glycerol	10 ml
Thymol 2.5 aqueous	10 ml
Water	140 ml

CARBOL FUCHSIN - *(Ziehl-Neelsen)*

Basic fuchsin	1 g
Phenol	5 g
Ethanol, absolute	10 ml
Water	100 ml

Dissolve fuchsin in alcohol; dissolve phenol in water and mix the two solutions.

CARBOL FUCHSIN, DILUTE - *(Gurr)*

Carbol fuchsin	10 ml
Water	90 ml

CARBOL GENTIAN VIOLET - *(Gurr)*

Genetian violet	5 g
Phenol	10 g
Ethanol, absolute	100 ml
Water	500 ml

Warm water to about 40°C stir in genetian violet and phenol; add alcohol and filter.

CARBOL PYRONINE METHYL GREEN - *(Gurr)*

Methyl green 2% aqueous	4 ml
Pyronine 5% aqueous	7 ml
Ethanol, absolute	2.5 ml
Glycerol	20 ml
Phenol 0.5% aqueous	66.5 ml

CARBOL THIONIN - *(Gurr)*

Thionin	1 g
Phenol	5 g
Ethanol, absolute	50 ml
Water	500 ml

Triturate thionin with alcohol, add phenol and water; filter.

CARBOL XYLENE - *(Gurr)*

Phenol	125 g
Xylene	375 ml

CARMALUM - *(Gurr)*

Carmalum	10 g
Acetic acid, glacial	5 ml
Water	200 ml

Gently boil for one hour, cool and filter.

CARNOY FLUID - *(Gurr)*

Glacial acetic acid	10 ml
Chloroform	30 ml
Ethanol, absolute	60 ml

CHROME HAEMATOXYLIN - (*Gormori*)

Haematoxylin 1% aqueous	50 ml
Chrome alum 10% aqueous	15 ml
Potassium dichromate 10% aqueous	1 ml
Sulphuric acid, conc.	0.1 ml

CHROME HAEMATOXYLIN - *(Hansen)*

Haematoxylin 1% aqueous	30 ml
Chrome alum 10% aqueous	30 ml
Potassium dichromate 10% aqueous	2 ml
Sulphuric acid, conc.	0.2 ml

COLOPHONIUM ALCOHOL - *(Gurr)*

Colophonium 2% in absolute ethanol.

COTTON BLUE-LACTOPHENOL - (Stock Solution) - *(Gurr)*

Cotton blue	1 g
Lactophenol	100 ml

For use dilute 5 ml of this solution to 100 ml. with lactophenol.

COTTON BLUE-MAGENTA-LACTOPHENOL - *(Gurr)*

Cotton blue 1% in lactophenol	4 ml
*Magenta 1% in lactophenol	2 ml
Lactophenol	50 ml

*Magenta; syn; basic fuchin

CRYSTAL VIOLET - *(Gurr)*

Crystal violet 0.5% aqueous

CYANOL STOCK SOLUTION - *(Gurr)*

Xylene Cyanol, (or Patent Blue V)	1 g

Water	100 ml

Dissolve and add -

Zinc powder	10 g
Glacial acetic acid	2 ml

Boil until decolourised and freshly filter before use; the solution keeps for several weeks.

DELAFIELD HAEMATOXYLIN - *Gurr*

(i) **Haematoxylin**	8.0 g
Ethanol Absolute	250 ml

Dissolve in water bath or electric hot plate, then cool and filter

(ii) **Ammonia Alum**.	16 g
Water	800ml

Dissolve by warming

Mix the above two solutions and add -

Glycerol	200ml

Mix thoroughly and leave exposed to daylight to ripen for at least 6 weeks; or ripen immediately by adding 0.4 g. potass. permanganate dissolved in 10 ml of warm distilled water.

EHRLICH'S TRIACID STAIN - *(Gurr)*

Orange G. satd. aqueous soln.	70ml
Acid fuchsin aqueous solution 20%	35ml
Ethanol, absolute	75ml
Methyl green aqueous solution 10%	62ml
Alcohol	50ml
Glycerol	50ml

Mix in the order given, using same measure unwashed; shake well after adding methyl green. Do not filter.

ELASTIN STAIN - *(Weigert)*

Weigert's elastin stain	0.75 g
Silver Sand	5 g
Ethanol, absolute	100 ml
Hydrochloric Acid, conc	2 ml

Place dry stain and sand in a mortar, add alcohol and grind for about five minutes. Transfer solution and residue to a flask, plug with cotton wool and boil in a water bath for 20 minutes. Cool, filter and add the acid. Note: The solution deteriorates in 3-4 weeks.

ELASTIN STAIN - *(Sheridan)*

Use Sheridan's dry elastin stain and prepare in the same manner as Weigert's. Note: The solution remains stable for at least 12 months provided it is stored in a well closed battle, kept in a cool place and away from strong light.

EOSIN - *(Gurr)*

Eosin Y water soluble	1 g
Tap water	100 ml

Add chloroform to preserve.

ERYTHROSIN - *(Gurr)*

Erythrosin	1 g
Tap water	100 ml

Add chloroform to preserve.

EAU DE JAVELLE - *(Gurr)*

A. Bleaching powder	50 gm
Water	250 ml

Shake well; then allow to stand for several hours in a closed bottle.

B. Potassium carbonate	37.5 g
Water	250 ml

Mix solution A with solution B; allow the mixture to stand for several hours; then decent or filter through glass wool.

FARRANT'S MEDIUM - *(Gurr)*

Gum acacia powder	150 gm
Water	150 ml

Warm the water to about 40°C sprinkle the gum on the surface of the water a little at a time and stir in with a glass rod; then stir in:

Arsenious oxide	0.5 gm

Leave overnight; then stir well before adding:

Glycerol	150 ml

Stir well; filter through surgical lint, or decant after standing for several days or weeks, if necessary.

FAST GREEN FCF - *(Gurr)*

Fast green FCF 0.5% in ethanol or water

FAT RED 7B (Fettrot 7B) - *(Gurr)*

Prepare saturated solution as Sudan Black (P.G. Solution).

FEULGEN SOLUTION (Schiff Reagent) - (Gurr)

Fuchsin, Basic	1 g
Water	200 ml
Potass. metabisulphite	2 g

Hydrochloric acid. pure	2 ml
Decolourising charcoal	0.5 g

Boil water and add fuchsin, when dissolved cool to about 50 degrees C. and add metabisulphite, dissolve and allow to cool to room temperature. Add acid, plug flask with wool and leave overnight. Add charcoal, shake and filter. The solution should now be colourless.

FEULGEN SULPHUROUS ACID SOLUTION - *(Gurr)*

Potass. metabisulphite	1 g
Water	200 ml
N/1 hydrochloric acid	10 ml

FIELD'S STAIN - *(Gurr)*

SOLUTION A

Methylene blue	1.6 g
Azure 1	1.0 g
Disodium hydrogen phosphate anhydrous	2.6 g
Potassium dihydrogen phosphate	2.6 g
Water	1 litre

Place all in a flask in a warm water bath for half an hour, leave to stand twenty-four hours at room temperature; filter.

SOLUTION B

Eosin Y	2.0 g
Disodium hydrogen phosphate anhydrous	2.6 g
Potassium dihydrogen phosphate	2.6 g
Water	1 litre

Dissolve in water and filter.

FLAGELLA STAIN - *(Casares-Gil)*

Tannic acid	20 g
Alum chloride hydrated	36 g
Zinc chloride	20 g
Rosaniline hydrochloride	3 g
Ethanol 60%	80 ml

Mix solids in mortar and add about 20 ml of the 60% ethanol; mix and stir in remainder of the ethanol.

FONTANA'S STAIN - *(Gurr)*

A. FIXATIVE (Ruge's solution)

Glacial acetic acid	1 ml
Formalin	2 ml
Water	100 ml

B. MORDANT. Tannic acid 5% aqueous

C. SILVER STAIN

Dissolve 5 g. silver nitrate in 100 ml. distilled water.

Dilute 5 ml. of concentrated ammonia solution (sp. gr. 0.880) with 45 ml of distilled water; add 35 ml. of the diluted ammonia to 90 ml. of the silver solution and continue to add ammonia dropwise until the precipitate formed is just redissolved. Now add sufficient of the remaining silver solution, dropwise until the solution again becomes slightly opalescent.

Warning: To minimise risk of explosion solution C should be kept in a cool dark place.

FOOT'S SILVER SOLUTION - *(Gurr)*

Silver nitrate 10% aqueous	12 ml
Sodium hydroxide 10%	9 drops

Add strong ammonia dropwise until the precipitate is just dissolved. Make up the volume with distilled water to 42 ml

FORMOL-ACETIC-ALCOHOL (F.A.A. fixative) - *(Gurr)*

Glacial acetic acid	5 ml
Formalin	5 ml
70% Ethanol	90 ml

FORMOL ALCOHOL (Alcoholic-Formol) - *(Gurr)*

Formalin	1 part
Ethanol, absolute	9 parts

FORMOL-CALCIUM (Baker 1944) - *(Gurr)*

Calcium Chloride	1 g
Formalin	25 ml
Water	225 ml

FORMOL-SALINE - *(Gurr)*

Sodium chloride	8.5 g
Formalin	100 ml
Water	900 ml

FORMOL SUBLIMATE FIXATIVE - *(Gurr)*

Mercuric chloride satd. solin.	90 ml
Formalin	10 ml

GALLOCYANIN - *(Einarson)*

Gallocyanin	0.15 g
Chrome Alum	5 g
Water	100 ml

Dissolve alum in warm water, add dye and boil for 10 minutes; cool. filter and make volume up to 100 ml with water.

GIEMSA STAIN - *(Gurr)*

Azur II eosin	3.0 g
Azur II	0.8 g
Glycerol	250 ml
Pure methanol	250 ml

Instead of azur II eosin and azur II, 3.8 g. of the compounded Giemsa stain powder may be used. Place all in a flask, plug neck loosely with cotton wool and warm on a water bath for 1 hour; cool and filter. Note: Gurr's Improved Giemsa R 66 solution should be purchased ready prepared.

GLYCEROL ALBUMEN - *(Mayer)*

*Egg albumen	100 ml
Glycerol	100 ml
Sodium salicylate	2 g

Shake well; allow to stand in a closed glass vessel for several hours; then filter if necessary; this takes a considerable time unless a filter pump is employed. *Four or five fresh eggs are required to produce 100 ml of albumen.

GLYCEROL JELLY - *(Gurr)*

Gelatine granular	100 gm
Water	650 ml

Warm the water, in a 2-litre beaker or an enamel bowl, to about 40-50 degrees C.; then sprinkle the gelatine on the surface of the water a little at a time, stirring in each lot with a glass rod; then add and stir in:

Glycerol	700 gm

Pour the mixture into screw cap storage jars, and add a crystal of thymol to each.

Keep the jars tightly closed.

GOTHARD'S DIFFERENTIATOR - *(Gurr)*

Creosote	50 ml
Cajuput oil	40 ml
Xylene	50 ml
Absolute ethanol	150 ml

GRAM'S IODINE - *(Gurr)*

Iodine	1 g
Potassium iodide	2 g
Water	300 ml

Add iodine and iodide to about 25 ml. water. when solution complete, add remainder of water.

GRAM'S IODINE - *(Kopeloff & Beerman)*

Iodine	2 g
N/1 sodium hydroxide solution	10 ml

After iodine has dissolved add:

Distilled water	90 ml

HAEMALUM - *(Mayer)*

Haematoxylin 10% in alcohol	10 ml
Sodium iodate	0.2 g
Ammonia or potash Alum	50 g
Chloral hydrate	50 g
Citric acid	1 g
Water	1 litre

Dissolve solids in the water and add haematoxylin solution.

HARRIS HAEMATOXYLIN - *(Gurr)*

Haematoxylin 10% alcoholic solution (freshly prepared)	25 ml
Ammonia alum	50 g
Water	500 ml
Mercuric oxide	1.25 g

Dissolve alum in water by warming: add haematoxylin solution: bring to boil, add mercury oxide and stir in while still boiling. Cool rapidly in a cold water bath; filter. Mallory recommended the addition of 5% acetic acid.

HAYEM'S BLOOD COUNTING FLUID FOR RED CELLS - *(Gurr)*

Sodium chloride	1.2 g
Sodium sulphate	3.0 g
Mercuric chloride	0.3 g
Water	120 ml

Dissolve mercuric chloride in warm water. add remainder, dissolve and filter.

HEIDENHAIN'S IRON HAEMATOXYLIN - *(Gurr)*

No. 1 or (A)

Iron alum	3 g
Water, freshly boiled	100 ml

No. 2 or (B)

Haematoxylin 10% alcoholic	5 ml
Water	95 ml

HELLY'S FIXATIVE - *(Gurr)*

Potassium dichromate 3% aqueous	50 ml
Mercuric chloride saturated aqueous solution	50 ml

Sodium sulphate	1 g

Immediately before use add 1 ml formaldehyde to each 20 mls

HISS'S CAPSULE STAIN - *(Gurr)*

No. 1

Basic fuchsin 10% in alcohol	5 ml
Water	95 ml

No. 2

Copper sulphate 20% aqueous solution

HITCHCOCK-EHRICH MIXTURE - *(Gurr)*

A. Malachite green	1 g
Water	50 ml
B. Acridine red	3 g
Water	150 ml

Mix one part A with 3 parts of B immediately before use.

IRON ACETO CARMINE - *(Belling)*

Aceto carmine	54 ml
Ferric oxide saturated solution in 45% acetic acid	2 ml

Histopathology Recipe FormularyS to Z

SANDIFORD'S STAIN - *(Gurr)*

Malachite green	0.05 g
Pyronin B	0.05 g
Water	100 ml

SCARLET R (Alcoholic) - *(Gurr)*

Scarlet R.	0.5 g
70% ethanol	100 ml

Dissolve in warm water bath; filter.

SCARLET R - *(Herxheimer)*

Scarlet R	1 g
Acetone	50 ml
70% ethanol	50 ml

Dissolve by shaking; filter.

SCARLET R (Glycol)- *(Gurr)*

Scarlet R.	1 g
Propylene glycol	100 ml

Dissolve in a warm water bath; filter.

SCHAUDINN'S FIXATIVE - *(Gurr)*

Mercuric chloride satd. squeous solution	100 ml
Absolute ethanol	50 ml

SHORR'S STAIN S3 - *(Gurr)*

(Single solution)

Biebrich scarlet, aqueous	0.5 g
Orange G	0.25 g
Fast green FCF	0.075 g
Phosphotungstic acid	0.5 g
Phosphomolybdic acid	0.5 g
Glacial acetic acid	1 ml
Ethanol 50%	100 ml

Dissolve by warming; then allow to cool and filter. or:

Shorr's Stain powder (Michrome)	1 g
Alcohol 50%	55 ml
Glacial acetic acid	0.5 ml

Dissolve the stain by warming with the alcohol; allow to cool; then add the acetic acid, shake well and filter.

SHUTT'S METHYLENE BLUE - *(Gurr)*

Ether	100 ml
Pure methanol	100 ml
Methylene blue	1 g
Hydrochloric acid	0.7 ml

The ether and methyl alcohol should be mixed together before the methylene blue is added. As soon as the dye is in solution the hydrochloric acid can be added. The solution is stored in a stoppered bottle and kept in a cool place when not in use.

SOLOCHROME CYANINE - IRON ALUM - *(Gurr)*

Solochrome Cyanine RS	0.2 g
Sulphuric Acid 98%	0.5 ml
Water	95 ml
1% aqueous from Alum	5 ml

Place dye in 250 ml. beaker and with a glass rod carefully stir in the acid; effervescence occurs and a thick creamy solution is formed. Add water and iron alum solution. Mix thoroughly by stirring, then filter. The solution keeps well.

STEVENEL'S BLUE - *(Gurr)*

A. Methylene blue	1 g
Water	75 ml
B. Potass permanganate	1.5 g
Water	75 ml

Mix the two solutions in a flask and place in a boiling water bath until the precipitate is redissolved. Allow to stand half an hour; then filter.

STIEVE'S FIXATIVE - *(Gurr)*

Mercuric chloride, saturated aqueous	76 ml

Formalin 20 ml
Acetic acid, glacial 4 ml

SUDAN BLACK - *(Alcoholic)*

Sudan black 0.5 g
70% ETHANOL 100 ML
Dissolve in a warm water bath; cool and filter.

SUDAN BLACK - *(P.G. Soln.)*

Sudan black 1 g
Propylene glycol 100 ml
Dissolve by heating to 100 degrees C. for a few minutes; cool and filter.

SUDAN BLUE - *(BZL Blue)*

Sudan blue 0.5 g
50% ethanol 100 ml
Dissolve in warm water bath; cool and filter.

SUDAN 111 - *(Gurr)*

Prepare as scarlet R.

SUSA FIXATIVE - *(Gurr)*

Mercuric chloride 45 g
Water 800 ml
Sodium chloride 5 g
Trichloracetic acid 20 g
Glacial acetic acid 40 ml
Formalin 200 ml

TARTRAZINE IN CELLOSOLVE - *(Gurr)*

Tartrazine 1 g
Cellosolve 100 ml

TOLUIDINE BLUE - *(Gurr)*

Toluidine blue 1% aqueous.

TOISON'S BLOOD COUNTING FLUID for white and red blood cells - *(Gurr)*

Sodium sulphate 8 g
Sodium chloride 1 g
Glycerin 30 ml
Water 150 ml
Dissolve and add:
Genetian violet 8% aqueous 3 ml

TURK'S BLOOD COUNTING FLUID for white blood cells - *(Gurr)*

Genetian violet 1% aqueous 1 ml
Glacial acetic acid 1 ml
Water 98 ml

VAN GIESON STAIN - *(Picro Fuchsin)*

Acid fuchsin 1% aqueous solution 2 ml
Saturated aqueous picric acid 100 ml

WATER BLUE-ORCEIN - *(Gurr)*

Water blue 1 g
Orcein 1 g
Glacial acetic acid 5 g
Glycerol 20 ml
Ethanol, absolute 50 ml
Water to 100 ml

WEIGERT IRON HAEMATOXYLIN - *(Gurr)*

A. Haematoxylin 5 g
Ethanol, absolute 500 ml
B. Ferric chloride (hydrated) 30% aqueous 20 ML
Hydrochloric acid, pure 5 ml
Water 475 ml

WRIGHT BLOOD STAIN - *(Gurr)*

Wright's stain powder 0.3 gm
Methanol, pure 100 ml
Place the methyl alcohol in a 250 ml. flask; add the powder and shake; plug the neck of the flask lightly with cotton wool; then heat on a water bath. Allow to cool; then filter.

ZENKER FIXATIVE - *(Gurr)*

Mercuric chloride 25 g
Potass. dichromate 12.5 g
Sodium sulphate 5.0 g
Glacial acetic acid 25 ml
Water 500 ml
Keeps badly; to avoid deterioration omit acetic from stock solution and add immediately before use.

ZENKER-FORMOL - *(Gurr)*

Prepare to Zenker recipe above but omit acetic acid and add 5 ml formalin per 100 ml immediately before use.

ZIKE'S FLAGELLA MORDANT - *(Gurr)*

Tannic acid 20 g
Chrome alum. satd. aqueous soln. (about 14%) 50 ml
Osmic acid 1% 15 ml
Gentian violet satd. aqueous soln. 10 ml
Water 80 ml
Dissolve tannic acid in water by warming, add chrome alum dropwise followed by osmic and gentian violet.

INDEX

H

I

R

S

T